CONTENTS

Session IV **Celebration Day Papers**

Chairman: *Mr A. L. Stuchbury, O.B.E., President of the Institution of Production Engineers*

Raising Productivity

Mr P. P. Love, Glacier Metal Co. Ltd

Manufacturing Process Research within the Company

Dr W. J. Arrol, Joseph Lucas Ltd, Research Centre

Session V **Production Research**

Chairman: *Professor D. S. Ross, Rolls-Royce Professor of Production Engineering, University of Strathclyde*

Trends in Machine Tool Development and Application

Mr C. F. Carter, Jr., Cincinnati Milacron Inc., Ohio

Superplasticity: A Contribution to Innovation in Forming

Prof. F. Jovane, Universita' di Bari, Italy

The Laser Cutting of Steel Rule Dies

Mr F. W. Lunau and Mr B. C. Doxey, BOC-Murex Research and Development Laboratories

Session VI **Decision Making in Design Manufacture and Marketing**

Chairman: *Mr J. D. Houston, Managing Director, Higher Productivity (Organisation and Bargaining) Ltd*

A Systematic Procedure for the Generation of Cost-Minimized Designs

Dr P. W. Becker and Mr B. Jarkler, Electronics Laboratory, Technical University of Denmark

The Design of a Production Control System: In Particular the Requirements Planning Segment

Mr A. W. Buesst, National Cash Register Co. Ltd

Decisions in Technical Selling

Mr D. J. Leech and Mr D. L. Earthrowl, Division of Industrial Systems Engineering, University College of Swansea

Session VII **Manufacturing Systems and Automation**

Chairman: *Mr W. Gregson, C.B.E., Assistant General Manager, Ferranti Ltd, Edinburgh*

PROCEEDINGS OF THE SECOND INTERNATIONAL CONFERENCE ON PRODUCT DEVELOPMENT AND MANUFACTURING TECHNOLOGY

University of Strathclyde, April 1971

PROCEEDINGS OF THE SECOND INTERNATIONAL CONFERENCE ON PRODUCT DEVELOPMENT AND MANUFACTURING TECHNOLOGY

University of Strathclyde, April 1971

Edited by Professor D. S. Ross, B.Sc., Ph.D., M.I.Mech.E., M.I.Prod.E.

Department of Production Engineering, University of Strathclyde

MACDONALD: LONDON

SBN 356 04124 7

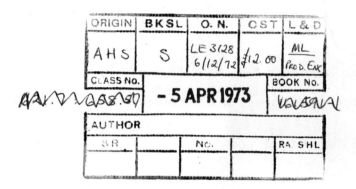

First published in 1972 by
Macdonald and Co. (Publishers) Ltd
49/50 Poland Street, London W1.

Printed by Balding + Mansell Ltd, London and Wisbech

PREFACE

One problem facing all conference organisers is whether or not they are going to have an adequate attendance, and invariably one cannot, under normal circumstances, have any assurance of this until delegates actually arrive. When circumstances are abnormal, as was the situation when the Second International Conference on Product Development and Manufacturing Technology was due to be announced, the organisers were faced with what could only be described as a 'cliff-hanger' decision – whether to proceed or to cancel.

In our case, postal communication throughout the U.K. was almost non-existent due to a strike, which very seriously curtailed our ability to make the Conference arrangements widely known as early as we would have wished. Rather than cancel or postpone the Conference, with the consequent inconvenience to authors, chairmen and others, we decided to proceed. We were gratified to have an attendance of eighty delegates, which although considerably lower than the attendance at the first PDMT Conference, was nevertheless sufficient to permit adequate and meaningfull discussions both during and between conference sessions. The countries represented at the Conference were: U.K., Canada, Denmark, Holland, Hungary, Norway, Poland, Sweden, U.S.A., W. Germany.

The Conference coincided with the Jubilee Celebration of the Institution of Production Engineers. To mark this occasion, the morning of the second day of the Conference was devoted to papers relevant to the growth of Production Engineering in the U.K.

The success of any Conference is due in great measure to the enthusiasm and support of its Organising Committee. In my role as Conference Director and Committee Chairman, I wish to thank the Committee members for their assistance and, in particular, the Organising Secretary – Mr F. I. Simpson – for his unstinted labours in achieving a successful event. Our thanks go, also, to the authors of papers presented, session chairmen and members of University staff for their willing cooperation; and to our publishers – Macdonald, London, for their invaluable assistance in producing this volume.

Professor D. S. Ross

SESSION I

Fulfilling the Market Need

Chairman: Mr J. W. Atwell, C.B.E.
Chairman
Engineering Division, Weir Group Ltd

THE IMPACT OF MARKETING STRATEGIES ON PRODUCT DESIGN AND DEVELOPMENT.

W. Hill
Clarke Chapman–John Thompson Ltd

SUMMARY

After a brief reference to the principles involved in formulating a strategy, the total concept of marketing is explained and its importance in formulating strategies to achieve the main objectives of a business.

The product is shown to be only one element (but nevertheless important one) in the overall strategy. Specific situations are analysed to illustrate the effect of profit requirements on the product.

A description is included of the methods employed to collect and analyse relevant data required for the formulation of strategies. Their evaluation and final selection by means of financial models is explained, models which simulate the cost effects of the complete design and manufacturing cycles of the product.

The paper concludes with descriptions of marketing situations for established products and one product with future potential.

1.0 INTRODUCTION

The word strategy comes from the Greek word 'strategus' meaning not only a general but also a civil governor with wide powers. Strategy can therefore be defined as the 'art' of the general or of the governor. For over two thousand years attempts have been made to formulate universal principles from the extensive case histories which presented themselves throughout the ages. In 500 B.C. the Chinese general Sun-tzu set forth 13 such principles which included the simple admonition to get there first with the most men. Napoleon listed 115 such maxims but modern thinking is content with about ten. Amongst these ten principles can be found the following:

- The objective
- Unity of command
- Economy of force
- Room for manoeuvre
- Security
- Simplicity

Although it is easy to overrate the analogy between business and waging war the disregard of the above principles in any marketing strategy can be very costly.

This paper is concerned with the impact of these principles on Product Design and Development.

2.0 DEFINITIONS

Discussions of certain subjects often suffer from either a woolly and unstructured approach or become sterile and trivial if reduced to purely mathematical symbols and relationships. In order to avoid both these extremes a number of definitions, assumptions and conjectures will be stated first so that further discussions can either question these assumptions or deal with their correct application to Product Design and Development.

- The main objective of a business is to stay in business.
- In order to stay in business it is necessary to satisfy the shareholders with an acceptable return on the capital they have invested.
- In order to pay the shareholders the business must accumulate profits and cash over and above that necessary for replacements, R & D new ventures and investments.
- Continuing to satisfy customer needs is one of several conditions for generating profits and cash.
- The identification of customer needs is an important marketing objective.
- The profitable matching of the design, manufacturing and management resources with customer satisfaction is the primary marketing objective. This is known as the total concept of Marketing and is particularly important in dynamic or changing circumstances.
- It is never the product as such which satisfies a customer need. This is accomplished only by the correct functioning of a number of interrelated products provided the customer need has been correctly identified.

3.0 CONJECTURES

- Customers' fundamental needs change little and are added to only slowly. The reasons why there is such a pressure to produce changes are two-fold and interrelated.
  (a) Competitive society
  (b) Desire to satisfy the same fundamental needs faster, more reliably and cheaper.
- Potentially the most rapid product changes can occur with vertical product chains where the ultimate customer satisfaction is subject to fashion trends.
- The number of truly new products is infinitesimal.
- The main scope for change lies in a better understanding and hence exploitation of the elementary build-up of matter and the natural forces controlling its interactions.

4.0 MARKETING STRATEGIES: The Objective

A definition of the total concept of marketing was given in section 2.0. From that point of view the product itself is only one link, albeit an important one, in the process necessary to satisfy a customer need profitably.

To investigate the effect of marketing strategies on the product raises the same problems as any analysis of the effect of the whole on a part.

We can, however, isolate specific instances which illustrate the effect of marketing ideas on the form, design or cost of a product. In theory at least, three distinct situations can be identified:

- Data have accumulated indicating a change in customer needs in a situation in which the existing product has in the past satisfied customer needs profitably.
- The required profit or cash flow objectives are no longer met by the present product.
- A change in the competitive position is forecast or actually happening owing to new materials, advance in technology, tariff restrictions, lack of essential raw-materials, need for more or less quality control, success of alternatives or even a simple withering away of customer need.

In practice, however, all marketing problems have some proportion of all three situations present at the same time. It is therefore important to define the objective in a clear and unambiguous manner leaving no doubt about the available time scale and making sure that all quantitative goals are labelled as maxima or minima.

In the situation in which there is evidence of a shift in customer need such as a demand for easier maintenance, cheaper or faster after-sales-service or higher reliability, the effect on the product design is largely determined by the following:

- The degree of urgency associated with the deterioration in competitive position.
- The need and cost of modifications to existing products either in customers' hands or in stock or of modifications to parts.
- The possible accumulation of other reasons for change. In a normally successful manufacturing business changes of this sort can usually be phased to coincide with new models, new manufacturing plant or methods. The point to remember here is the need for good communications which allows records of these changes in customers needs to be stored and retrieved at the appropriate time.

The situation where an existing product does not meet the required profit or cash flow objectives may require action ranging from a complete suppression of the product to the establishment of a new factory on a greenfield site. In general, however, less drastic remedies will suffice, such as Value Analysis of the product, rationalisation of the product range, examination of the overhead costs associated with the product and other actions which can be summarised as good housekeeping. An important aspect of the role of marketing in this process is in establishing the scope of these investigations in such a way that a particular course of action has a chance of achieving the objective.

Potentially the most dangerous inroads into product profitability, or even into the viability of whole industries are those changes caused by the third situation where new materials and technologies can completely alter a product concept. This is also the area where national or political considerations expressed as tariff restrictions or reductions, stock-piling or dumping can alter radically the competitive situation of an existing product line. It is potentially the most dangerous situation because a successful defence depends on a timely anticipation or correct forecasting of the situation coupled with the necessary investment in R & D. Conversely the opportunities for achieving very high rates of return on capital are exactly those where new technologies or ideas make the very same inroads into competitors' product profitability. As so often in the last 5000 years taking the offensive may be costly but can constitute the best defence. A very important point which any discussion of market anticipation or forecasting must make concerns the definition of what is a correct or incorrect forecast. Apart from the situation where a forecast can be made a self-fulfilling prophecy it is more important to predict correctly what people will do than to predict what would be the right course for them to follow. Considerations of this nature emphasise once again the role of the product as a means of penetrating and dominating the market, but only for such a time as will allow its competitive advantages to produce the required return on capital.

5.0 THE SELECTION AND EVALUATION OF MARKETING STRATEGIES

Lasting success in business, politics or warfare is rarely, if ever, achieved without a penetrating understanding of the circumstances, the personalities and the driving forces involved. Such understanding must be based on a thorough examination of all the data which can be compiled. Sound judgement and experience will help considerably in selecting the relevant areas from which to collect data. It is clear, however, that the ability to learn from the data and modify the plan of collection in mid-stream, so to speak, can be a valuable characteristic which can shorten the time necessary for decisions.

Market research is the customary and powerful method available for the systematic collection of data necessary for any formulation of marketing strategies. There are very simple and very sophisticated methods of market research. Cost-effectiveness evaluations are necessary to decide the degree of accuracy required. Very often in the industrial field it is only the order of magnitude which is required for a 'Go' or 'No-go' decision. Amongst the relevant details which must be established for the selection and evaluation of marketing strategies for a particular product or product line are the following:

- Total Market for the product in various geographical areas and to different industries.
- Actual sales of this product in the same geographical areas.
- Details of competitors' products in these areas.
- Analyses of competitive advantages of relevant products such as price, delivery, after-sales service, ease of maintenance, reliability.
- Attempt to correlate competitive advantages with actual sales.

A critical examination of such details will produce a product profile which can be regarded as almost as distinctive as a finger print. Conclusions, however, are more difficult to draw with certainty. What is important is to segregate product attributes from external factors as responsible for affecting the ability of the product to satisfy customer needs. A particularly interesting example of such a situation existed when an examination of sales of electric motors to the company manufacturing cigarette making machines fell off dramatically. Probing into the reasons appeared to produce the answer that the electric motors were running hot, that is were not complying with the respective British or International specification. As no manufacturer has an entirely clear conscience on compliance with such limits, care was taken to supply a sample batch, carefully selected and tested to be within the permissible limits for a customer evaluation. To the surprise of the supplier and the customer these samples were rejected out of hand as too hot after only a short trial. A few weeks and several management levels later it transpired that the hot air convection currents from the surface of the motor distorted the very delicate mechanism of the machine. The motor had been moved a small distance compared with the previous designs where this effect had not been harmful. The remedy was, of course, very simple.

With the establishment of the product profile it is possible to conceive a number of different courses of action all designed to exploit the profit potential of the 'product' which can, and normally does, have different detailed specifications for different courses of action. It is sometimes useful to formalise all possible courses of action into a 'decision tree' and then decide to analyse the effect of a certain number of them in depth. As the object of such an analysis is the selection of a strategy maximising return on capital this analysis must therefore be expressed in financial or plain money terms. The most effective method for this financial analysis is the construction of a 'model' corresponding to every one of the possible courses of action. In general a model is built up from the point at which raw materials are bought, stored and handled. The costs associated with the proposed manufacturing methods and processes, the floorspace and lifting capacity, the demands for heat-treatment and quality control must be represented in the model to the extent to which they would apply in reality if that particular course of action were chosen for implementation.

From such a complete operational model the financial results can be estimated for any given volume of sales. Judgement must then be exercised to determine what volume of sales or market penetration is reasonable for that particular set of product attributes. Evaluation of the financial results of all possible strategies by way of these models will enable each decision in the decision tree to have at least two values associated with it.

The two values are:

- Investment Cost
- Return on investment.

This is usually as far as a Marketing Strategy can take the analysis, as the final decision between several courses of action may depend on the availability of capital, alternative uses of capital and the whole area of risk analysis. A case can be made out that even these areas are logically within the total concept of marketing.

It may be as well to state here that the correct application of marketing policies can in the last resort only prevent competitive disadvantages. If it is assumed that either of two competitors is likely to have access to the same market data and the logic of neither is faulty, then they must arrive at similar market strategies.

There are other areas where the financial model described above can be of considerable help. The model may indicate that for a given percentage share of the market an adequate return on the capital required is only possible by relying on some competitive advantage which commands a higher than the existing market price. Such a competitive advantage could be a rapid world-wide spares service always appreciated, say, in the marine field. Such a model will also throw up what might be termed the 'Honda' principle. Although originally implicit in the concept of the Model T Ford motor car, it was used in launching the 'Honda' motorcycle. Basically, it illustrates the fact that unit costs of a product can be reduced almost continuously as the numbers produced increase, provided an increasing amount of mechanisation, automation and sophistication is used in the manufacturing and production control processes. If, in addition, the reducing unit cost permit of the tapping of an ever increasing market, say, as the result of encompassing lower income groups or smaller firms, then the Honda principle states that there is a number, however astronomical which optimises profits or return on capital. The computer industry is somewhat undecided whether to sell a few mammoth computers to a small number of large firms or to concentrate on smaller units which a very much larger number of smaller firms could afford.

A further valuable aspect of such model operating budgets is the help it can give in tactical decisions such as are needed to decide whether to buy-in design under licence or develop the product in-house. It also illustrates very convincingly the danger of running out of cash when a new product is so successful that profits, even if high by modern standards, are insufficient to finance the cost of materials for the production of an increasing number of units.

6.0 BRIEF DISCUSSION OF OTHER PRINCIPLES OF MARKETING STRATEGIES

6.1 *Unity of command*

The effect of the unity of command on product design is one of the most important aspects of profitable trading. A marketing strategy must be conceived with the stark realities of the industrial life firmly represented including the manufacturing background and its limitations. Conversely it is necessary for works management and sales policies to adhere to the details of the marketing strategies finally adopted. Unilateral disregard of agreed product specifications by one department can easily negative an important competitive advantage such as reliability, low cost or the ease of maintenance. Similarly disregard of selling into specified industries or areas may well destroy the volume basis on which the product was expected to show the expected rate of return. Unity of command will ensure that a common purpose permeates the relevant sections involved and that necessary changes and modifications do not destroy the basis on which success was predicted. In an industrial organisation, unity of command means a line organisation capable of achieving this common purpose.

6.2 *Economy of force*

The Principle of the 'Economy of Force' is closely related to that of industrial good housekeeping. You do not advertise on television when a few direct mail shots to the relevant consulting engineers will reach 95 per cent of all possible customers. You do not involve the T.U.C. when a few words with the shop steward will be as effective. You do not have a committee when it is possible for one person to encompass the collective experience.

6.3 *Room for manoeuvre*

This is such a classic principle that few examples are necessary. In product design and development the switch to alternative materials in case of shortages, rising prices or failure on test is well documented. To have enough room for the replacement of a spot weld by a nut and bolt may be an elementary precaution.

6.4 *Security*

Security will become increasingly important where industry is investing in designs and products having advanced technological competitive advantages which require years to perfect. If you are trying to steal a march on your competitors, keep it dark. Alternatively, do not assume your competitors are inactive, it may just be the result of their good security.

6.5 *Simplicity*

Developments in the design of a product tend towards greater simplicity. The electrical connection eliminated by a better lay-out cannot cause trouble. The deliberate search for greater simplicity in product design can therefore anticipate future developments and produce a competitive advantage today rather than prevent a disadvantage in the future.

7.0 DISCUSSIONS OF TYPICAL SITUATIONS WITH REFERENCE TO PRODUCT DESIGN

Tracing the effect of marketing strategies on the design and development of the electric motor over the last 70 years would be a very instructive exercise indeed. As their initial price has always been a small proportion of the price of any installation marketing strategies in the early days tended to build up an image of indestructible rugged dependability worth paying a high ring price for. Motors were large in diameter, short in length and barely hot to the touch and extremely wasteful with what are now scarce raw materials. When this image started to be questioned and costs began to matter, the product changed its shape within a few years. The advent of self-ventilation changed the shape from the large diameter machine to the small diameter long carcase one, with a saving in material of approximately 50 per cent. High prices to customers were still defended on the basis of a highly technical advisory service before purchase of a motor and a life-long after sales service which included help with commissioning and trouble-shooting.

There was also an extreme willingness by the supplier to manufacture special variations at little or no extra cost. Standardisation of fixing dimension in those days meant standardising on the products of one supplier with its consequent captive market.

Since 1948 the emergence of international competition, the phenomenal increase in world trade, the availability of synthetic insulation, the power of the multi-national and supra-national companies buying in several markets at once, all combined to alter the climate in which electric motor manufacturers operated. Few, however, appreciated the changed circumstances and re-examined their marketing strategies. When price fixing became illegal in 1959 few manufacturers had taken the trouble to analyse the new situation. The present competitive position is one whereby a manufacturer must produce an electric motor within a standard dimension envelope at the lowest possible price with a permissible temperature rise of twice that permitted a few years ago. The size of motors has declined again drastically but unit costs are still too high for a reasonable return on capital. This can only come from a large increase in manufacturing output from a single plant in which unit costs have been reduced considerably. Attempts to boost the return on capital by other than the correct marketing strategy have included a reduction of the active material of the smaller motors below a reasonable value taking account of voltage and load variations with the predictable result of customer resistance. Another strategy was to penalise very heavily any demand for non-standard features. Most of these manufacturers, if not all, are, of course, no longer with us.

New product development is very much affected by the strategy of whether to design and test a new product or become a licensee. Evaluation of the optimum strategy by means of analysis of financial models has been mentioned, but there remains the problem of assessing probable time-scales and hence the cost of development.

Two situations can occur and it is useful to treat each separately.

- The design of the product is subject to a patent held by the licensor.
- The design of the product or of a product line is the result of many years of experience and product development by the licensor.

If a product is the subject of a patent, it is often useful to investigate the following:

- Is the patent still in force? It is surprising, how many patents lapse after only four or five years because the statutory fees due each year after the fourth have not been paid.
- What exactly is the specific claim made in the patent and is there a better or different way to achieve the same result? If there is, apply for a patent citing the first patent as prior art of which the new application is an improvement.
- Was the original patent applied for after such a product had already been sold for money by the original patentee? If so, the patent is invalid. If the patent is sound, valid and uneconomic to circumvent, then only exceptional circumstances will make it profitable to manufacture under licence.

Under normal competitive circumstances there is never enough money for both profit and royalties. In high technology areas like nuclear energy, electron beam welding or others, manufacture under licence usually means a virtual monopoly in a given geographical area. The same reasoning applies to the manufacture of a product like shell boilers the present design of which is the result of many years of experience.

Unless the normal competitive pressures are suspended by protective tariffs, import quotas and state monopolies there will again be too little money for both royalties and profits.

An interesting marketing situation can arise from the examination of a product, which although not necessarily new, has properties which due to development demands in other products suddenly become of value.

Such a product is silicon nitride, a ceramic substance capable of high strength at temperatures up to 1400°C. If silicon nitride were used as a simple replacement of, say, steel where steel is used now, the selection of the best marketing strategy would differ little from those already discussed; in fact, it can be shown that such a replacement would have few competitive advantages to off-set the present high cost.

If, however, silicon nitride is used in such processes and in such applications where materials used to-day limit the permissible temperatures to 750°C or so then we have a new marketing situation. With apologies to J. W. Dunne ('An Experiment with Time') this could be called a serial marketing problem. As there can be no existing market data it is necessary to extrapolate into the future the likely development of products which to-day operated under temperature limitations. Obvious examples are die-casting machines, continuous-casting apparatus, gas-turbines, boilers, I.C. engines, fuel cells and others.

In these days of pre-occupation with pollution, for instance, a higher operating temperature will make the essentially low pollution causing gas-turbine even cleaner, particularly as the oxides of nitrogen are burned more completely. The use of gas turbines to replace I.C. engines, however, depends on the provision of a reliable heat exchanger capable of withstanding gas temperatures of 1000°C or more. The production of silicon nitride heat exchangers is possible but a profitable marketing strategy for silicon nitride heat exchangers depends entirely on the successful implementation of a profitable marketing strategy for the replacement of diesel engines by gas turbines.

There is, however, a limited marketing strategy possible for silicon nitride which allows a build-up of future applications whilst providing less sophisticated products like thermo-couple sheaths and liquid metal pipes.

Several aspects of the impact of marketing strategies on product design and development have been illustrated above and in conclusion it may be appropriate to indicate that there is a converse effect.

If, as the result of a general marketing strategy, certain products are manufactured as, for instance, turbo-alternators or ship's deck machinery then the pace of technical achievements in these fields will impose its own pace and direction on subsequent market strategies. Unit outputs of turbo-alternators have increased in workload, the size of tankers has increased by factors of four and yet a 200 000 ton tanker still only requires one anchor windlass. All these changes tend to produce a large surplus of manufacturing capacity for such products with profound effects on the room for manoeuvre in marketing strategies. Fundamentally, however, this problem is analogous to that of a control system the parameters of which are unsuited to respond to the frequencies of the signals imposed on it. As some of the important system parameters are not determined by companies or even industries this type of problem has solutions provided the system boundaries are drawn in such a way that they include all decision making elements.

8.0 ACKNOWLEDGEMENTS

Permission to publish this paper has been given by the Board of Clarke Chapman–John Thompson Ltd. All views and opinions expressed in this paper are entirely those of the author.

CASE STUDIES IN PRODUCT DEVELOPMENT

S. S. Carlisle
Sira Institute

SUMMARY

Two case studies on projects involving inspection techniques for industrial products are outlined.

The first, involving the development of equipment for the automatic detection of surface defects in aluminium strip production. The same basic technology is applicable in other fields. The second project led to the successful development of a 'speck counter' which detects the presence of impurities in powders, or surface defects in solids, using a flying field scan system.

The problems associated with the development of these specialist instruments on a commercial basis are indicated.

1.0 ELECTRO-OPTICAL APPARATUS FOR INSPECTION OF SURFACE QUALITY OF MOVING FLAT PRODUCTS

Around 1965/66, Sira started development of a flying field scanner system for automatic detection of surface blemishes in metal sheet and strip at speeds of up to 500 ft/min and for widths of about 1 metre. The design objective was to be able to detect automatically all those defects which influence the classification of the finished sheet or strip and so provide an automatic system of classification to replace the present labour intensive inspection process in sheet mills.

The project was undertaken principally for the following reasons:

(1) The current manual inspection process was subjective and liable to inconsistency.
(2) Production speeds were rising to make visual inspection less practicable.
(3) Quality control was becoming tighter.
(4) Labour costs were rising.
(5) Sira had expertise in advanced electronics and optical systems development and design and these appeared to be the techniques predominantly required for solution of this problem.

Initially it was thought that there would be combined support from the ferrous, non-ferrous and paper producing industries all of whom attached growing importance to the problem of continuous inspection of the finished product. However, initial financial support was obtained only from the aluminium, copper and brass strip producers. From the beginning the development was done in close co-operation with aluminium sheet producing plants.

Fig. 1. The flying field scanner and associated electronic set-up in the laboratory for inspection of the surface quality of aluminium sheets moving on a conveyor belt.

Figure 1 shows the essential features of the system developed. It employed a rotating lens drum scanner which had been developed previously by a company for paper inspection. Our work concentrated on design of suitable illumination systems to give maximum resolution of the defects to be detected and particularly on the electronic signal processing systems. Circuit techniques were established to give constant defect resolving power against normal changes in surface reflectivity, against small variations in sheet flatness and against reasonable variations in lamp brightness and optical window transparencies. Techniques were established for counting numbers of defects which produced defect signals above a level which is adjustable in initial setting up of the equipment. A novelty of the system is the 'video print' which displays on a storage tube an 'electronic photograph' of the defects actually registered during a sheet inspection for the particular sensitivity setting used.

Altogether about four years of work was done in getting to the stage of a well engineered system suitable for extended works trials and with a power to detect most of the defects considered significant. A greater portion of the time and the cost of this project was concerned with intermediate works trials each of which involved considerable preparation to get an adequate number of typical sheets available at the plant for running through the inspection equipment and the co-ordination of manual inspectors' observations with the quality assessment made by the instrument.

This work highlighted the particular difficulties of developing an apparatus for objective quality appraisal to replace a subjective appraisal system. The basic difficulty is that the criteria used in subjective assessment cannot be adequately described by the human inspector to enable a satisfactory specification to be written for the automatic inspection device. The result is that as the performance of the apparatus progresses, the user's specification continually extends to want more or different defects to be recognised. This is a type of development exercise to be wary of and where one is concerned with what might be called a 'creeping specification'.

The present position is that the apparatus is considered by the aluminium strip producer to meet most of his needs for automatic detection of the specific defects. However, this performance is only obtainable on oil-free sheets, and sheets in such condition are not provided at the required inspection station in the present mill practice, although we had expected this would be so at this point in time. The adoption of the instrument in such an aluminium sheet producing plant thus depends on future development of sheet mill practice. When this happens it will probably be associated with much higher strip speeds and so on, this particular inspection problem our activities must be directed toward.

(1) Finding aluminium sheet or other non-ferrous sheet production situations in UK or abroad to which the current level of our technology would be applicable.

(2) Raising the speed of inspection capability to five or ten times the present level.

The accompanying tabular presentation of this project case history shows numerous other product inspection problems that have been tackled using the same basic technology. This widening of the potential area of use was done purposely to increase the probability of ultimate reasonable return on the relatively high R. & D. costs incurred in pursuing such an innovation. Early in the development for aluminium inspection the requirement for automatic inspection for glazed ceramic tiles was identified and this led to the conducting of a major project for an integrated automatic inspection, classifying, stacking and packaging system for tiles with a capacity of handling up to 9 units per second. Operational studies conducted with the cooperation of a leading UK tile making company showed economic justification for quite an expensive system if it could be provided, and thus considerable support for development was

THE TECHNOLOGY AND ITS PROGRESS.

1965 Commenced development of flying field optoelectronic scanner for detection of surface defects – diffuse or specularly reflecting.

Resolving power improved.

Sensitivity of electronics to electrical noise reduced.

Automatic correction for variation in surface reflectivity

Self adjustment for different strip widths

Video print developed to assist setting up of sensitivity.

1970 Well engineered system available for works use at line speeds up to 200 metres per minute.

The 'Speck Counter' variant of the system developed for high resolution of very small defects.

1971 Very high speed laser beam scanner system demonstrated in Laboratory and capable of detecting 1 mm. defect in 1 mm wide strip at 3,000 metres per minutes.

Aluminium sheet and strip

Problems:

Defining defects considered most significant.

Changing location of works trials.

Electrical interference from plant.

Co-ordinating plant and sheet availability for trials with apparatus availability.

Correlation of instrument readings with human inspector assessment.

Oil films on sheets varying with works practice.

1971 Result

Reliable apparatus for works use detects most significant defects at speeds up to 200 metres per minute provided sheet is oil free

but

to date situations not available in UK where sheet can be inspected in-line, oil free. Later there probably will be such situations but higher speeds of inspection will then be required.

Future

Requirement expected for the very high speed inspection system.

Glazed ceramic tiles

1967 onwards

System shown in laboratory capable of detecting most significant surface defects, both colour spots and indentations.

Operational research studies showed need for a complete inspection and handling and stacking system to make automatic inspection economic.

1969 System installed in a tile factory for trials.

Result Inadequate consistency of performance in routine use – particularly in detecting edge chips and tile curvature. Showed promise of ultimate solution but more development expenditure required. Project discontinued due to decline of business in the tile making industry.

Future Requirement for single colour tile inspection not now likely because of major switch in market to patterned tiles for which the system is not appropriate.

APPLICATIONS

Paper	*Steel industry*	*Polymer powder*	*Other industries*
			Several developme contracts undertak or being negotiatec solution of specific problems in other industries includin textiles, rubber fab and tobacco.
1969 Requirement arose for inspection of pinholes and opacity of paper for computer punched tapes. Variant of systems developed for transmission measurements and to cope with different paper types.	*1969* Apparatus demonstrated in Laboratory showing capability of detecting most significant defects on typical samples of cold reduced steel sheet.		
	1970 Although apparatus could replace human inspectors at inspection stations in a sheet selection line industry decided not to install because changing practice in sheet and coil production puts much more emphasis on need for the higher speed inspection system.	*1970* Variant of the scanner system but with same electronic signal processing system developed for detecting small (75 microns) specs in polymer powder known as 'Speck Counter'	
1971 Engineered equipment ready for installation at a paper mill. *1971* Use for high resolution 'speck counter' identified on very high quality papers. 'Speck Counter' tested and proven and under offer to client.	*1971* Major German steel plant considers the present system meets their needs and contract under negotiation. *1971* Present system installed for trial for in-line inspection of coated steel sheet. *1971* UK steel industry considers backing of the very high speed inspection system under development.	*1971* Now under trial in polymer plants.	
Future Probably a successful market outlet for existing systems and need for the new high speed system under development.	**Future**Probable significant markets in Europe for existing systems and pronability of home supported development of the high speed system.	**Future** Very promising sales in polymer manufacturing industry and probably in other powder industries.	

won from the UK tile making industry in the 1967/69 period. The development of this rather complex system was quite rapid, and a system was installed in a tile factory involving four scanner heads to detect discoloration and geometric surface defects and two special units for edge chip detection, all associated with a new design of high-speed tile diverter system for classification and stacking. After some three months' trials in routine use the system was considered promising but insufficiently consistent in behaviour. It was reckoned that we were 75 per cent of the way to a successful system. However, just at this point when it was estimated that a further £20 000 would be required for development to completion, the economic situation of the UK tile industry had deteriorated and plants were running seriously under capacity. Thus, the development had to be discontinued.

More recently some investigation of this inspection requirement has been made in the tile making industries in Italy and Germany who are major producers, and this has provided some evidence that the need for automatic inspection is quite strong but principally for the inspection of pattern glazed tiles which are now becoming the market leader. Plain coloured tiles to which our system is applicable and which were (and still are) the major UK market are of decreasing interest in the advanced European countries. Thus, although this particular project might well have paid off if the UK tile market had not deteriorated, it now shows little promise of success because of change of fashion which the product has to adapt iself to meet. An outlet may, however, appear for the work in the automatic inspection of tiles before glazing as this seems to be a growing requirement in the modern high volume production plants.

So far the story is disheartening but publicity given to this work in recent years raised the interest of numerous other industries. Sira was becoming recognised as a centre for innovation in automatic inspection technology. Demonstration of the surface inspection system on cold reduced steel sheet showed a defect detection capability which raised the interest of the British Steel Corporation. However, after detailed analysis they concluded that they would not wish to install the presently developed low-speed system (200 metres/min.) but would be interested in a much higher scan-speed system which we now believe we can achieve with laser beam scanning. So a new development project is being formulated. However, one of the largest German sheet and strip producing plants has come up with a strong expression of interest, and their representatives have seen a demonstration of the apparatus in its present form and they state that in this form it would meet a requirement in their plant. The market is thus being tested for this current range of equipment in other European and Scandinavian steel sheet plants. Furthermore, the steel industry explorations have brought to light an immediate application for this inspection system on coated steel sheet and strip products.

The performance achieved with this inspection system on metal sheet raised interest of some manufacturers of special purpose papers. The case in point is paper used for preparation of punched-tape for computer inputs. To ensure high-reliability in reading holes or not-holes, it is essential that there be a reasonable uniform and high level of opacity in the paper itself. Our system of inspection is readily modified for transmission measurements of this nature at high-scan speeds. As a result one of our systems is now commencing trials at a paper mill.

2.0 SPECK COUNTER

A quite different inspection problem was put to us in 1969 by a polymer manufacturer. It was to detect the presence of a few discoloured particles (as small as 75 microns) in an otherwise white polymer powder. The current practice is to spread the powder on a flat surface and examine visually. We conceived that our surface inspection system was basically capable of doing the job automatically and this led to a development programme out of which has now evolved an instrument known as the 'Speck Counter'. It adopts a flying field scan system, and although the scanner head is of different design to the general purpose scanner equipment in order to give the very high resolution for extremely small defect areas which is required, the electronic signal processing system is built of the same modules as the surface inspection equipment.

Figure 2 shows this apparatus which is now fully engineered and undergoing acceptance trials in two plants.

An interesting outcome of this 'Speck Counter' development is that it offers a solution for the inspection of the surface of very high quality papers which are used for documents to be subjected to optical character recognition reading systems. Thus another potential market outlet is opened up.

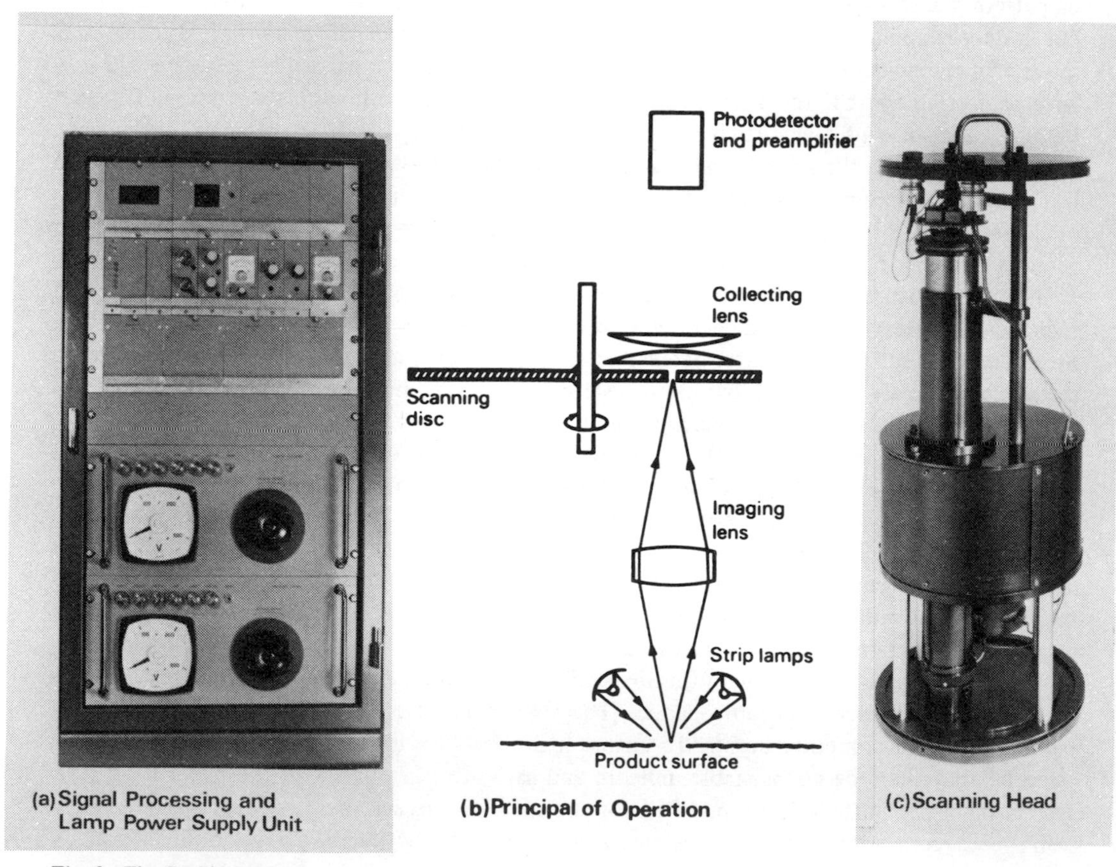

Fig. 2. The Speck Counter.

3.0 OBSERVATIONS

I chose this project for a case study in innovation not only because it has been one of our highest growth rate projects in winning industrial sponsorship for research and development but because it also shows a number of very interesting features and pitfalls in its development from which lessons can be learnt.

- Although after the initial Sira investment in technical development, the aluminium industry and the ceramic tiles industry were the earliest to come in with financial support (to which large Government grant aid was applied), and have facilitated most of the development of the apparatus technology, other industries are most likely to be the first to adopt the technology.
- Failure to adopt in the two initiating industries after what might be regarded as a reasonable time (six years since initiation) and after reasonable technical success was achieved, is attributable to two different reasons in these cases. In the case of ceramics it was due to an unforeseen deterioration of the market for the product to be inspected, and in the case of the aluminium sheet inspection, due to insufficient precision in initial specification of the inspection problem and in forecasting changes in the industries process technology.
- The successes which now seem likely in exploiting the technology in other industries, for example steel sheet, coated sheet, paper, polymer powder, are due to publicity given to the earlier technical developments before they were industrially adopted. The fact that this publicity could be given was due to the work being done under grant aided industrial sponsorship.
- When developing radically new industrial instrumentation involving relatively large speculative investment, it is essential to have intimate contact and co-operation with the potential user industry, but even with this market forecasts can go wrong and so it is equally important to keep doors open to application of the technology in as wide a spread of customer industries as possible.
- Specifications have a habit of 'creeping' and if the sponsor continues to finance the further development required, an R. & D. organisation may not be too unhappy. But, if the organisation is concerned with not only getting support for R. & D. but in actually achieving profitable innovation, as is the case with Sira Institute, then one must adopt a different attitude. This is to identify a state of development of the new product at which there is a reasonable prospect for a market and mount a sales drive for a product to that specification in as wide a market area as is appropriate to it. Further development to meet this 'creeping' specification may be continued if properly financed as a separate exercise.
- Obviously one has to be wary of the 'one customer' or 'one industry' requirement for an innovation. We have also found it wise to review and compare the requirements of an industry operating in different European countries. Sometimes the position and attitude in markets for the industries product differ considerably in the different countries and this can influence that industries policy in adopting a particular innovation.

SESSION II

Exploiting Innovation

Chairman: Mr D. McLean
Director and General Manager,
Aero Engine Division,
Rolls-Royce (Scotland) Ltd

THE NEW L. M. ERICSSON KEY PAD

Leif Branden,
ERGA Division, L.M. Ericsson,
Telefonaktiebolaget, Stockholm

SUMMARY

This paper describes the design of a new key pad to be used primarily in telephone sets where space is at a minimum. Reducing the overall dimensions of the key pad required the development of a new wire spring-contact sufficiently small to be incorporated within the pushbuttons. The spring contact has a long wiping action for self-clearing the contact surfaces from dust contamination.

For the proper functioning of central office equipment a contact duration of at least 40 milliseconds was required. This was obtained by incorporating a slight snap action in the push-button, thereby pursuading the user not to release the button too quickly.

1.0 DESIGN OBJECTIVES AND REQUIREMENTS

The key pad should be provided with 12 buttons arranged in four rows and three columns according to the C.C.I.T.T. recommendations.

There should be two individual electrical make contacts associated with each button as well as one common springset to be operated when depressing any of the buttons. The actuation sequence should be such that the individual make contacts close before the common springset is fully operated. It was also decided that, if possible, the operating characteristic of the push-buttons should influence the user not to release the button very fast in its bottom position in order to obtain a contact duration of at least 40 milliseconds.

As the key set in the first place would be used in fancy telephones, as for instance the Ericofon, its overall dimensions should be as limited as possible.

2.0 SPACE CONSIDERATION

To comply with the requirement for restricted dimensions it was considered that, if the individual make contacts could be made to operate within the button itself, this would be a way to minimize the overall dimensions of the key set for a predetermined distance between the buttons. The separation of the buttons may not, however, be arbitrarily chosen for an optimal user performance. Human factors associated with the keying of telephone numbers have been subject to fundamental studies by R. L. Deiniger[1] at the Bell Telephone Laboratories. Among the results of these studies it was shown that considerable differences in keying times and errors owing to the arrangement of buttons and the button spacing were obtained. With essentially the same arrangement as required for the present design, the buttons could be spaced $\frac{5}{8}$" between centres without significant change in performance in favour of a somewhat larger spacing. For the key pad to be used in the Ericofon a button spacing of 15·8 mm was chosen. Also, to avoid interference with adjacent buttons when keying, the size of the button tops may not exceed certain limits. In the new design a top size of 9·5 X 11 mm was found to be convenient.

3.0 INDIVIDUAL CONTACTS

For the reasons outlined above it was obvious that small sized individual contacts had to be developed.

3.1 *Movable contact springs*

A movable wire spring designed as shown in Fig. 1, for co-operation with a stationary contact, was shown to have the desired physical properties. The spring has two cylindrical spiral coils. The upper coil, 1, besides taking part in the spring action as a whole may also be used for mounting the spring in a suitable working position. The lower coil, 2, provides the desired contact force when the spring is actuated and also serves as the actuation member for the pushbutton. At the end of the one arm the spring is provided with a cylindrical contact sleeve, 3, for working against the stationary contact and the other arm, 4, is connected to the conductors of a printed circuit board.

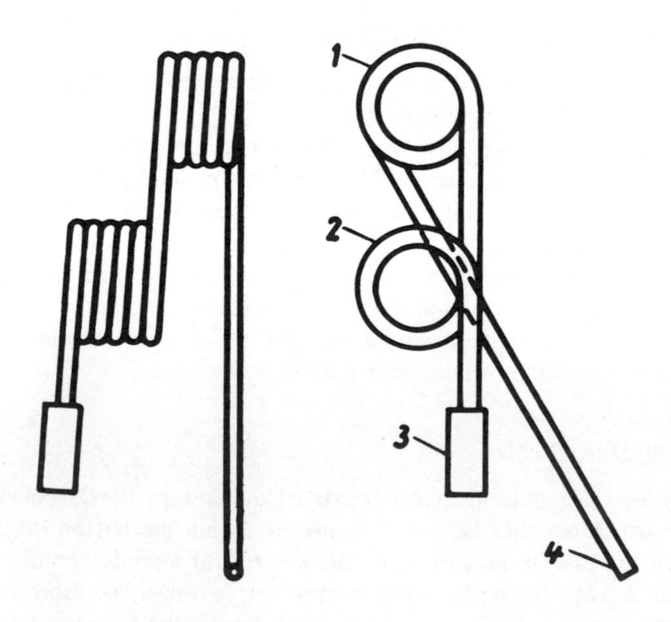

Fig. 1. Movable contact spring.

3.2 *Stationary contacts*

The stationary contacts are made in the form of bars designed for four contact points each, as shown in Fig. 2. In order to ensure a reliable twin contact the stationary contacts are coined

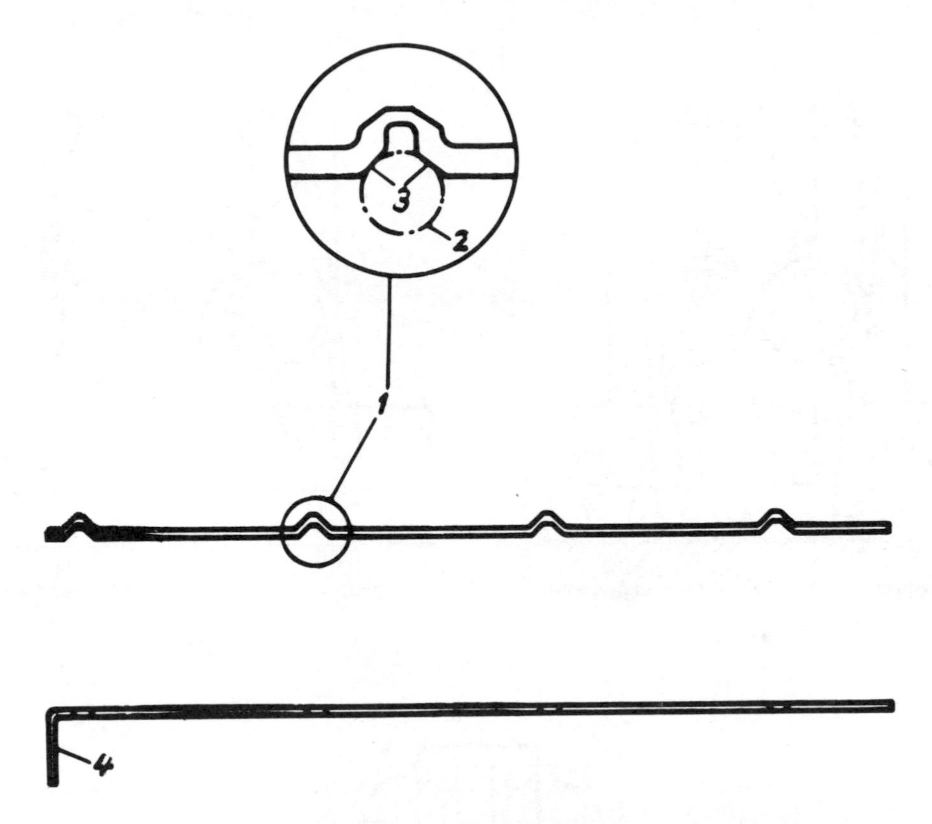

Fig. 2. Stationary contact bar.

as V-shaped indentations into the bar, 1. The contact sleeve, 2, enters this indentation perpendicularly in such a way that one contact surface, 3, is formed against the branches of the indentation on each side of the sleeve. The bar has at the one end a bend, 4, for connection to the circuit board.

3.3 *Arrangement of individual contacts*

Fig. 3(a) shows schematically the arrangement of the two individual contacts in a single key mechanism. The wire springs, one being a mirror image of the other, are mounted on a support, 1, projecting from a base plate, 2, and are retained in their working position by means of the upper spring coils, which fit into a recess, 3, in the support. The recess has an opening upwards that is somewhat narrower than the outer diameter of the spring coil, which may, however, be snapped through the opening as the support is made of an elastic material.

The straight arms, 4, of the springs are when mounted inserted in holes through the base plate so that they may be soldered to conductors, 5, of the circuit board, 6, and thus be incorporated with the electrical circuitry. The spring arms carrying the contact sleeve are centred into a

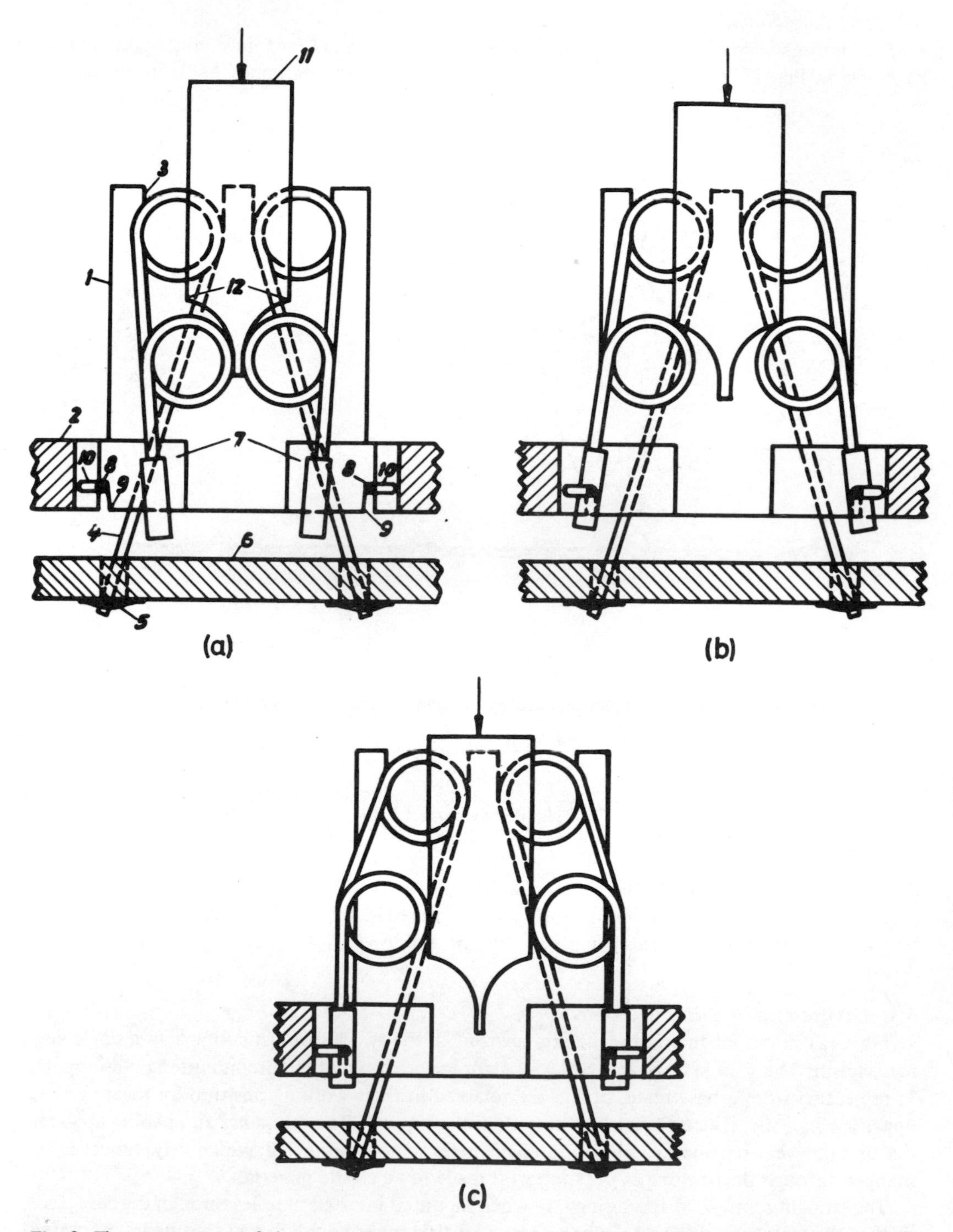

Fig. 3. The arrangement of the two individual contacts in a single key mechanism. The operation of the pushbutton is shown

slot, 7, in the base plate in such a way that they may move freely in a lateral direction when the spring is actuated. The contact bars, 8, are pressed into grooves, 9, in the base plate and cross the slots perpendicularly so that one contact indentation will be located at each end of the slot. See also Fig. 4.

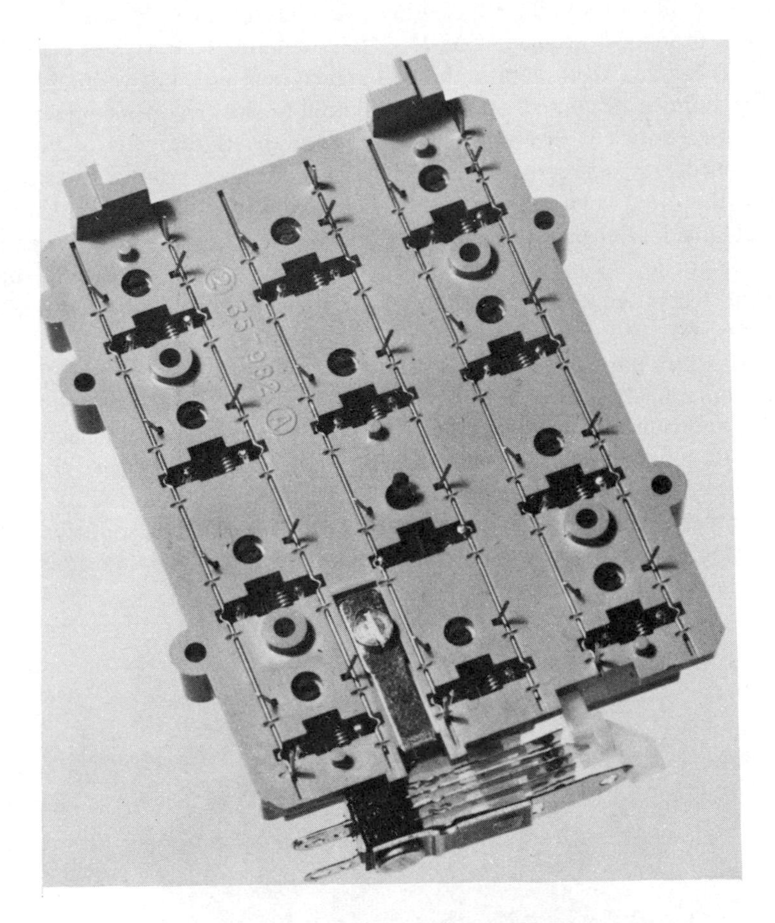

Fig. 4. The underside of a partially assembled key pad. The stationary and movable contact springs can be seen clearly.

3.4 *Actuation of contact springs*

The actuating member of the pushbutton is designed in the form of a wedge, 11, which in Fig. 3(a) is shown in a position that corresponds to the released position of the button. When the button is pressed, the wedge actuates the lower spring coils so that these are displaced laterally as seen in Fig. 3(b). Thereby the contact sleeves follow a circular path until they make contact with the stationary contact. A continued downward movement of the button increases the separation of the spring coils, which now become loaded due to the torque produced by the reaction force of the stationary contact against the sleeve [Fig. 3(c)]. This reaction force constitutes the contact force and attains its final value as soon as the button travel has proceeded so far that the spring coils press upon the parallel side surfaces of the wedge.

3.4.1 *Contact force and contact wipe*

As the contact force is built up by stressing of the spring coils a sliding movement called contact wipe takes place between the contact members. The contact wipe is of great importance for self clearing the contact surfaces from dust contaminations by operation. The contact force also contributes to the self clearing effect but less than the wipe.

This process is thoroughly studied by M. M. Attalla and Miss R. E. Cox of the Bell Telephone Laboratories[2]. These investigators have shown by analytical methods verified by experiments that the average number of contact operations required to clear an open contact due to foreign matter on a single contact is inversely proportional to the third power of the contact wipe, when the contact force is kept constant, while the contribution of contact force is but inversely proportional to the force when the wipe is held constant. It was thus shown that the change in cleaning rate obtained by changing the wipe by a factor of 2 can also be obtained by changing the force by a factor of 8. The experiments were carried out for a range of wipes between 0·06 and 0·25 mm at a constant force of 10 gf and for a set of forces between 5 and 80 gf at a constant wipe of 0·06 mm.

In practice it is not unusual that telephones are installed in dusty localities. The atmosphere may also be contaminated with corrosive gases or vapours that make electrical contacts deteriorate. The individual contacts of a key pad carrying weak alternating currents are particularly sensitive to such contaminations, which may interrupt the proper functioning. These contacts in the new LME key pad are therefore all gold plated.

In order to prevent excessive wear of the gold layers, a relatively moderate contact force of 12 gf on the single contact has been chosen. On the other hand, to secure a high clearing rate, the contact wipe is made as large as 0·3 mm.

Fig. 5. The underside of a pushbutton. The wedge for actuating the contact spring is at the rear.

3.4.2 *Contact bounce*

The disturbances originated in the rebounds of the individual contacts are eliminated as mentioned in 5.0.

4.0 PUSHBUTTONS

Fig. 5 shows the pushbutton seen from below. The wedge for actuation of the individual contact springs appears in the rear. In the front part the button has a rectangular raised portion that functions as a bottom stop. The shaft projecting from the bottom stop fits into a hole in the base plate, which hole serves as a guide for the shaft. A second guiding means is provided by one of the square holes in the cover plate through which the button protrudes. The cover plate is shown in Fig. 6 left. The top of the shaft has a ring shaped enlargement which may be snapped through the guide hole when assembling the button to the base plate. When this operation is carried out the helical restoring spring for the pushbutton, inserted in the hole adjacent to the shaft, becomes tensioned against the base plate. At the assembly, the shaft top ring functions as a provisional back stop to prevent the shaft from being pulled out of the guide hole by the spring force. Fig. 6 right shows the pushbuttons thus retained by the provisional back stop.

Fig. 6. The cover plate (left) and a partially assembled key pad (right).

5.0 COMMON SPRINGSET

The common springset is composed of four flat contact springs and one restoring spring. The purpose of the springset is to perform one break action and one make-before-break contact operation necessary for the proper functioning of the electronic part of the key set. There must also exist a correlation between the operation of the individual contacts and the ones of the common springs so that the individual contacts close before the make-before-break-operation of the common springset takes place, thus preventing the disturbances originated in the rebounds of the individual contacts to be transmitted.

Fig. 7. Linkage system for operating the common springset.

As is shown in Fig. 7 the common springset is located at the one end of the base plate. A displacement of any of the buttons is transferred to the springset via a linkage system consisting of four rods and one slide. The rods are supported by bearings in the rim of the base plate and have four levers each, three for co-operation with the buttons in one row and one operating lever perpendicular to the others for actuation of the slide. The slide has four rectangular holes, one for each operating lever and one cross-piece which functions as a back stop for the slide in the unoperated position.

Actuating one of the levers by pressing a pushbutton, the corresponding rod describes a rotary movement about its end bearing. This movement is translated by the operating lever into a linear movement of the slide, whereby the common springset is actuated.

When the pushbutton is released, the slide is pulled back to the unoperated position by the restoring spring incorporated into the common springset and is stopped in this position by the cross-piece against the rim of the base plate as illustrated in Fig. 7.

6.0 FORCE-DISPLACEMENT CHARACTERISTIC

When displaced the pushbutton is subject to a counteracting force caused by spring loads and frictional forces within the mechanism. This counteracting force has to be overcome by the actuating force supplied to the button top by the user's finger. Force versus displacement based on essentially static measurements may be graphically represented by a curve called force-displacement characteristic. Fig. 8 shows the force-displacement characteristic of the new LME key mechanism. The abscissa represents pushbutton travel through the maximum button displacement of 3·2 mm and back again. The ordinate shows the actuating force which is seen to have two values for each value of the displacement. Consequently the curve has two branches, the upper one representing the force on the downward movement and the lower one showing the force on the back stroke. The area between the curve branches represents the mechanical hysteresis or friction losses in the mechanism. The frictional force at any point of the button travel is one-half the difference between the two ordinates for this point.

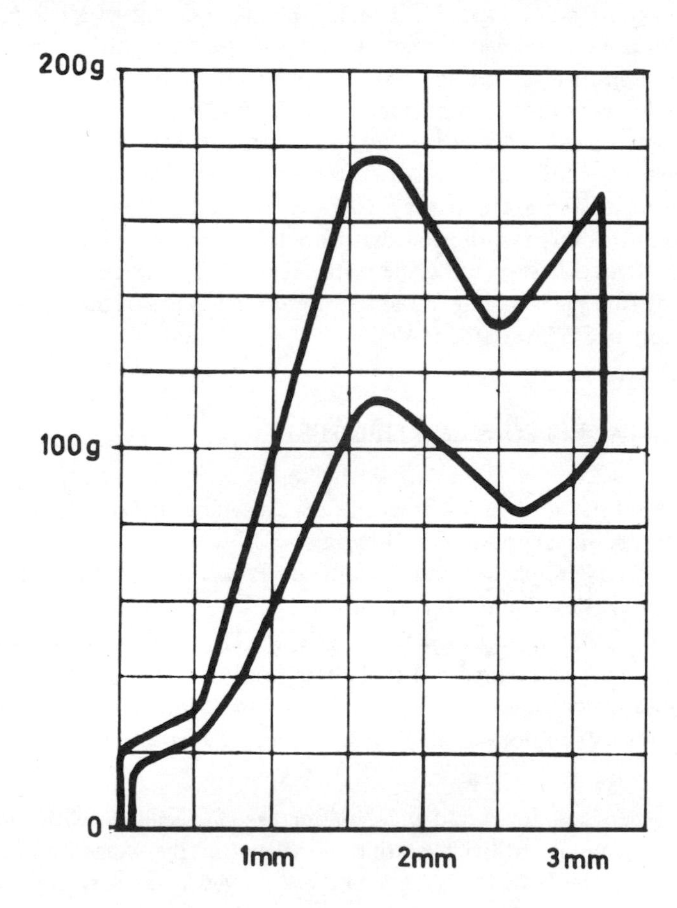

Fig. 8. Force-displacement characteristics of the LME key mechanism.

Following the upper curve branch it is seen how the force increases until the button travel has proceeded so far that the lower coils of the individual contact springs pass over the edges, 12, Fig. 3(a), of the button wedge and begin to slide on its parallel side surfaces [see Fig. 3(c)]. At this moment the upward-directed force component, acting on the pushbutton due to the load of the contact springs, suddenly disappears and the actuating force on the button top required for further depression drops down to about 130 gf. From this point the force increases with the displacement due to the tensioning of the common springset and the restoring spring of the pushbutton and attains a final value of about 170 gf when the button has completed its downward stroke.

The rapid change in actuating force causes the feeling of a slight snap action, which provides for a distinct bottoming and an agreeable touch of the pushbutton. Moreover, it was assumed that the snap action rapidly followed by a distinct bottoming would influence the user in the desired way to spontaneously retard the release of the button in its bottom position in order to obtain a somewhat longer contact duration.

In a subsequent investigation carried out under the management of Mr S. E. Magnusson[3] it was shown that this assumption was correct. Two pushbutton telephone sets called A1 and A2 respectively were subject to a special test to investigate how the contact duration was influenced by the force-displacement characteristic. A1 had a key set with an almost linearly increasing force-displacement characteristic while A2 was provided with the new LME key set. Fifty randomly selected LME employees used these two instruments with respect to number keying. Each individual was asked to key 15 listed seven-digit telephone numbers on each instrument. The result of this test showed that a contact duration of less then 40 ms was three to four times more frequent when using the instrument Al than when keying on A2. Furthermore, 75% of the persons participating in the test preferred the use of the instrument A2.

The assembled key pad is illustrated in Fig. 9.

7.0 ENVIRONMENTAL EFFECTS AND LIFE TEST

7.1 *General*

All products are subject to environmental stresses caused by climatic and mechanical factors during handling, storing, transportation and life span.

Knowing the actual environment and requirements as to function reliability one establishes the environmental properties required for the product.

To test the properties of a product it has to pass through a series of standardised environmental tests. Here it must be pointed out that these tests are often accelerated and the limits applied do not always correspond to the actual environment. Therefore the tests do not correspond to the actual conditions in which the product may have to work.

7.2 *Environment*

The key pad is designed for world-wide indoor use, in localities with inferior heat and moisture insulation without artificial heating anywhere in the range from cold to damp-tropical climate zones. The LME test programme has proved to be very good with respect to actual conditions encountered.

Where international norms are known, e.g. I.E.C. recommendations of standards, our test methods are based on these recommendations.

The quality of the key pad will be in accordance with the following tests and requirements:

Fig. 9. The assembled key pad.

Environmental tests

Temperature (I.E.C. 68–2–1, cold, I.E.C. 68–2–2, dry heat)

Temperature range for correct function	–10 to + 40 °C
Temperature range for function	–25 to + 60 °C
Temperature range for non-destruction	–40 to + 70 °C

Change of temperature (I.E.C. 68–2–14)

Correct function after exposure to temperature change of ± 15° C within ½ hour

Relative humidity (I.E.C. 68–2–3 56 days, long time exposure, and I.E.C. 68–2–4 6 cycles, accelerated test)

Correct function within the humidity range of 10–100% RH (more than 90% RH only during 60 days a year)

Air pollution

Salt atmosphere (A.S.T.M.: B 117–64, 2 days)

This test method is unreliable and is therefore performed as a parallel test together with a reference product.

Sand and dust test

Since contact reliability is one of the most important factors for the correct functioning of a key pad, a lot of attention has been paid to the design and testing of the contact itself and its functioning.

Together with the design of the new key pad we have developed a test equipment for dust (Fig. 10).

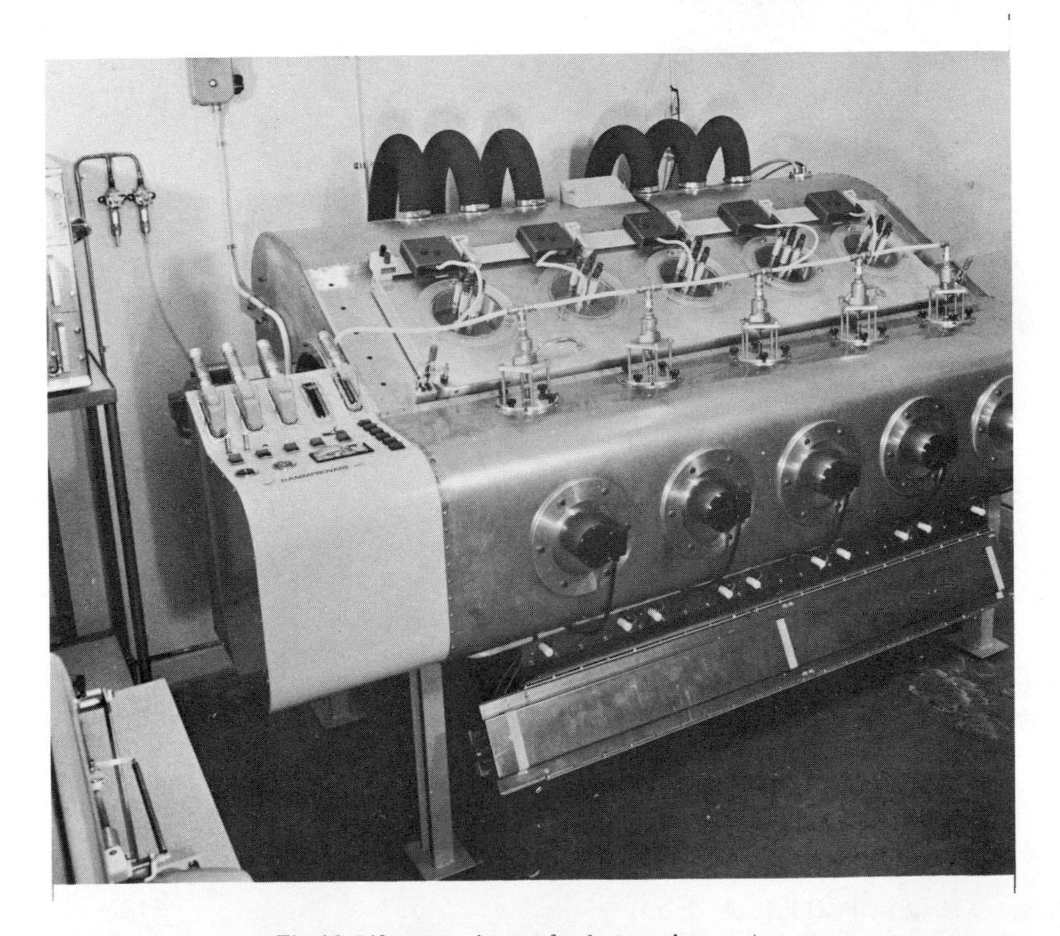

Fig. 10. Life test equipment for dusty environments.

Normally, standards recommend sand as a medium, since sand is the main contributor to wear of moving parts. Dust is more dangerous with respect to contact failures due to its tendency to absorb moisture thereby remaining on the contact.

For our test we use a mixture of sand and textile lint as a medium. The test is carried out in parallel with tests on a reference product.

Mould (I.E.C. 68–2–10)

Mould appears very frequently in humid climate

Drop test

The key pad placed in the actual instrument shall tolerate 800 mm drop test.

Vibration (I.E.C. 68–2–6)

The key pad placed in the actual instrument shall tolerate 10–55 Hz, movement amplitude 0·75 mm, 2 hours in each direction.

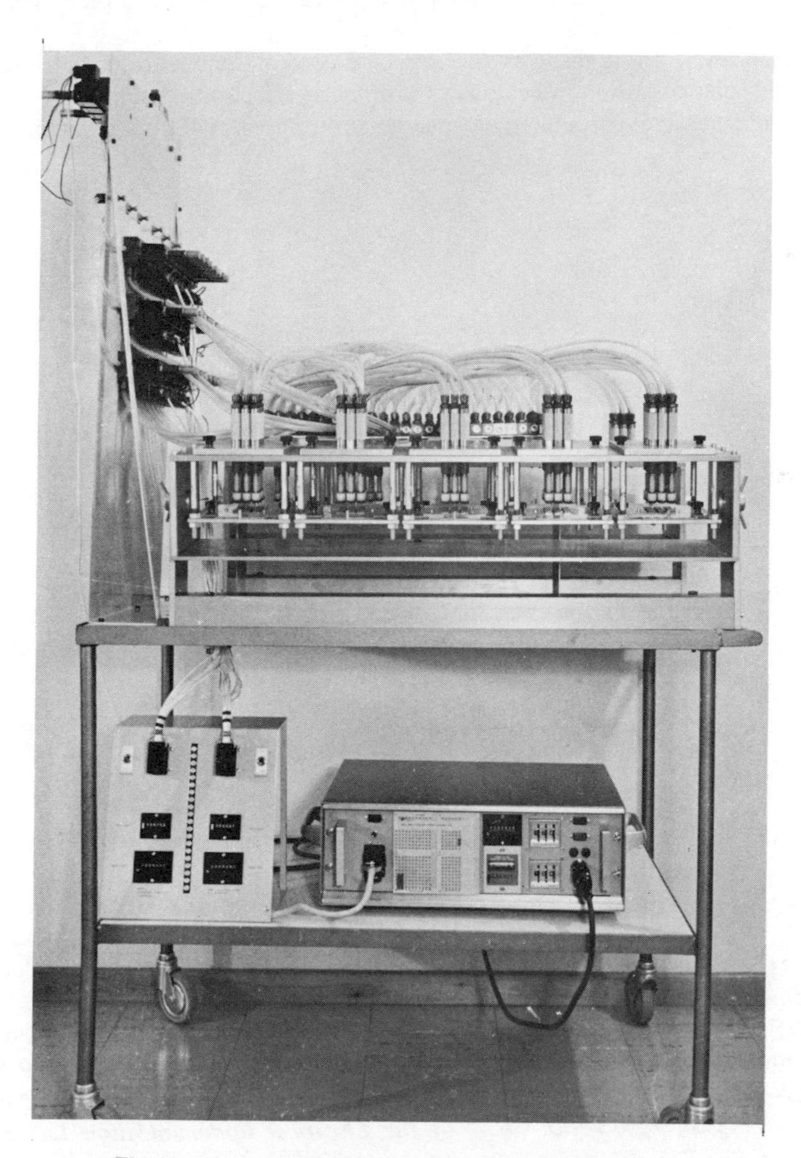

Fig. 11. Life test equipment for normal climatic conditions.

Bump test (I.E.C. 68–2–29)

6 × 1000 bumps with the acceleration of max. 30 *g*.

7.3 *Life test*

The key pad shall withstand 200 000 operations per key, corresponding to an approximate lifetime of 25 years under normal environmental conditions.

The test equipment allows the key pad to be tested in different environments. Fig. 11 shows the life test equipment for normal climatic conditions and Fig. 10 shows the life test equipment for dust environments.

It is important that the speed at which the keys are actuated is limited to prevent excessive acceleration forces in the mechanism.

According to an investigation made by the Nippon Electrical Communications Laboratory[4] the maximum attainable operation speed on keys with toggle effect is 800 mm/s.

The test equipment used in the actual case operates at the speed of 400–600 mm/s.

REFERENCES

1. DEININGER, R. L., 'Human Factors Engineering Studies of the Design and Use of Pushbutton Telephone Sets', *The Bell System Technical Journal,* 985 – 1012, Jul. 1960.
2. ATTALLA, M. M., COX, R. E., 'Theory of Open-Contact Performance of Twin Contacts', *The Bell System Technical Journal,* 1373 – 1386, Nov. 1954.
3. MAGNUSSON, S. E., 'Human Reaction to Push-button Sets with respect to Geometry and Pressure Characteristics', Telefonaktiebolaget L. M. Ericsson, ERGA Division, Stockholm,
4. MATSUZAKA, Y., TAKIMA, J., HASHIMOTO, S., 'Studies on Pushbutton Telephone Systems', *International report No. 3438 of the Electrical Communication Laboratory*, Japan.

INDUSTRIAL DESIGN IN THE CAPITAL EQUIPMENT INDUSTRY

J. H. Hilton
Mather & Platt Ltd, Manchester

SUMMARY

There are two popular misconceptions about industrial design. One is that it is an activity peculiar to this day and age. The other is that it is some kind of cosmetic process which is applied to 'tart-up' a product.

In fact, what we now mean by industrial design has been practised with varying degrees of aptitude since man first concerned himself with the visual and utilitarian aspects of his environment and sought to combine function and purpose with form and colour. Today the industrial designer does not seek to imitate the past by the use of contrived ornamentation. The aesthetic content of his work, particularly in the field of capital equipment, is a derivative of present-day technology and methods of production, and, like engineering design, his contribution must possess commercial validity.

This paper expands this theme and propounds a design philosophy which has been applied successfully to a wide range of engineering products.

1.0 THE BACKGROUND OF INDUSTRIAL DESIGN

Industrial design concerns itself with the environmental factors of design, within the meaning of that word in its broadest sense. In that role it has been accepted as a valid factor of design in those industries where ergonomic and visual attributes can be significant to sales. The industry which makes most use of the industrial designer, and has done for a considerable period of time, is the consumer goods industry. In this sector function, utility, and appearance must go hand in hand if the aim is to stay in business in a highly competitive market.

The same aim now applies in the field of capital equipment. In the world at large there are many more customers than ever there were who expect more for their money than a collection of working parts. Nowadays the discerning customer seeks a greater totality in design, and whilst he may not regard appearance, for example, as a primary reason for purchase, there is enough evidence in existence to prove that it is one of the factors that can influence choice. If one is to be honest, performance, price and delivery are the first criteria which must be met, but it is now bad business to ignore those factors which can complete the totality of a product to match to the mood of the market. Total design has become a 'must' in the capital equipment industry just as it is in the consumer goods industry.

2.0 AESTHETICS

Of the two disciplines associated with industrial design, i.e. ergonomics and aesthetics, perhaps the latter is least understood in the context of capital equipment. The part it can play, and the influence it can bring to bear on methods of manufacture and the reduction of manufacturing costs, is not always appreciated by those who have only taken a cursory glance at the subject. Further, the mention of aesthetics in the same breath as engineering design has been found sufficient to create antipathy which has only been dispelled by proof positive as to its commercial usefulness. Proof of the required kind has accumulated of which some examples serve to illustrate this paper. The aesthetic content within the design of these machines will be seen to differ radically from the classical forms of visual design which a traditional upbringing and a national background has subliminally imposed and sustained. The replacement, for instance, of a curved line by a straight line in the configuration of a product is a piece of aesthetic field-play where the whistle would have gone for a foul a decade ago. Tortuous field tactics expend energy, likewise complex shapes in the design of a machine expend cash.

Every tangible object which man has the sensitivity to perceive has two fundamental characteristics, function and form. From the days when he first changed flesh for fur there is evidence of the glimmerings of an aesthetic sensibility which superimposed itself, if not intentionally, then instinctively, on the articles he produced to satisfy the needs of his environment. Time and experience taught him how to develop an aesthetic theme and to formulate rules for its use. The popular example of his first attempts to apply his singular gift of aesthetic sensibility are the cave drawings which communicate and record his activities. Later he embellished his civic and domestic architecture, his pots and pans, his chariots of war and his artifacts of peace. This he continued to do in very much the same way, using the same idioms of ornamentation on an ever increasing scale of elaboration, until comparatively recent times. In so doing he has left striking memorials to his periods of elegant sophistication and some interesting examples of his orgies in the field of ornamentation. Too few of the former remain. Maybe too many of the latter survived their day to confuse and detract from the best of the past.

3.0 INFLUENCE OF MASS PRODUCTION

Prior to the industrial revolution the architect and the designer/craftsmen were the arbiters of visual taste in the manufacturing industries of that time. The objects which they produced, if their creative stature was significant, were notable for vitality of form, elegant interpretation of classical modes of design, and for their sheer practical utility. In the field of architecture the creative giants, Wren, Inigo Jones and John Webb applied a genius unequalled except, perhaps, by Ictinus, who in 432 B.C. designed and built the Parthenon. The advent of mechanical power on a scale previously unavailable made the process of replication possible and allowed a greater number of people the opportunity to possess articles which heretofore had been available on an individual and not a collective basis. Many designer/craftsmen, seeing, no doubt, an easier and quicker road to prosperity, exploited the art of duplication and established themselves as manufacturers. In so doing they remained loyal to the tenets of their original calling and continued to give of their best both in the functional and aesthetic design of their products. The articles produced at this time show a high regard for quality and craftsmanship, and whilst some of them may be regarded as curios in this day and age, they possess a design integrity nothing short of admirable. The deterioration in refinement of form which later infiltrated into the mass-produced articles of the mid and late Victorian period was on the whole absent. But it was not long before the technical problems associated with the desire to reproduce, in cast metal, designs which had originally been conceived for timber or stone, became evident and had to be resolved. The problems were indeed solved, generally by modifications of form and proportion, and the result was often a grotesque sham of the original design. The Great Exhibition of 1851 housed many typical examples of this approach as applied to a range of products both domestic and capital. It is unfair, however, to be too unkind in the criticism of this period of design for it can be said that the concept of total design was accepted with a conscientious understanding not only from a commercial but also from an ethical aspect, which countenanced pride in workmanship and skill in its execution.

The styles of design following the Victorian period were experiments in aesthetic taste entered into with a spirit of adventurous revolt against classicism. Many of them emanated from the continent of Europe. But in Britain, ever the stronghold of conservatism in all things, little impact was made capable of rocking the foundations of in-bred preferences for Graeco-Roman and Gothic styles. 'Art Nouveau' came and went, but not without leaving its mark, particularly on furniture and textile goods. It was also at this time that the designers of capital equipment began to shed the residual pieces of architecturally inspired decoration which had adorned this category of machinery since the advent of the beam engine. The configuration of capital equipment such as textile machinery, machine tools, and other machinery having the same principles of mechanical design and manufacture, assumed an appearance which was to be retained, with little modification, until the middle of the twentieth century.

4.0 THE INDUSTRIAL DESIGN CHALLENGE IN CAPITAL GOODS INDUSTRIES

Came the end of World War II and the prospect of lush markets in a brave new world. One attitude which tempered the action of some in their efforts to satisfy demand was centred on the belief that all that was required was to pick up the threads and begin again where pre-war marketing had left off. The idea that post-war markets would be avid for this type of revival soon proved illusory. British capital equipment became associated with such descriptions as 'old-fashioned', and 'comparatively expensive'. The word 'comparatively' used in this context meant that machinery was expensive compared with that of Continental origin, the sale of which was being conducted on the lines of a commercial blitzkrieg in markets accepted as a traditional British monopoly, i.e. the Empire. These machines were notable for their 'new-look' as well as for technical characteristics and their other sales attributes – delivery, price, availability of spares etc.

The fact which was clear was that aesthetics had been formally wedded to engineering design and the prime purpose of this partnership was to improve the saleability of the product. And it proved to be good business.

The same kind of match in Britain was more of a shot-gun affair. There were, however, some companies which saw the commercial potential of embodying in their designs improved control and maintenance facilities, more up-to-date appearance, and colour schemes a little more imaginative than black, bottle green, or battleship grey. There were others who tried a different tactic to maintain a competitive degree of sales initiative. The basis of this was to festoon an existing design with the latest technological components rather as one hangs gifts on a Christmas tree. Price and performance being equal the customer bought the tidier looking, better arranged model.

It was at this juncture that the 'industrial designer' entered into the field of capital equipment design. With very few exceptions he came from the consumer goods industry or the architect's office. He was a 'stylist' dealing in the art of creating, and co-ordinating, modes of visual design at a far more subjective level than is thought to be appropriate today. His masterpieces were to be found in the motor-car industry – particularly in the United States – and in a range of domestic appliances, furniture and the like. There was a fetish for 'streamlining' and polished chromium plating, and the tendency, at first, was to apply these idioms to capital equipment as the be-all and end-all of up-to-date machine appearance. This was the approach that led to the practice of hiding the working parts of a machine under sheet iron cladding of the removable – and not so removable – type, and the use of chrome trims. It was the era when appearance was added as a contrivance to sugar-coat the pill of utility.

The industrial designer today shuns the superficial cosmetic approach and seeks to bring into his part of product design a professional contribution that means much more than a face-lift. He tackles his job by looking into the function and purpose of a machine, studies its relationship with the people who will build, operate and maintain it, investigates the possibilities of new materials and methods of manufacture. This he does as a member of a design team, not as some kind of artistic hermit living apart from the realities of cost and compromise. If he is worth his salt his contribution will be seen in the achievement of *total design* in terms of the combination of function, ergonomics, and up-to-date appearance, with the two latter factors built into the machine, not added as expensive afterthoughts.

5.0 INDUSTRIAL DESIGN IN PRACTICE

The thesis which the foregoing submits is that the base on which industrial design practice must stand today is one of full integration with the engineering disciplines associated with product design and manufacture. This can be illustrated by a simple diagram thus:

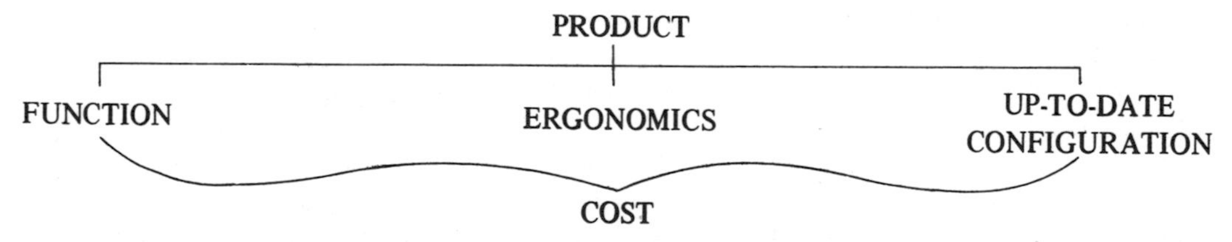

Here the concept postulated is in fact the theme recurrent in this paper, i.e. that if a product is to be significantly commercial its design must not only satisfy the demands of function, it must also incorporate the best possible ergonomic and aesthetic features, and in aiming to achieve this integration in a practical way it is imperative to link the endeavour to the controlling parameter – COST.

FUNCTION in the above context means all the engineering disciplines which contribute to the satisfactory fulfilment of the technical specification.

ERGONOMICS means those features which must be incorporated so that the machine can be built, operated, and maintained as conveniently and safely as possible.

UP-TO-DATE configuration means the utilisation of appropriate aesthetic concepts which will provide this attribute as a reciprocal of practical design and production techniques.

The element of compromise has to be accommodated in design activity and the ruling factor is usually one of cost. It follows, therefore, that in the design of many types of capital equipment most of the money available has to be spent on engineering essentials leaving little, if any, for flights of fancy in the sphere of configurative design. Far from being restrictive, cost limitation can be helpful. It can provide that measure of restraint which leads to better and more valid aesthetic solutions, and can at the same time ensure that a real contribution to cost reduction is made.

6.0 CASE STUDIES IN 'TOTAL DESIGN'

This, then, is the backbone of today's concept of *total design* as the author of this paper sees it. The industrial design unit which he leads is engaged in the day-to-day practice of this philosophy, the implementation of which is planned to achieve the following aims:

(1) *Simplification* of mechanical design to reduce manufacturing costs.
(2) *Standardisation* of as many parts as possible within each category of equipment.
(3) *Up-to-date configuration* based on simple geometric shapes derived directly from the aesthetic possibilities of engineering materials and methods of manufacture.

The various machines which have undergone redesign according to these aims are notable for their simplicity of form when compared with the machines they replace. Ergonomic features have also been improved, and the cost factor, which in machines of traditional construction is more likely to increase than decrease, has been curbed.

The configurations which have emerged are totally different from traditional designs, as reference to the 'before and after' examples illustrated here will show. (Figs. 1 to 6.)

The industrial design unit, mentioned earlier works in close association with the engineers concerned in a design project and its staff are accepted members of the design team. The unit is expected to apply its creative capacity in a more worthwhile manner than merely 'tarting-up' a design. It is, therefore, very much engineering-orientated, and believes that industrial design in the capital equipment industry is an engineering discipline, and that engineering design has its own aesthetic which is now far removed from the classical forms associated with artifacts of the past. In carrying out its work the unit is called in at the early stages of design. This is vital if full use is to be made of its services. The industrial designer must be involved from the beginning right through to the end of a project and it is also important to properly co-ordinate his part in the programme. This is best done within the terms of a written brief which emphasises the time factors relevant to the overall time scale for the project. The techniques used to communicate the ideas which the unit submits can take the form of perspective drawings, scale models, or full-scale mock-ups. The latter are most favoured and made use of since they are capable of conveying the right impression and relationships ergonomically and visually.

The configurations which the unit promotes are derivatives of primary geometric forms. Their visual characteristic is rectilinear as opposed to curvilinear, the latter being far more expensive to produce, the former having been found to be more economical, particularly in the case of non-mass-producers where batch sizes may range from one-off to say 50-off.

Over the past five years the unit has helped to apply the foregoing design philosophy to textile finishing machinery, electric motors, food processing machinery, package boilers and other items of capital equipment. In every project the industrial designer has been part of the design team and has been able to prove that industrial design alongside engineering design can play its part in the saleability and profitability of a product, and this is an important contribution, for without the achievement of these two aims there can be no prosperity in any commercial enterprise.

Fig. 1. A flameproof motor with a cast-iron casing of traditional design. (*Mather & Platt Ltd.*)

Fig. 2. The same motor as seen in Fig. 1 after redesign. The casing is built on the rectilinear concept and is of welded steel construction – saving in labour costs and a marked reduction in total weight were achieved. The price was reduced by 20%.
(Mather & Platt Ltd.) (Photograph by courtesy of the Council of Industrial Design.)

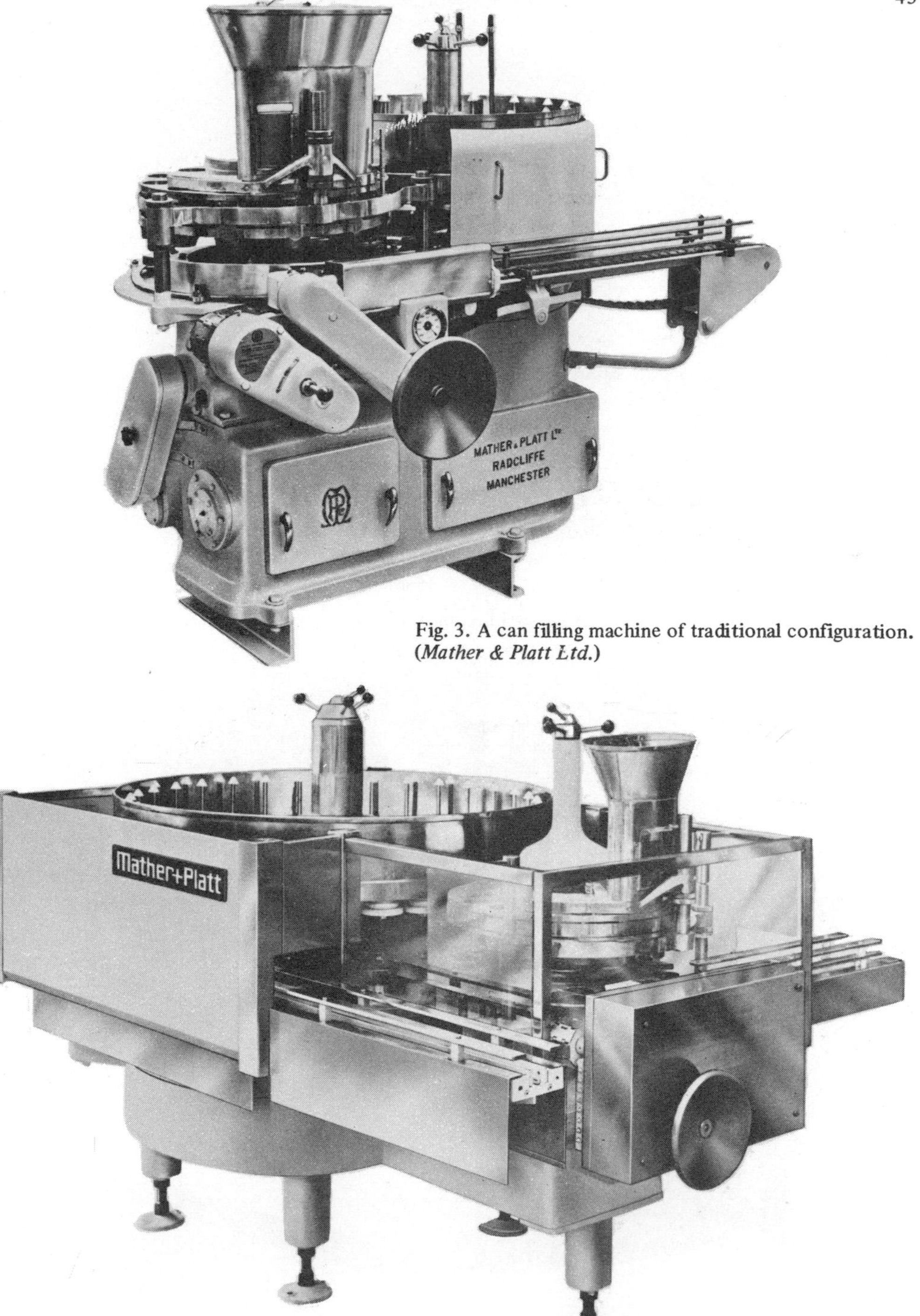

Fig. 3. A can filling machine of traditional configuration. (*Mather & Platt Ltd.*)

Fig. 4. A can filling machine after redesign. The main structures have been simplified and 'clean-lined'. Ease of cleaning and accessibility have been improved. (*Mather & Platt Ltd.*)

Fig. 5. A package boiler of conventional design. (*Joseph Adamson & Co. Ltd, Hyde.*)

Fig. 6. A package boiler with an up-to-date configuration. The structure has been simplified and the rectilinear casing also serves as a heat insulator.
(*Joseph Adamson & Co. Ltd, Hyde.*)

DEVELOPING AND MARKETING INNOVATION – A CASE HISTORY IN THE MAKING

L. S. Barrus
Prosper

SUMMARY

For any worthwhile new product concept to be of value, there must be linked to the concept the development of a sound and economic design, an effective marketing arrangement, adequate finance for development, and an efficient management of all these resources. The author of this paper is a director of a company whose purpose is to forge together these necessary links in the chain of successful development and marketing of innovation. Case histories are cited covering the company's own new products and inventions, those of individual inventors, and those of other companies. Examples are also given of successful and unsuccessful development, some of them still in the balance. An analysis is attempted of the prime causes of success or failure. Suggestions are made as to how, in an inherently high-risk field of endeavour, the chances of success can be improved and failures assessed before they become costly.

When Professor Ross approached me concerning this discussion he said in effect 'You have formed a company, the purpose of which appears to be to bring new products to the market place. Tell us how you do this.'

We will do our best.

You start with some essential beliefs – and with some essential limitations, determined by the skills and finance available. In our case these beliefs include the following:

- We believe that with respect to design, the simplest solution to a given problem is generally the best.
- And we wish to the limit of our ability to produce designs for products which have functional excellence. Or in other words, which work well.
- We believe, with respect to people, that it is a proper function for management to do all it can to bring to all employees the satisfaction which comes from working at the limit of their innate and acquired skills.
- We believe that, not only is the quality of a product important, but equally important is the quality of the life of the people producing that product. We do what we can, therefore, to heed this in the design of the products we work on.
- We believe that, while in our society to-day the trend is to ever-larger enterprises, this also increases opportunities for smaller companies to provide personalized or selective services or products with an efficiency not achievable by the large corporation. We try to keep our eyes open for 'niches' left uncovered by the big boys.

- We believe that successful innovation results from the intensely *personal* involvement of an individual. Central to it is the *will* to succeed. Such a drive can only be generated and sustained in the right working conditions. And these involve *no* politics. But man is a political creature. We argue, therefore, that innovation can flourish only in small groups. If these small groups are part of large organizations, then the administrator must act as a lightning conductor. His major function must be to earth political currents.
- We believe that worthwhile innovation is to-day, as much as it ever was, often the product of individuals or very small groups of individuals, working with very limited technical and financial facilities. It is our observation, however, that in to-day's society the dice are increasingly loaded against the lone innovator ever getting his product to the market place. We see activity in this area as a particular niche which we hope to have the wit to fill.
- We have observed that daily production problems can clog up the most creative mind. We resolved, therefore, to organize to avoid this. We decided that as much as possible development and manufacturing would be geographically separated.
- We also try to use the machinery and production facilities of others as much as possible; restricting ourselves to fitting, assembling and testing. This policy we felt would have the additional merit of leaving us entirely free to licence any product we happened to be producing, if this seemed desirable – we would have no fixed investment dictating a different policy.
- We believe that creative engineering is one skill, and creative management another. We believe both skills can be exercised in partnership, with the individuals concerned enjoying comparable status and remuneration.
- Technically we feel quite deeply, that where it is feasible to form rather than cut metal, and other materials, this generally makes sense. It makes sense often in cutting costs, and improving quality. But also such developments help in their small way to preserve our limited mineral resources, and to avoid, not only waste, but the problems connected with the disposal of waste. So, while by no means intending to be exclusively involved in what has come to be known as 'chipless machining', we do wish to become quite deeply involved.

1.0 ORGANISATION OF ACTIVITIES

These then were the beliefs which, combined with limited financial resources, and our available skills, dictated the form of our organization, and our areas of activity. Over the two years plus of our existence the organization has come to look like this. (See Fig. 1.)

Our activities are classifiable in four segments:

(1) Metal forming developments.
(2) Development of products of our own invention other than metal forming.
(3) Product and process consultancy to industry.
(4) Technical, marketing, and management assistance to individual innovators.

1.1 *Metal forming development*

This has been our most ambitious development. We have designed and built a *production prototype* form-rolling machine.

This has a three-roll configuration – the only one at present offered in this country – and the combination of machine design and roll design constitutes a novel, and we believe patentable, approach. The prime objective of this innovation is to accomplish metal deformation with much lower pressures than those required by conventional equipment. For both technical and

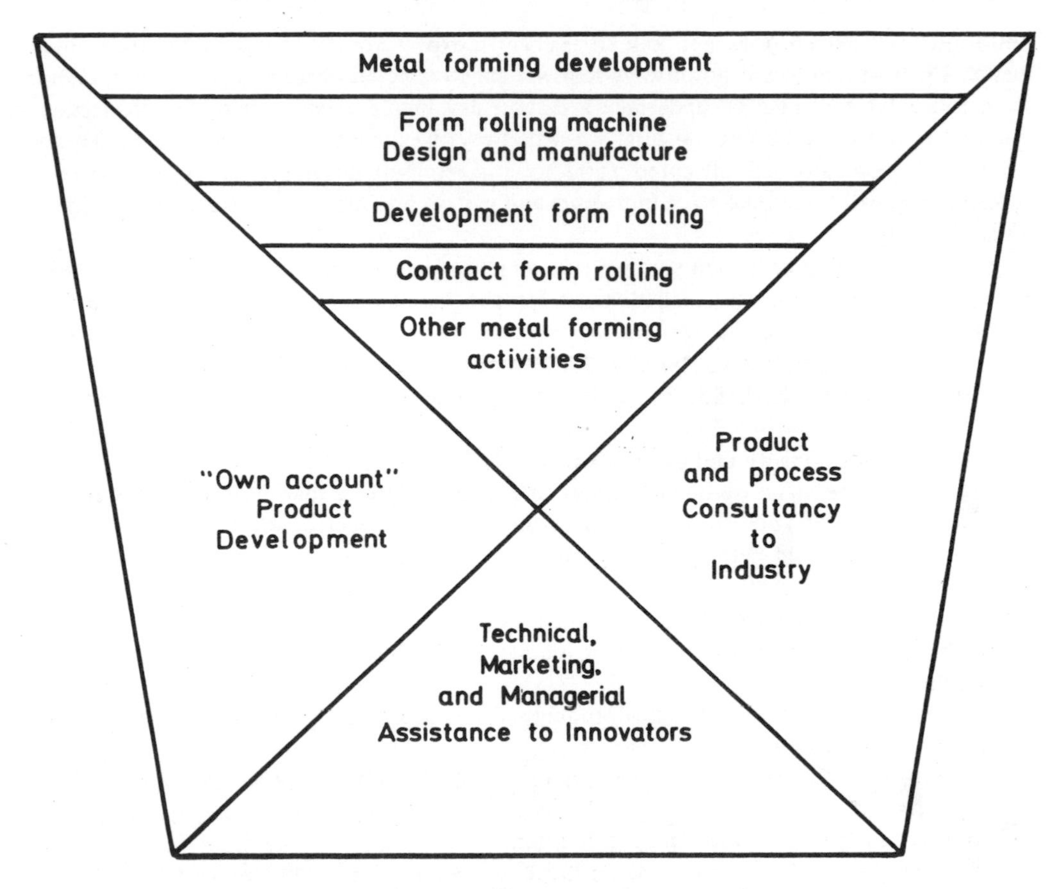

Fig. 1. Organisation of Prosper.

financial reasons, our work in this area is then divided as follows:

- The design and manufacture of purpose-built form-rolling machines.
- We undertake contract form rolling. This gives us needed production experience, and it also helps us to generate some income. Also it enables us to get a client off the ground on a project before he receives his own machine.
- We do development work for clients on more far-out projects. These have included very-fine-pitch finned tube, ball bearing inner race rolling, ball screw rolling – to cite three.
- In addition to the form rolling machine, we have designed other metal forming machines. One of these substitutes a hydraulic forming operation for a conventional cutting operation. Another substitutes metal forming for a welding operation.

All of the above products or services are marketed through two agents in this country, and overseas by direct contact.

1.2 *Development of products of our own invention other than metal forming*

In the category of new product development outside of the metal forming area, we have designed and developed a damping mechanism such as might be usable on a door closer, or a similar item.

The damping medium in this case is not oil, but another substance of great stability and uniformity and which does not leak so easily. We expect always to have some such developments. Creative minds will not always follow a single selected channel; and where it does not seem too great a gamble to undertake a certain development, and where the chances of an eventual return do not seem too slim, we will always be inclined to have a bash. In this case, a major U.K. company has allocated £500 for further joint development. If this meets with success, they will undertake to manufacture and market a jointly developed product, paying us a small royalty.

Also in this category, it will surprise you, almost as much as it did ourselves, to find that we have established a separate small division for the manufacture of a 'family' of hand-made dressed teddy bears. Their design was the work of the accomplished wife of my associate and fellow director, originally undertaken with no commercial thoughts in mind. Later, subjected to value analysis by her husband, the teddy bears became a production item following the logic that.

- They meet our quality and design standards.
- They could be made by older ladies of considerable skills in their homes, in Stewarton and district, thus providing for them a worthwhile activity, as well as additional income.
- We felt they would help provide a modest income base, and thus assist in the financing of our longer-range developments.
- Technically it provided us with first-hand experience in the press cutting of cloth. This is a technique little used in the garment industry, but which we believe to have considerable potential.
- In line with our philosophy, this production operation is geographically separate from our development centre.
- Finally, and incidentally, teddy bears have become small but growing $ earners.

1.3 *Product and process consultancy to industry*

That segment of our business which has to do with product and process consultancy to industry, has proven a most interesting development. 'Hardware' – as distinct from 'software' – consultancy would appear to be almost unknown in the U.K. in the mechanical engineering field. We believe that we have found yet another niche in offering this consultancy, and that this activity is destined to grow. We have only been at this particular function for about a year, and because it is both a new concept and one proposed by a new and unknown company, it has taken a while to get established.

- We find that the 'hardware' consultant can function as a catalyst. By challenge and query he forces more aggressive thought by the client's own personnel. Action and problem solution is as often instigated by them as it is by us. Either way the client benefits.
- In performing this work we like initially to spend perhaps a week in the client's plant working on a few selected, specific problem areas. This provides an intensive introduction to the whole 'feel' of the plant situation. And this is invaluable. One cannot divorce a product or process development from the whole environment of product, personnel, policy and philosophy which is peculiar to each plant.
- For much the same reason, we also like to 'stay with' a job after it is apparently finished.
- We are free to admit that we tend to adopt a rather personal approach to this work. We do this because we *like* to work this way. But it would not surprise me to find that this approach may also be pretty much a 'condition' for good creative engineering consultancy.
- Another interesting aspect of this work is the application of what may be called 'Transfer

Technology'. For example, we have found developments originally undertaken in the garment industry to be applicable in the manufacture of confectionery. We have found press and die design experience in the metal-working field applicable in the garment industry. We have worked on the design of car-safety belts, and found that press and die design knowledge proved helpful.

- Fascinating marketing problems are posed by this activity. The creative mind at work is not by nature commercially oriented. It is all too easy to give a would-be client sufficient information in the course of enquiry and discussion – and never hear from him again. The curtain comes down and one never really knows why. And how do you set a proper value on consultancy aimed at modifying a client's product design or one or other of his processes? The results are frequently difficult to forecast in concrete terms. You know that they are there, but how do you quantify them accurately? There are no 'set' answers. We feel the best approach is to earn sufficient confidence from the client for an annual retainer to be agreed. The consultant is then completely free to give of his best.
- Finally, we like to do a certain amount of this work because it keeps us on our toes, and in touch with new products and processes in a wide spectrum of industry. It can help us avoid inbreeding. We believe there are many companies who cannot justify their own full-blown design development teams, who can benefit from such a product and process engineering consultancy service.

1.4 *Technical, marketing, and management assistance to individual innovators*

Last, but far from least, we come to that segment of our work which has to do with offering assistance to creative individuals, who have developed to one stage or another a concept or a product which we feel is worthwhile, and which we believe it may be within our ability to help bring to the market place.

The great problem is that the finest new design of mouse-trap – or any other product – is valueless until the product is brought to the market place. And to do this one has to link together:

- A working prototype.
- A production prototype which is low-cost and works.
- An effective production facility.
- An effective marketing facility.
- The finance and management necessary to ensure that all of the above work harmoniously together. The lack of any one of these elements means total failure.

Our endeavour is to provide for the innovator whichever of the above elements he cannot provide himself.

This then is one more 'niche'. Most industrial concerns prefer to take on new products only when they have been perfected and, preferably, marketed. Conventional wisdom in larger companies to-day is to acquire new products, either through internal development or through acquisition of a company or new product line at the point when it is already started on the fast upward slope of the classic 'S' growth curve. So the lone inventor has a rough time getting even the most worthwhile innovation to the market.

2.0 ASSISTING THE INVENTOR

If the inventor, unsuccessful with his approach to industry, goes to Government agencies such as the D.T.I. or N.R.D.C., he will find that they are quite frank to admit that they generally are *not* set up to help. They tend to work with the larger established industrial concerns. Agencies such as T.D.C. or the merchant banks similarly, while having funds available, have no means of providing the other links in the chain. So many worthwhile innovations never do get to the market place. How do we go about trying to fill this niche? To enumerate the *less* obvious items.

2.1 *First we look at the invention*

- Is it the simplest way we can conceive to achieve a given purpose?
- Is the innovation *primarily* the result of a clever and creative mind at work? Or is it primarily an effort to meet a serious need felt by a significant and identifiable group of customers? We are wary of the former.

2.2 *Then we look at the inventor himself*

- We permit ourselves the luxury of working only with people we find 'sympathique'. I will not attempt a more scientific description. We suspect that this may be commercially justifiable, as well as personally more enjoyable. I must say that this is not a very fine screen. We tend to find most innovators interesting. Often fascinating people. Amongst our friends we count a university lecturer, a Free Church minister, a former circus performer, and a London-based Ghanaian who has now founded a company in Ghana to make the food processing machine we jointly developed.
- We find out early whether the inventor has what we consider a realistic idea of the worth of his invention. Some do not: they underrate all that is involved between the concept and the final marketable product.
- We have come to the belief that the innovator who seeks our aid should make some financial contribution to the work involved, even though this may in some cases be small, and our reward might come only in the event of success. This precept is partly a screening process, to test the degree of confidence the innovator may have in our ability to help. Partly it is economic necessity. We just cannot afford too rich a mixture of projects, however worthwhile, for which the work and cost involved is *now* but the pay-off is in the distant future. This is our single greatest problem at this time. The affluent innovator is a rare bird indeed.

2.3 *Next we decide strategy*

Theoretically the best policy is often to develop a new product, to market it in a limited way, perhaps in a limited area, and then, once it is established, seek to licence it to a larger company which has the finance and the marketing organization to do it justice. In practice this is often not possible. The economics of volume can make it impossible to market a new product competitively.

Ideally a company such as ours should have several satellite production and marketing companies covering various product groups. We are just not that far along to date. Our approach generally, therefore, is to make and market if we feel we can successfully, and then attempt to find a licensee from a position of comparative strength. If this is not possible, we tend to go for a provisional U.K. patent, and then try over the next 12 months to find a manufacturer who is at least willing to undertake a joint development, with a promise of a royalty arrangement, given success.

3.0 CASE HISTORIES

I will cite two cases by way of illustrating the dynamics involved.

3.1 *A food processing machine*

The first of these has to do with a machine developed to produce a food which is the staple diet of several West African nations. The basic foods used are yams or plantanes of many varieties. These are cut into chunks and boiled and are then traditionally pounded with a mortar and pestle by a team of three for up to an hour.

This must be done twice daily, since the resultant food, which goes by the name of Fu Fu, does not last in the tropical conditions. A young Ghanaian some years ago had the concept of mechanizing this operation. Obviously the hours saved would be of national significance in a country where everyone eats Fu Fu twice a day. Of equal importance, however, was felt to be the question of hygiene. This young man, not an engineer, spent several years and several thousands of pounds, attempting to prototype a suitable machine. He had the concept, but not a functionally effective machine. When approached we undertook, for a fee, to develop a properly functioning production machine and to provide him with all cost and supplier information for the purchase and manufacture of the required components, if he was able to obtain the necessary backing.

He returned to his native land, obtained the backing, and went ahead, following exchange approval, with the development. We are also providing him with a modest amount of what is essentially management consultancy and advice. This product is in its final development stages at the time of writing. We feel that to date it has been correctly organized. Needless to say, the risk involved in such an undertaking is of a very high order. The innovator, however, we judged to be a man of high intelligence and proven determination. He, in turn, is counting on the integrity of our design and development of a simple, but robust machine. Unquestionably, there will be many problems to be overcome, but we feel that the essential elements of success are present.

3.2 *A copy milling attachment*

The other case has to do with a copy milling attachment. Briefly, the concept involves a wholly independent copying device able to be mounted on any vertical milling machine and capable of fully automatic two-axis operation. The inventor, in this case, had spent many thousands of pounds, much in international patent fees, and some five years, in development to date. At the time he was referred to us he had pretty well run out of finance, and had had no success in finding an established machine tool company interested in taking on the product at that stage.

In this case, the inventor knows a great deal more about hydraulics – the discipline primarily involved – than we do. Our assistance, therefore, is less technical. Rather is it to provide a location for him to complete his development, a modest amount of loan funds and an undertaking either to make and market the machine ourselves, or to find a licensee. The machine at the time of writing is not yet perfected. We still believe it stands a good chance of being so. We like the essential simplicity of the design.

A necessarily limited analysis of the market suggests to us that there is a niche in the small tool and die shop for a machine such as this, which should be able to be sold for around £1000. The lesson one learns with a relatively sophisticated machine such as this is that it can take many years to bring the device successfully to the market place.

4.0 CONCLUSIONS

Through working with these projects and with other men and women we find an initially unthought-of bonus. Their talents become available to us. In helping them, we strengthen our own organization.

There is now emerging a synthesis of our activities. On the one hand we have alert companies looking for new products, on the other we have a growing band of 'associates' as well as our own design staff. The possibilities are as thrilling as they are obvious. Can we not tie-in inventive minds with felt market needs – and provide the prototyping facilities and engineering skills to make a theoretically interesting concept a practical, commercially viable, reality?

It's going to be great fun trying!

SESSION III

Innovation Through Materials Science

Chairman: Professor C. Timms
Head of Machine Tool Branch,
Department of Trade and Industry,
Visiting Professor,
Mechanical Engineering Group,
University of Strathclyde

INNOVATION IN MATERIALS

H. B. Locke,
National Research Development Corporation
London

SUMMARY

Following a brief history of the evolution of product design in relation to properties of materials, some recent developments in materials, including composites, are considered together with potential applications of these in manufacturing. Some implications for management in the use of new materials, are indicated. A technique for selecting a specific material for a given application precedes a brief discussion on cost implications in introducing new materials.

1.0 INTRODUCTION

Materials are either taken-for-granted–or expected to do the impossible. Scientific interest and the needs of the more difficult design problems led to the concept of the interdisciplinary 'Materials Science'. Today there are new materials with improved values in almost any desired property, and this makes us think again from scratch because the balance of properties in a new material will usually be different from that to which we have become accustomed both in design and in use. In addition, new production techniques are often required, and high property composites even require the material itself to be designed. When physical design is at its limit, the only way of extending performance of either machines or structures, is by using materials developments.

2.0 MATERIALS AND PROPERTIES

2.1 *The past*

Originally man presumably made things somehow, and adjusted shapes by evolution so that required functions are fulfilled. If something broke it was made thicker next time; if it did not, then urge to economy tended to thin it down until it did break–and then the limit would be known to have been reached. This way throughout history, walls of buildings for example, could become thinner in relation to their height, and window areas could increase. New techniques of design and of fabrication were continuously developed for materials that did not themselves change greatly: the various components of materials-using evolution served performance and economy with ever increasing efficiency.

In buildings this is illustrated by the progression from Greek temple (stonework without cement) via Roman villa (stonework and also brick, with cement) and then mediaeval churches, to the office block or cross-wall estate-type housing of today. Of structural materials brick is interesting in having been developed early on in technological history, so that there was little opportunity for further improvement until recently. Also bricks needed cement, or mortar, for buildings of any size; and a bonded stone or brick structure is itself a sort of composite in which the joints can 'give' when required. Roman buildings withstood earthquakes better than those of the Greeks partly because of this (and partly because their designs as they evolved, precluded the swaying of heavy walls). Stone was used without improvement, but usually cut according to accumulated experience along the most satisfactory planes: timber though carefully grown to the sorts of shapes needed for ship construction, was nevertheless limited to the properties imparted by careful seasoning: and metals must have had very variable performance except for the highly specialist products of a very few master craftsmen. Of non-structural materials, textiles were mainly linen, wool and other animal fibres (though silk was highly developed early on); the only plastics were the naturally occurring leather, parchment and horn; and glass was really a speciality, as also such materials as papyrus, tree barks etc.

A more recent example of design evolution is the motor car where the space and comfort per passenger as well as performance, safety and availability of gadgetry are usually far greater in today's (UK) cars than in their larger and heavier predecessors. In this case there is much evidence of the continual incorporation of materials improvements as soon as they became economic.

2.2 *Properties and design*

Strength is the property that usually comes first to mind, but it alone does not determine mechanical design. Distortion under load can be equally important. Elastic modulus is in fact a specific component of basic beam and shaft design theory, although it tends to become hidden within the methods, techniques and parameters used in everyday work. Elementary design tables, for example, are based upon such formulae as the following:

$$\frac{M}{I} = \frac{f}{y} = \frac{E}{R} \quad \text{for beams}$$

$$\frac{I}{J} = \frac{f}{y} = \frac{C\theta}{l} \quad \text{for shafts}$$

(where E is the tensile modulus and C the shear modulus; M is bending moment and T the torque; I and J are the moment of inertia and the polar moment of inertia for cross sections of beam and shaft respectively; R is the bending radius of the beam, with θ the twist angle and l the length of the shaft; while f and y are the surface stress and distance from neutral axis (or radius) in each case.)

The elastic modulus of the material in such an expression introduces the 'cause' of deflection, the value of which then depends on the stressed length involved. For this reason, Rolled Steel Joist tables, for instance, limit 'safe' loads to those within some predetermined maximum deflection. This deflection limit (1/325th of span is typical) is purely for convenience in use, even though a beam would in fact withstand far greater loads without failing. The next stage after this is found in the use of rule-of-thumb factors that take practical properties into account too. An example is Buckling Factor (length between notional pin joints divided by least radius of gyration) for columns, where a common maximum value might be, say, 180 for simple sections of ordinarily used steels, or up to 250 for certain made-up structures (often also called slenderness ratio).

Design in stone is based upon virtually no deflection, but with metals people have become accustomed to the small distortions normally encountered. A hundred years of designing in this way means that loading deflection is always assumed, and its magnitude tends to be intuitively understood in relation to the overall strength of a structure–by the user as well as by the designer. When a material is used of much higher modulus in relation to strength, the entire basis of experience in use is changed, and the designer must start again from scratch.

Strength and stiffness, although important, are only the obvious beginnings to considerations of properties in materials. Yield point, elongation and proofstress, for example, are all of consequence in practice. And when aluminium first came into engineering industry hardness also was seen to be a factor that previously had tended to be assumed. Another example of assumption lies in low temperature properties, which showed themselves to be significant when metal failures began to be a problem in a quite new field of use, in early cryogenic applications.

Where one property is markedly improved it is wrong to 'blame' other properties either for not keeping pace, or for changing in type. Carbon-fibre reinforced plastics are a good example. The fibres are of value because of their high specific stiffness. Their deformation is almost entirely elastic–so the stress/strain diagram is virtually straight from no-load to rupture, with, of course, a steep slope. So it is no use lamenting that, for instance, the material ruptures at 0·5 per cent longitudinal strain (for the type of carbon fibre with UTS in the region of 300 000 p.s.i. and Young's Modulus around 60.10^6 p.s.i.). It must be remembered that the purpose of using such a reinforcement is to make a light-weight article stiff, and so almost undeformable, but still capable of withstanding a reasonable total load. If more strain at rupture is required, then a lower-modulus carbon fibre should be used–and it will, incidentally, have a higher UTS.

When designing with any new material it is necessary first to consider the properties needed in the article or component, and to ensure that the value of each is sufficient. Secondly, a suitable balance of these essential properties must be achieved–for example having dealt with both loading and deflection, what about behaviour before failure? or locked-up energy that will be released on rupture? or the mass of the component from a kinetics point of view? Thirdly, other properties must be taken into account–some may be useful and others possibly deleterious: will the corrosion characteristics of the system be a problem and how? will any consequential friction loads be too high, or too low? Fourthly, it will, of course, be necessary to consider the user of the article, and whether or not any of his habits may have to be changed, often to his ultimate advantage; and finally the manufacturing processes will be taken into consideration, sometimes even to the extent of 'designing' them too. With a new material the body of experience must be created that already exists for traditional materials. Table 1 lists some of the more common properties of materials – properties that are either used or else assumed to be suitable in the objects that are designed, made, and used in everday life.

2.3 *Property improvement*

It will readily be seen that with such a long list of properties (which nevertheless does not pretend to be complete) a marked change in one important property may completely upset established values. Plastic packaging provides a current and perhaps extreme example of this. Until the war 'tin' cans, glass bottles, board and paper in the form of bags and newspaper, plus a relatively small tonnage of 'silver' paper and cellulose film comprised virtually all packaging materials. Natural processes of decay such as the effects of wind, sun and rain, biological degradation, and rust served to break up, eat through, or otherwise dispose of most used

TABLE 1 SOME GROSS PROPERTIES OF BULK MATERIALS

Mechanical

Tensile strength (and specific strength)
Elastic modulus (and specific modulus)
Yield point
Elongation at break
Flexural strength and flexural modulus
Compressive strength
Impact strength
Hardness
Low temperature departures from above
Effect of increased temperature on above
Creep data
Fatigue limitations
Abrasion resistance
Damping capacity

Other Physical

Specific gravity
Thermal conductivity
Coefficient of expansion
Specific heat
Refractive index
Coefficient of friction
Velocity of sound
Colour and texture
Optical transmission
Restitution coefficient
Lubricity
Specific surface area

Chemical and Similar

Corrosion resistance
Resistance to solvents
Resistance to radiation
Internal chemical stability
Resistance to surface deterioration
Internal structural stability
Diffusion rate–of one component through another
Taste
Smell

Electrical and Magnetic

Resistance/Conductivity (in the three dimensions)
Dielectric constant
Dielectric strength
Specific inductive capacity
Power factor
B/H relationship and hysteresis
Electronic characteristics

Thermal and Similar

Glass transition point
Softening point
Heat distortion temperature
Weight loss on heating
Thermal shock resistance
Regain
Porosity
Phase changes
Melt index
Coefficient of expansion (in the three dimensions)
Fire resistance

If Liquid

Viscosity
Flash point
Surface tension
Dissolved solids
Suspended solids
Emulsion characteristics
Dissolved gases

For All Materials

Cost
Specification
Delivery

packaging within a relatively small number of years, and rubbish centrally collected could easily be burnt. Now, however, plastics are much used on account of ease of processing, appearance, impermeability, cost—and a great deal of work has gone into making them remain in better and safer condition for longer periods than the materials they have so considerably replaced. Today civilised countries are becoming worried about the longevity of plastic litter over the country-side, and also about the corrosion caused (mainly by hydrochloric acid in the flue gases) when PVC is burnt in incinerators.

So this innovation—more a revolution—in materials that has been so highly successful from many points of view, has decided disadvantages from certain other standpoints. It is no solution, of course, just to follow one line of attack, and try for instance merely 'to make plastics biodegradable': then they might break down, perhaps only in part, before the intended time. A partly degraded vacuum pack of ham, say, could be most dangerous. So, further work must go on to see what sorts of solution might be technically and economically feasible, and not, in their turn bring further and possibly even more serious problems.

Materials scientists have been hard at work, particularly since the war, and have made possible considerable advances in properties of all kinds. Such advances derive partly from direct improvement in materials properties themselves, and partly from new understanding of imperfections and modes of failure under different conditions.

2.4 *Materials developments*

Table 2 shows some recent examples of materials development and the properties that have been improved. While it is far too early to be certain, it seems at present reasonable to imagine that the following may one day come into relatively large-scale and widespread use: superplastic alloys (motor vehicle bodies); carbon fibre reinforced plastics and metals (aerospace and high-performance engineering); concrete reinforced with alkali-resistant glass fibres, and also reinforced gypsum (building and construction industry); new textiles (as distinct from felts) not wholly knitted or woven (clothing, furnishing, packaging); and rigid-foam mouldings and extrusions (building and general industry). All these, it will be noted, are in fact quite new materials with entirely new properties, and not mere improvements upon conventional materials.

Although not as likely to see large-scale use, certain specialities should also be important in the future. In such cases the volume of direct business would be only a small fraction of the consequential business so stimulated. For example, there is a whole range of electronic crystals of high purity and structural perfection, especially doped in order to achieve specific effects—magnetic, optic, acoustic, mechanical, electrical. Expensive though such crystals are, a radar microwave generator, or radiation detector for example, would contain only a minute 'chip' —and the cost of material would be very small compared with the value of the complete equipment made possible by the crystal fragment itself Other examples of speciality materials of likely future consequence include: silicon nitride for furnace parts, investment casting moulds (lost wax), etc.; vitreous carbon for biomedical uses; special glasses; high temperature polymers; carbon materials for accurate cheap casting moulds; and carbon fibres incorporated into plastics or metals as bearing materials—and some of these, albeit specialities, may expect to achieve considerable tonnage production.

2.5 *Composite materials*

As well as making a homogeneous material with improved properties, it is also possible to combine various constituents so that each contributes more of some particular property than

TABLE 2 SOME EXAMPLES OF MATERIALS DEVELOPMENTS

Metals	Higher strengths	Maraging steels TRIP steels Metal powder components Aluminium bronze variants Dispersion hardened copper, tungsten, lead, nickel and cobalt alloys
	Corrosion resistance	Clad composites and coated metals Improved stainless steels Weather-resistant low-alloy steels Chloride resistance–Hastelloy, increased use of Titanium
	Super conduction	Niobium alloys with titanium, tin, chromium, aluminum etc.
	Hard edge propensities	Mixed carbides of tungsten, titanium etc. Coated boron carbide Metalliding
	Easier fabrication	Superplastic alloys Memory alloys
Ceramics	Thermal shock resistance	Silicon nitride
	Chemical inertness	Vitreous carbon
	Thermal anisotropy	Pyrolytic graphite
	Production developments	Hot pressing variants
	Glass	Bendable glass–strained composite Nucleated strain-reduced glasses
Constituents of Composites	Reinforcing filaments	Carbon fibres Boron fibres Steel fibres Silicon carbide whiskers Alkali-resistant glass fibres
	Light weight space fillers	Nylon paper honeycomb Aluminum foil honeycomb Plastics, ceramic and metal foams
	Matrices	Thermally improved polymers, polyimides Cement/polymer developments Metal incorporation developments
Reinforced Composites		Carbon fibre reinforced resins and thermoplastics. Alkali resistant glass fibre reinforced cement products Glass-fibre reinforced gypsum building panels Carbon moulds for castings
Some Recent Products		Non-woven textiles Electronic crystals Synthetic leathers Thermally triggered metal fastenings Polyolefine cordage and sacking Carbon fibre bearings Foamed plastic rapid antiques reproduction

would be otherwise obtainable. Strength, hardness and 'lightness' are the major properties that can be enhanced in this way, along with thermal and electrical conductivity, corrosion resistance, colour, lubricity and other characteristics required in specific applications.

Present-day lightweight strong materials are partly the descendants of the successive box-girder, RSJ, castellated-beam evolution. With plastics particularly, a gas-filled rigid foam can fulfill the same function between stress-bearing 'flanges' as does the web of an H-Section beam. The foam has to withstand crushing and shear loads but its main function is to fill space. Another way of doing the same thing is to use honeycomb with the cells axially normal to the principal stressed surfaces. Aluminium alloy, and special nylon paper impregnated with phenol-formaldehyde resin, are commonly used–and also end-grain balsa wood.

Fibre reinforcement is another technique. Carbon fibres, or, less conveniently, boron deposited on fine tungsten wires, are much stronger and stiffer, weight for weight, than metals such as steel. They need, however, to be incorporated into a matrix before they can be used for most of the purposes for which a solid material is required. Resin matrices are much used, following upon glass-reinforced-plastics practice, and there is some experimental work involving metal matrices. Such a composite material is anisotropic (that is its properties along the three dimensional axes are not all the same) and will, for example, be stronger, and stretch less, along the line of the fibres, than at right angles to them. The principle of designing is to align the fibres along the lines of stress, making suitable arrangements to transfer load to and from the reinforced object.

When fibre-reinforced materials are combined with hollow formations, then the advantages of both techniques can be realised. Whole composite structures are made up, particularly in the aircraft industry: a common example is where sheets of parallel carbon fibres cured in resin are separated by aluminium foil honeycomb fixed to both faces by adhesive, and incorporating any necessary aluminium alloy sections or inserts for transferring load to or from the structure. In another technique, a hard resin skin can be made to form the surface of an object made by filling a mould with foaming rigid plastic–polyurethane for example. Such formations can be very light and yet possess 'strength' normally associated with solid objects.

Another way of constructing composites is applicable in metallurgy, where throughout history advantage has been taken of the hardness imported by cementite suitably dispersed throughout steel. Modern developments include dispersion hardening of metals with particles of alloy, or of oxide; and it is even possible by directionally solidifying certain alloys to 'grow' strong eutectic fibres within the structures of the object. Finally, with all composites already often anisotropic, the properties can also be made to vary along one direction where it would be useful. This 'property grading' can be either gradual, or abrupt according to need. Understanding of timber as a material is well established, of course, but high-property anisotropic materials are only just beginning to be used, and it will take some time before their advantages can be fully realised.

3.0 MANUFACTURING TECHNOLOGY

Table 3 shows some of the main production processes

In the main they are applied to bulk materials, and will apply to new developments too, where the properties have been improved without significant involvement of the material structure. It will be seen that some of the production processes involve cutting or deforming the material as it stands, while other processes alter the shape of the material while it is in a soft condition ordinarily resulting from the application of heat.

TABLE 3 SOME PRODUCTION PROCESSES FOR BULK MATERIALS

Material	*Size Reduction*	*Shaping*	*Cutting*	*Building-up*
Plastics	spinning-various extrusion	moulding-cold, – compression, – transfer, – injection casting thermoforming – blow, – vacuum	as little as possible: standard tools drilling grinding	welding adhesion screw threads &c. lugs, rivets &c.
Metals (bulk)	rolling drawing extrusion	casting pressing forging spinning	shearing grinding turning milling drilling eroding-e.d.m., – shot blasting dissolving-e.c.m., – chemical milling thermal-oxygen, – laser &c.	electroforming welding adhesion electron beam &c. screw threads &c. rivetting, bolting &c.
Timber		steam pressing	sawing planing drilling	adhesion screws, nails &c. load spreader plates
Concrete		casting extrusion pultrusion	grinding splitting	pouring around extending reinforcement
Reinforced Plastics		moulding (as for plastics) filament winding pultrusion casting	as for plastics	adhesion
Ceramics		casting moulding hot pressing turning while soft	grinding	adhesion
Metal Powders		pressing sintering forging rolling	as little as possible; standard tools drilling grinding	as for bulk metals

When new materials are used, care must be taken not to alter the special characteristics involved, as well as to ensure that the most economic processes are fitted to the new circumstances. Also, some conventional processes may just not 'work' when the material changes. In the same way that, say, turning practice for cast iron is different from that for bronze, high property reinforced composites need still different treatment. Boron/resin composites may be virtually unturnable, and some carbon fibre reinforced plastics may require the use of diamond tipped tools. Fibre reinforced metals will present different problems still: while they may be bendable by the application of sufficient force, to bend them beyond the rather low limiting strain of the reinforcement would turn an advantage (high stiffness) into a disadvantage and break the fibres, while leaving the unstrengthened matrix useless for the job.

To use conventional processes indiscriminately on unconventional materials may be to waste production opportunities. The aim is always, of course, to do as little work on materials as possible, and plastics help because moulding, thermoforming, filament winding and other techniques can often, in one operation, turn raw material into very nearly final finished form.

4.0 MANAGEMENT AND MATERIALS INNOVATION

In most manufacturing there is always the question–whether to continue using material within the current specification, or to use a cheaper material to a less onerous specification. The issues are just the same over the possible use of materials of improved properties, perhaps enhanced corrosion resistance, for example, or high-properties mechanically. The important question is, what material will enable the job as a whole to be done for the lowest overall cost. Aspects to be considered may include the life, performance, reliability and appearance of the article to be made, along with subjective factors in sales; the production costs–which will entail consideration of the number of operations required, and on what types of machine, shop overheads, tool life, complexity and cost, inter-machine transport, etc; design complications and uncertainties of performance; any purchasing factors, reliability of supply, storage problems, and locked-up investment.

Reliability of properties is a continuing problem in some branches of engineering, using conventional bulk-produced materials. Many new materials by their very nature offer opportunities in constancy, once satisfactorily developed, as an additional advantage beyond the property advantages that were the purpose of development.

Materials innovation opens up new possibilities, some of which are presumably deliberately sought by management. As always it is income from sales that pays for production; and production that is the aim of the sales/design team that conceived the manufacture, and of the designers and production engineers that made it possible. Markets need to be created. In parallel, there will be consequential changes and opportunities that require the whole of management thinking to be flexible–also receptive to the possibility of business in new directions and applications as well as those of cheaper current production. Table 4 shows some of the implications

5.0 PRODUCT DEVELOPMENT–AND THE FUTURE

5.1 *Products*

One can perhaps make three main divisions of materials usage:

(1) Highly sophisticated, for example aerospace, with considerable technical requirements upon low-weight materials; where it is important to anticipate as many as

TABLE 4 SOME MANAGEMENT ASPECTS AND IMPLICATIONS OF NEW MATERIALS

Aspect	*Simple Replacement By Cheaper Similar Material*	*Higher Property New Material Displacing Conventional*
Economics	Reduction in prime cost. Sometimes minimum economic throughput. Possibility of variation in other costs.	Probable increase in prime cost. Opportunity to simplify fabrication and reduce wastage. Use only where properties required. Use criterion of cost per unit of performance. Evolve new make-up of production cost.
Sales	Emphasis on price reduction for adequate performance.	Sales where performance more important than any initial cost increase. Create entirely new markets. Economics of the overall job.
Design	Need to avoid any consequential inferiorities.	Capability to design to greater duty. Need to redesign component from scratch. Need to consider whole design of machine or structure. Need to design fabrication processs Need to optimise properties from scratch. Often a need to 'design' the material e.g. in case of composite. Capability of designing for applications previously impossible.
Production	Need to avoid any consequential inferioritites.	Develop techniques, making use as far as possible of existing production techniques. Fabrication speed with expensive new materials less important than with conventional. Components in production machinery may drastically raise maximum product-ion rates. Different shop floor layout. Check that the process is the best in the circumstances.
Research & Development	Search for technical feasibility.	Technical feasibility under conditions that perhaps no materials had previously so performed under. Close contact with sales department required. Need to create testing procedures.
Management	Ensuring no consequential losses to company – in terms of finance, reputation or expertise.	Encouragement of innovation. New approaches to design, fabrication and use. Need to create new markets and new technology. Helix of material, opportunity, design, market, production, and growth, with feedback at all stages.

possible of the otherwise 'unexpected' adventitious demands (qualification testing); and where one material or component may have multiple functions, for example stressing of aircraft skin.

(2) Average, for example most of engineering where the industry has become used to wider tolerances upon properties–leading to much larger factors of safety than in aerospace even though duties are generally much less severe.

(3) Rough, for example building, where tolerances are wider still in many instances (though some, for example for sand-lime brick dimensions, are stringent); the environment is dirty, wet and sometimes antagonistic; and where nevertheless a long service lifetime is expected.

Most developments help in raising standards, and the building industry, for example may change considerably in the next few decades. Plastics and finished metal components are being increasingly used, but are still liable to damage in an operation that also includes bricks, cement, tiles, sand, concrete and boots. Prefabricated concrete components are also used, but they tend to be heavy–often more to ensure adequate strength to resist damage in transit, than because the job the component has to do actually requires such weight and strength. Perhaps the big change can come when building evolves into on-site assembly of works-manufactured wall sections and other units, finished on all faces, and only requiring manipulation under clean conditions on already completed foundations. This would be a 'step' change: lightweight reinforced cement products and plastics should have increasingly interesting parts to play in making it happen. The average and sophisticated uses may be expected to evolve steadily along lines already apparent. Greater constancy in properties, plus improvement in properties, will be the aim. Whenever a new materials possibility arises it will be necessary to consider two aspects:

(1) Is there a worthwhile use of the new property made available? – in overall terms of basic cost, manufacture and use.

(2) Are the improved properties balanced in the same sort of way as those in the conventionally-used material?–in which case the same basic design philosophy can apply. Or is one property so far in advance of the others that quite new design principles require to be derived *ab initio*?

Table 5 illustrates an example of one version of morphological approach to a simple problem (another approach might be to consider the passage of the fluid from the source to destination). Here some likely materials have been considered from a property point of view in the light of known needs. It might be an interesting exercise to consider, say, a reinforced composite for the same purpose. The purpose of the exercise would be to select the material with the greatest overall advantage (usually to the user though not always) under the prevailing circumstances–which may well vary from one manufacturer and one sales climate, to another.

5.2 *Economics*

Since new materials are new there is no immediate established market for them. Demand will presumably grow as a result of use in trials either first hand, or reported. So an exponential growth would be expected, with increase in demand at any one time related to the then current usage.

$$d = f_t\,(e^u)$$

where d is demand, u is usage, at time t.

TABLE 5 AN EXAMPLE OF MATERIALS SELECTION FOR A PRODUCT

Possible Materials For a Beer Mug	*Metals*	*Glass*	*Ceramic*	*Synthetic Resin*	*Thermo Plastic*
Properties Involved					
Strength	surplus	enough	enough	enough	poor
Elastic modulus	surplus	surplus	surplus	enough	poor
Corrosion resistance	some suitable	surplus	surplus	surplus	surplus
Fabricability to give volume tolerance	enough	enough	enough	enough	enough
Bounceability	surplus	poor	poor	enough	surplus
Freedom form taint	some suitable	surplus	surplus	some suitable	some suitable
Favourable economics	many suitable	many suitable	many suitable	surplus	surplus
Properties Involved					
Colorability	surplus	surplus	surplus	enough	enough
Subjective factors	many suitable	many suitable	many suitable	poor	poor
Isotropic	yes	yes	yes	yes	yes
Examples of Properties Unused or Irrelevant					
Hardness	√	√	√	√	√
Thermal conductivity	√	√	√	√	√
Coefficient of friction	√	√	√	√	√
Refractive index		√			
Thermal stability	√	√	√		
Electrical conductivity	√				
Electrical resistance		√	√	√	√
Extra Properties Beneficial	polish/ patina surface finish luxury possible	polish translucency cheapness of complex shapes luxury possible	glaze decoration cheapness of complex shapes luxury possible	cheapness if habits change	unbreakable cheapness if habits change

Usage will tend to be related to need, price and advantages, that is some complex function of properties and cost

$$u = f'_t\,(p,c)$$

where p is properties, c is cost, also at time t. Cost will tend to depend upon the extent of development of the manufacturing process, raw materials prices, labour rates and return on investment, which itself may have a power relationship to output.

$$c = f''_d \left\{ m + l + \left(\frac{i_d}{i_{d-1}} \right)^{\frac{2}{3}} \right\}$$

where m is materials, l is labour costs, d is demand, i is investment, related to plant size so related to demand.

6.0 CONCLUSION

Innovation in materials includes many highly sophisticated developments, certain of which will undoubtedly be of enormous value to industry as time goes on. There are also much more mundane developments amongst accepted materials that will also be important–and in areas often overlooked. What is more necessary than dentistry for instance?–with a whole range of materials uses and developments upon which most of us may depend rather critically. Developments here are fascinating.

Sometimes, too, quite humble basic materials can be combined to achieve far reaching innovation. Paper and resin suitably processed become a real aristocrat–the decorative wall-board that seems to be about the only inexpensive means of standing up to football crowds on trains. Other developments go in the opposite direction–for example the once highly specialised polyethylene, and the p.v.c. (now around forty years old) that are now ubiquitous both in industry and domestically.

Materials innovations come in many forms, and have many effects, all worth considering seriously so as to use each in its most effective way. Production engineers must continuously be on the look-out for the opportunities that materials scientists are making possible. Can traditional materials be used in new ways? Can traditional materials be combined in new ways? Can a new material developed for one purpose be adapted for another one? If a materials problem seems insoluble can the materials scientist suggest an entirely different approach? Can a mechanical or structural design problem be so formulated as to suggest what sort of new material might be needed? Progress in design and performance starts by asking questions about materials, their fabrication processes, and the job to be done.

Above all–how can the purpose in hand make the best use of the properties that can be made available? and, what opportunities are opened up by new materials possibilities?

BIBLIOGRAPHY

COTTRELL, A. H. and KELLY, A., 'Strong Fibrous Solids', Proc. Roy. Soc. A, Vol. 319, No. 1536, pp. 1–143, 6 October 1970.
KELLY, A., 'Strong Solids', O.U.P., London 1966.
HOLLIDAY, L. (Ed.), 'Composite Materials', Elsevier Baring 1964.
LOCKE, H. B., 'Composites in Progress', *Composites*, September 1970, p. 261.
'Materials Developments & Planning', *Long Range Planning*, September 1969, pp. 67–73.
'Working with Carbon Fibre Composite Materials', Prodecon '70, PERA, 1–3 December 1970.
'The Development of New Processes & New Materials', Achema Triennial Chemical Engineering Exhibition-Congress, Frankfurt 1970.
PLASTICS INSTITUTE CONFERENCE, 'Carbon Fibres, Their Composites & Applications'—some fifty papers on most aspects, London, 2–4 February 1971.
GORDON, J. E., 'The New Science of Strong Materials', Pelican 1968.

FUTURE TRENDS IN PLASTICS

C. Menges and W. Dalhoff
Institut fur Kunststoffverarbeitung (IKV), Aachen

SUMMARY

The future growth in the use of plastics is predicted, and the need for education and training of designers and engineers in the effective utilisation of plastics is emphasised. Various production factors are considered, including composite materials and fillers, as also the potential use of computers in process control. Research work at Aachen in this field is briefly reviewed, and a high pressure plastification unit is described. Future trends in plastics processing methods are indicated.

1.0 INTRODUCTION

1.1 *Future trends in plastics*

Our matter of fact attitude towards plastics makes us easily forget that the plastics industry is still very young. Its main development started only in the thirties. Despite this short period these new synthetic materials became so important that we almost can't imagine life without them. The main reason is their big share in things making our life comfortable. From 1950 to 1968 the world production and consumption of plastics has increased fifteenfold. In the same period the world production of crude oil–an indicator of industrial progress–grew scarcely fourfold. The share of plastic materials going into the plastics processing industry (production of semifinished materials and finished parts with the exception of rubber) can be estimated at 60% of the total consumption. This means that in 1968 approximately 4630 million lbs of plastic materials were processed by the German industry. The total world production amounted to approximately 53 000 million lbs. In 1980 it is likely to be 176 000 million lbs[1,2] (Fig. 1).

According to recent forecasts[2] in 1973 the annual production of plastics in the U.S. will exceed by volume that of steel and nonferrous metals with 20 million m^3 to 16 million m^3. According to this source in the eighties, world production including fibres and synthetic elastomers is expected to be about 60 600 million lbs. This quantity would mean the top position even for a comparison by weight.

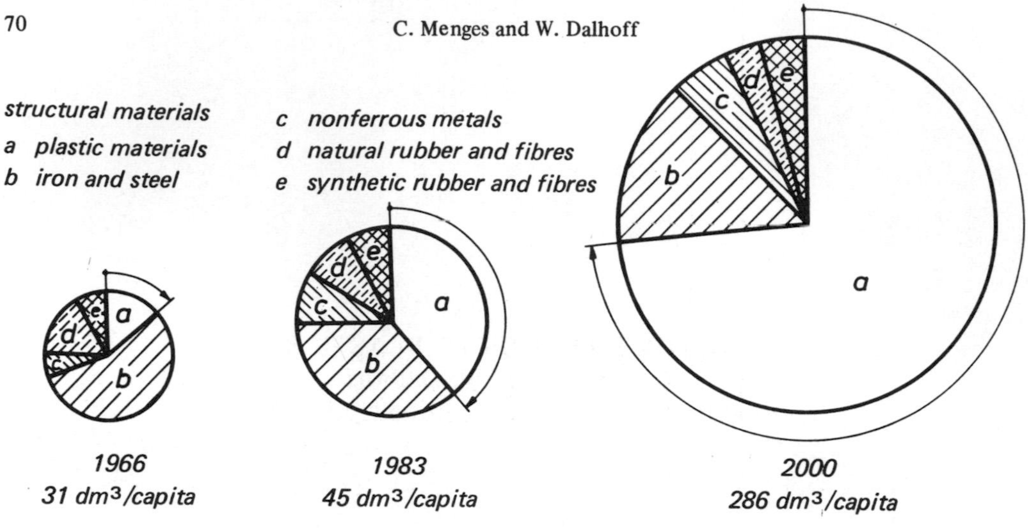

structural materials
a plastic materials
b iron and steel
c nonferrous metals
d natural rubber and fibres
e synthetic rubber and fibres
a
b
c
d
e
1966
31 dm3/capita
1983
45 dm3/capita
2000
286 dm3/capita

Fig. 1. Estimation of the world consumption of structural materials per capita[1].

The consumption of plastics per capita can be looked upon as a criterion for the technical progress and economic strength of a country (Fig. 2).

kg / capita
60
50
40
30
20
10
1960
62
64
66
68
70
W.-Germany
Japan
France
USA
GB
Italy
UdSSR

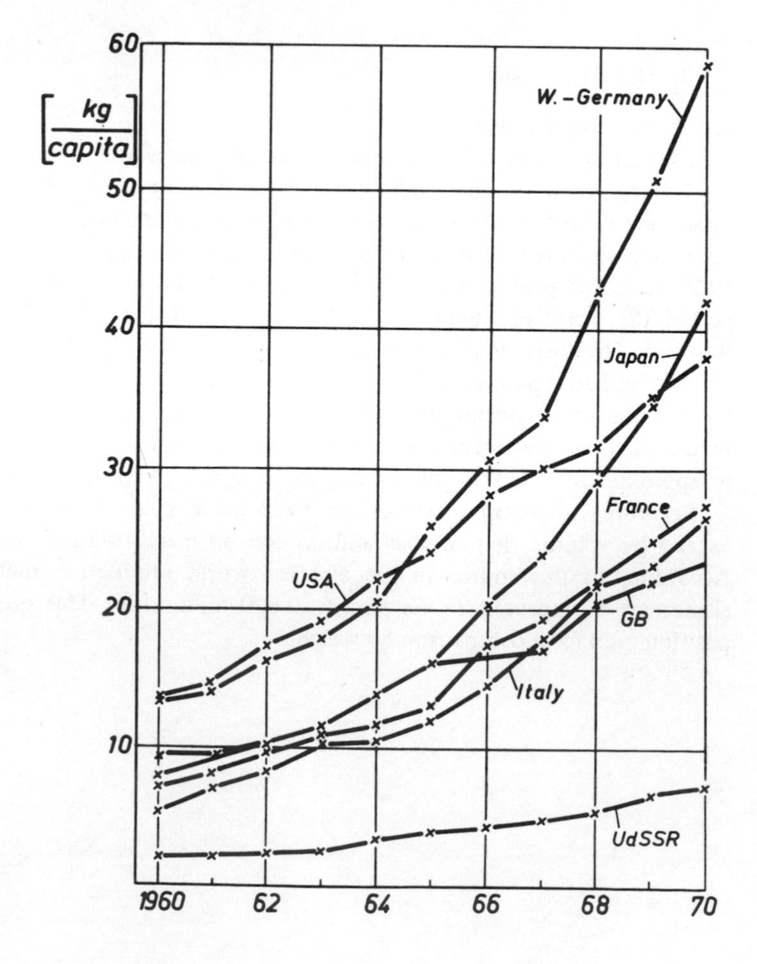

Fig. 2. Plastics consumption per capita of several countries.

The plastics producing chemical industry is prepared for this trend. The necessary large investments require financial strength and international markets. The big chemical companies cope with both conditions by acquiring smaller producers as well as by establishing new plants for promising markets. The integration of interesting, that is economically important successive production as fibres, films, paints, etc. also has to follow.

If, however, the plastics producers expect extremely high investments (Fig. 3), a similar development of the processing industry should be observed. This applies only to big trusts. The question arises if the medium sized plastics processing companies can cope with this tendency. All foreseen developments could be destroyed by lack of capital and by inability to finance investments by issuing shares. In addition, too many branches and high taxation might leave only small funds for reinvestment. However, it can be observed that other big industries take considerable interest in plastics processing even if it is not yet profitable for them. This situation can be expected to change soon and by this the present medium size of the German processing companies will alter.

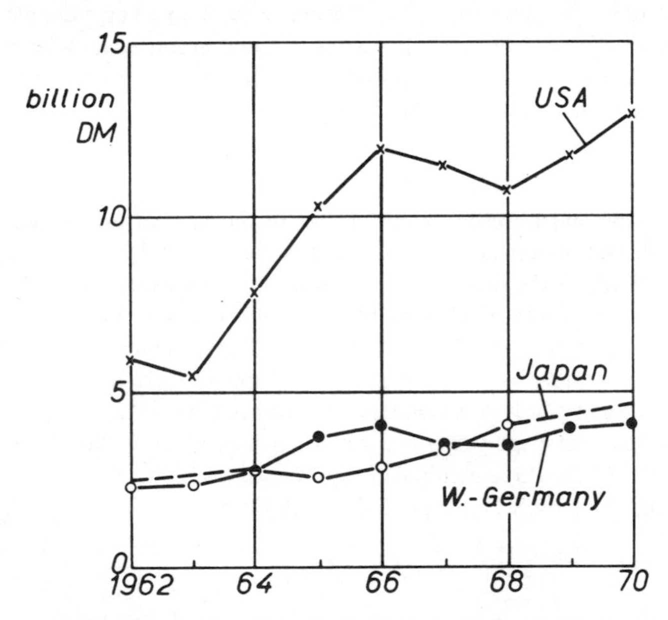

Fig. 3. Chemical investments of some industrially important countries.

Financially powerful machine-building companies or trusts prepare even more for the future. Mergers are daily news. It is not surprising that this tendency is stronger here than for the processing industry as the German plastics machine-building industry accounts for approximately 40% of the world output in this field[3]. In 1969 the top-ranking company produced plastics processing machines equivalent of 250 million German marks.

1.2 *Education and training requirements*

None of these plans would be feasible, however, without people prepared for the corresponding tasks. For the plastics producing, processing, and machine-building industry we need leading personnel whose qualifications will be decisive for the realisation of the forecasts and the role West Germany will play in this field. We especially need engineers for our plants. But the customers must be prepared, too. As sales will shift more than at present to technical purposes, it will be vital that engineers and architects are acquainted with these materials through their education. They must know the properties of these materials to make successful use of them.

Until now in West Germany it has been possible for mechanical engineering students to specialise in plastics technology at technical colleges and universities by taking this as an optional subject. The first step towards the education of plastics engineers was made by the technical colleges in Aalen, Bielefeld, Darmstadt, Paderborn, Rosenheim, and Ulm, where special branches for plastics technology were set up. Now the university of Aachen has decided (as the first university in Germany to do so) to establish on our instigation a branch of plastics technology. The educational work started during the winter term of 1970/71. The necessity for this branch of study was acknowledged by professors and assistants as well as by the students. The schedule of studies was set up by team work. Our aim is to educate mechanical engineers by teaching the application of engineering taking plastics technology as an example.

There is, however, a considerable lack of teachers trained in materials science for colleges and universities for civil engineering and especially for architecture. In this most important technical field of construction, engineering should be done much more.

2.0 PRODUCTION

The biggest share of the plastics production will be, in the near future as well as today, the cheap mass produced plastic materials polyvinyl chloride (p.v.c.), polyolefins, and polystyrene which amount to two thirds of the present total produced in big quantities are the main reason for this fact. Of special importance, also, is the easy and economical processability which cannot be obtained for the mass production of finished parts from other materials. Their properties, which can be excelled by far by other more expensive plastic materials in terms of strength and heat resistance, are only of secondary importance (Fig. 4)[4].

The polyolefins and especially polyethylene which reached recently the top position, have the biggest rate of growth of the three material groups mentioned.

It is unlikely that this situation will change in the next years[5].

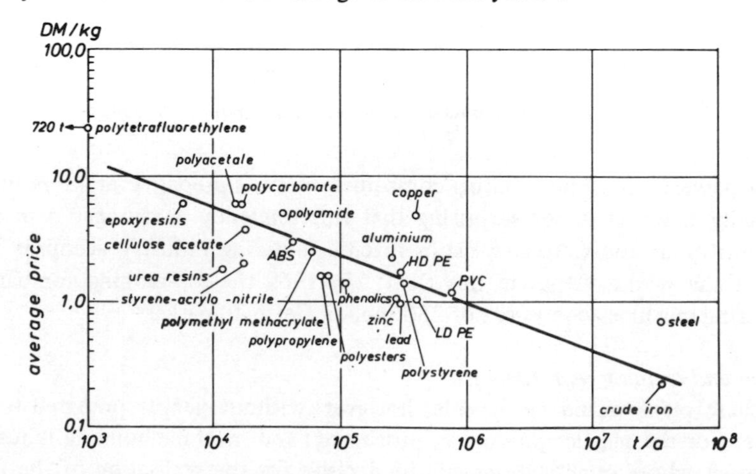

Fig. 4. Prices of several materials related to annual production (W. Germany 1969).

A considerable reduction of producers' prices for the presently mass produced plastic materials can hardly be expected in spite of increasing rates of growth as the provision with raw materials will be looked upon by the petrochemical industry no longer as a secondary, but probably as a principal line of business[5,6]. A further price reduction for plastic materials is, in principle, possible only by new applications leading to bigger production units and by rationalising and automating the processing of semifinished materials and finished parts. If the inevitable rise in the prices of other materials caused by increasing wages and raw material costs is taken into account a further relative price reduction for plastic parts will result, especially because these are well suited for rationalised mass production[7]. The same applies to all the thermoplastic and thermosetting materials following the top group as polymethyl methacrylate, cellulose acetate, polyamide, polycarbonate, polyacetat, etc. and phenolics, polyester, and epoxy resin. Especially for plastic materials of excellent strength and heat resistance a final break-through will come, not by producers' price reductions, but by simplifying and economising in methods of fabrication. New plastic materials developed in the last years, for example high-strength and more heat-resistant materials as polysulfones and polyphenylenes, will gain in importance. Furthermore a rather strong trend exists towards intentional variations of materials. Already copolymers of the mass produced plastic materials such as terpolymers of styrene, (e.g. ABS, and ethelene vinyl copolymers) are very important. Concerning the more heat-resistant and high-strength resins it must be considered that these properties make processing more difficult and expensive. The temperature level of metal processing is reached. Because of this there will be an inducement to use these materials only in cases where their properties are absolutely necessary for the application. This concerns mainly electrical applications, if big markets are taken into consideration.

2.2 *Composites and fillers*

The modification of existing plastics by blending with other polymers, but also with fillers, especially with fibres, is growing fast. At present glass fibres are the bulk, which are used not only for low pressure thermosetting resins but also increasingly to improve injection moulding materials from polyamide to polyethelene. The biggest advantage can be seen in reduced mould shrinkage and higher stiffness of parts manufactured from these materials.

Of course there is often talk of super fibres as whiskers, boron-tungsten fibres, steel and carbon fibres. Certainly exceptional properties can be obtained with composite materials consisting of suchlike fibres and high-strength resins. This is shown by the result of creep tests up to temperatures of 300°C with reinforced polyimides of which several types are being investigated in our institute (Fig. 5). Some of these composites can already be manufactured at reasonable cost. Nevertheless their production, and this applies to all these developments, will be restricted to special purposes although already there are prices of 10 DM/kg for SiC-whiskers and of 50 DM/kg for carbon fibres in view. The price seems to be only a question of the quantities produced so that this level is believed to be feasible in about 5 years' time.

Only in this field of high-strength and highly heat-resistant resins and super fibres is there little activity in West Germany and Western Europe, if the cyanate resin of Bayer withstanding continuous loads at 200°C and the easily processable bis-maleinimid of Rhone-Poulenc are not considered.

At present there are strong learnings in the U.S. aerospace industry towards the synthesis of highly temperature-resistant polymers with temperature limits up to 1000°C. The Narmco Materials Div. of Whittaker Corp. considers ladder polymers on the basis of aromatic-

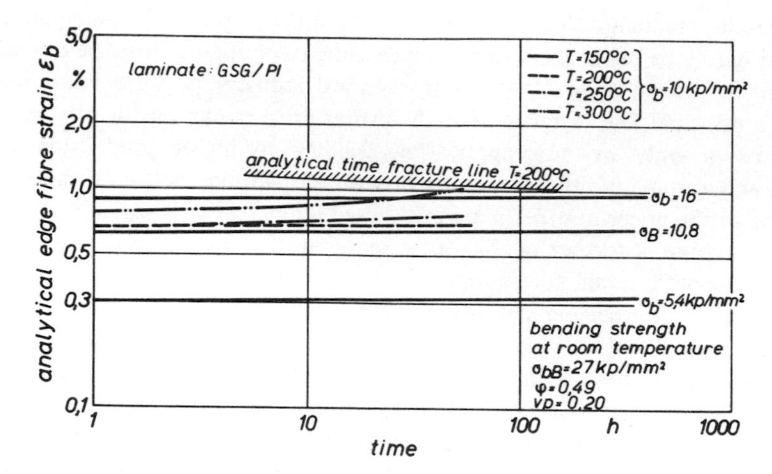

Fig. 5. Creep of fabric reinforced polyimides

heterocyclic systems to be the way to these exotic plastic materials. Some test materials on the basis of boron-nitrogen-monomers resist temperatures up to 2200°C.

Inorganic long-chained molecules on the basis of silicon-nitrogen and phosphorus-nitrogen also yield high thermal stabilities. However, there are no methods to process these materials. The high heat-resistance and also the processing difficulties are caused by considerably higher bond energies due to ion binding.

Until now there is no reason to be afraid of the future. A technological gap does not exist in plastics[8]. But the high consumption per capita must not lead to complacency as developments today which appear visionary can be of importance tomorrow. We must participate in the new plastic materials, too. Consequently, experiments with super fibres and exotic resins are made not only at the IKV, but in many other places too.

3.0 PREPARATIONS

Most raw materials are in one or more steps provided with fillers and improving additives and blended. There are attempts to simplify these operations for bulk materials more and more, for example simple mixing of pvc compounds in fluid mixers, or to integrate them into the processing machine. Attempts are made to introduce additives as plasticisers into the material during the polymerisation[9].

This tendency is developing strongly for polyolefins. The task here is to get a material with a particle-size distribution favourable for processing[10] and to combine a basic stabilisation and the processing in one operation. The most promising developments appear to be those which add fillers of any kind through masterbatches. In this case the processing machines remain simple and only a part of the raw material (about 10 per cent) has to be prepared. Furthermore, there is the economic advantage of using large amounts of the same basic material. This cuts down not only the costs of purchasing and storekeeping, but facilitates also inner- and inter-departmental transport, for example by pneumatic conveying (Fig. 6). A steady growth of the variety of masterbatches offered in many cases by new producers–mostly big trusts–can already be observed.

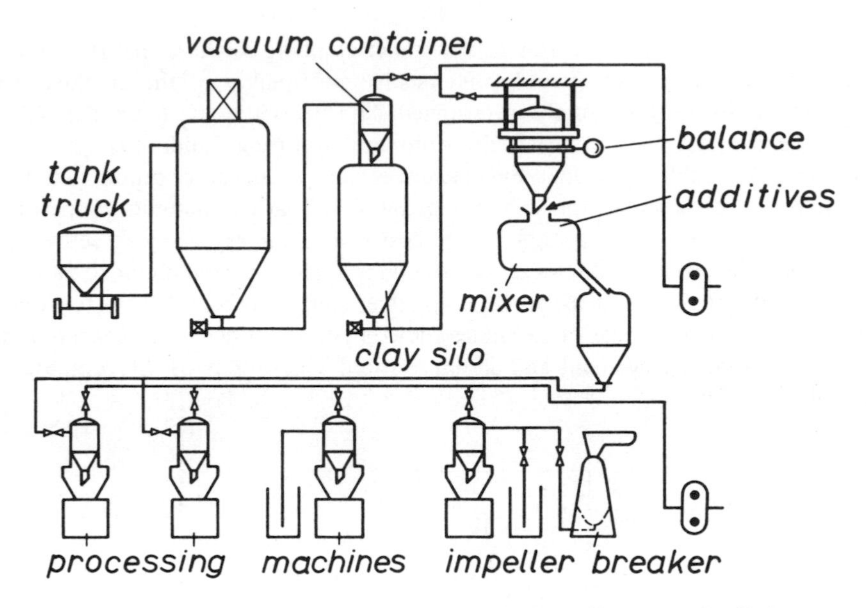

Fig. 6. Example for fully automatic material supply (according to von der Ohe)

4.0 PROCESSING

4.1 *Computer process control*

The assumption can be made that industrial processing developments are finished to a high degree or are at least known and under test. Completely new methods are hardly to be expected. However, a strong trend towards rationalisation is growing. The most interesting of the new developments aim in this direction.

The most important changes can be expected from the use of electronic data processing. By means of computers it is possible, not only to automate extensive and complicated processes, but also to control automatically complete production plants having many machines.

In ten years' time it is envisaged that complete industrial plants with a hundred or more machines for injection moulding, compression moulding, foam moulding, bottle blowing, extrusion, film blowing, calendering, etc. will be controlled by computers. An enormous wage explosion accompanied by considerable shortening of working hours will lead to a general limitation of manpower to some control operations.

Computer process control of manufacturing finished parts by injection moulding seems to be possible in the near future. In our institute a project in the scope of a research programme 'Process Control of Manufacturing Techniques' is being carried out, which will include the experimental computer control of an injection moulding machine in the next year. There are already several machine models suitable for connection to a computer. The most interesting development of this kind has a programmable electronic control which allows rapid mould changing[11].

Also the first inspection of a mould on an injection moulding machine—a very expensive operation—can be carried out considerably faster by means of a computer, to save expensive operator hours.

The present research work related to this aim attempts to establish mathematical models for the interactions of machines, mould, melt quality and quality of the moulded parts. A programme set up by experimentally determined data gives the computer the opportunity of choosing the shortest way to reach the optimum operating conditions quickly and reproducibly. This is possible even when the machine settings of an earlier production with the same mould have to be altered, for example because of varied material properties. As with numerically controlled machine tools, where cost savings are especially effective for small and smallest series, the economical lot size of injection moulded or blow moulded parts will become smaller. Lower costs result furthermore from fewer staff and from less waste. Apart from the actual control of a multitude of machines the computer can take over background work, for example concerning management and administration. The objects of the computer control are accordingly (Fig. 7):

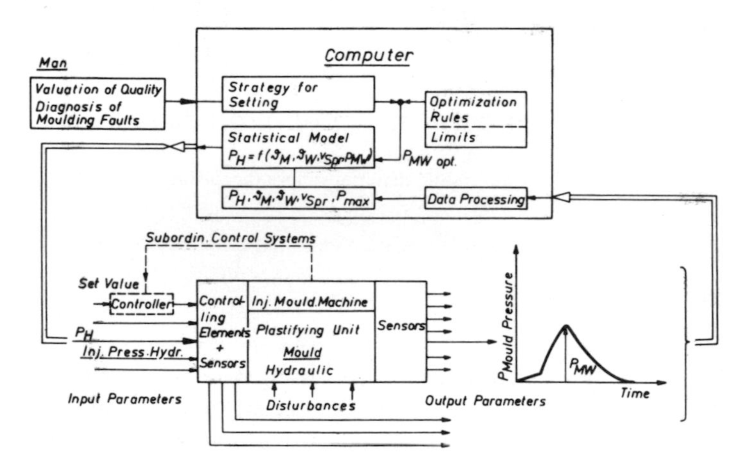

Fig. 7. Principle of computer control of an injection moulding machine with injection pressure as example. PMW melt pressure in the injection mould.

(1) Start-up control, that is fast building-up of stable and optimum operating conditions;
(2) Optimisation of product quality and production time;
(3) Supervision of the production process;
(4) Recording of the course of production and by this filing of data;
(5) Cost accounting and administrative duties as background work.

To reach these aims injection moulding machines must be adapted to suit a computer control. For the continuous evaluation of the particular conditions of machine, melt, mould and cooling, suitable sensors are necessary, for example pressure transducers and temperature transmitters for the plastifying unit as well as for the mould. The recorded signals are fed to the computer for evaluation. Then the computer can control the single operations of the injection moulding process exactly and reproducibly by means of suitable adjusting elements. If, for example, the course of pressure in the mould for an optimum cycle is proven, it is the task of the computer to reproduce this pressure course constantly by influencing the injection pressure, injection speed and melt temperature (Fig.7)[4,9].

Machine motions of high speed and large forces must be controlled within close tolerances.

For this purpose electronically adjustable correcting elements are necessary for the rotary as well as for the translatory motion of the screw.

Despite the high degree of automation attainable by use of computers there are economic limitations for the processing of plastics with conventional injection moulding machines. The low heat conductivity of plastic materials–almost three powers of ten lower than that of metals–is a severe impediment to reductions in cycle time. For example, long production times in plastics processing are one reason why glass bottles are still keeping up competition with plastics. Glass bottles have become thinner, and as a result cheaper and faster to produce. Methods of overcoming this disadvantage in plastics processing could be warm forging and high-pressure plastification.

4.2 *Warm forging and high pressure plastification*

Especially thick-walled parts can be produced by warm forging, a kind of drop-forge technique. Parts produced by this method from semi-crystalline plastic materials have a very low mould shrinkage if the forging is done at temperatures near crystallisation and if a few seconds are spent as holding time.

To economise on the plastification and shaping of plastics by heat input through adiabatic compression and energy dissipation, a high-pressure plastification process was developed. The disadvantages of present single-screw machines, including high-speed machines which have a growing success on the market, are the still rather long dwell times of the material in the plastification cylinder and also feeding problems which grow with increasing screw speeds. Furthermore conventional screws become uneconomic, not only because of poor filling of screw-channels at high screw speeds. They also discharge an inhomogeneous melt because of decreasing dwell times of the material[12].

By high-pressure plastification, for example by means of the plastification unit developed by the 'Institut fur Kunststoffverarbeitung' (IKV) at the Technical University of Aachen (Fig. 8), plastic materials can be molten at pressures of about 150 000 p.s.i. and more[13].

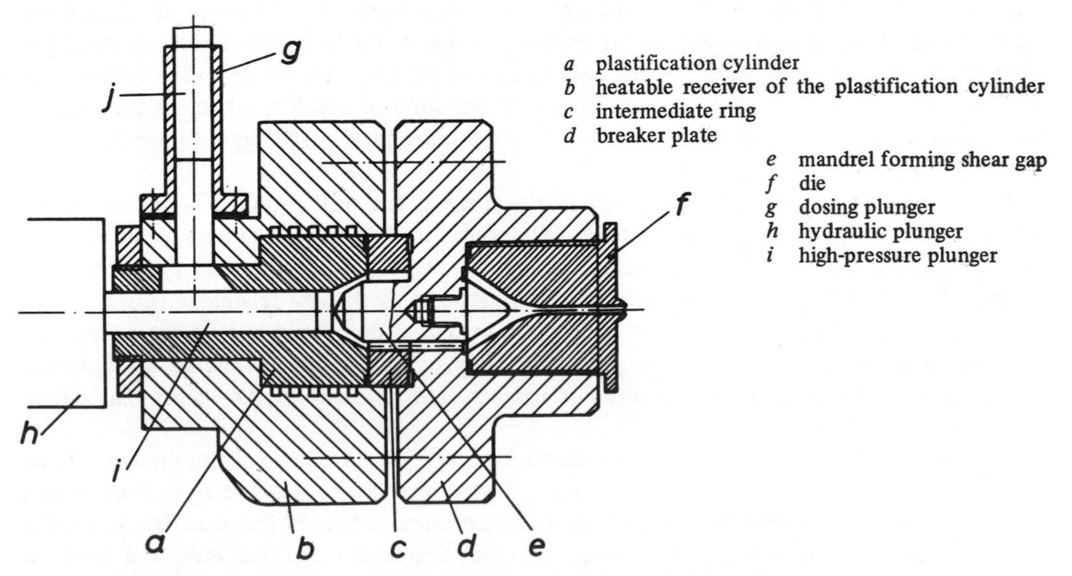

Fig. 8. Cross-section through a high-pressure plastification unit developed by the 'Institut fur Kunststoffverarbeitung' (IKV) at the Technical University of Aachen.

The special advantage of adiabatic heat input is the generation of heat in the material itself within a fraction of a second. Exceptionally short dwell times can be reached which permit the forming and cooling even of rapidly decomposing materials, before damage by the degradation reaction occur. The actual plastification cell consists of the cylinder *a*, which has an opening in the first third and widens funnel-shaped to the end. The cylinder is shrink-fitted to the receiving plate *b* with a system of channels which makes a liquid-tempering of the cylinder for protective heating possible. In front of the receiving plate an intermediate ring *c* and a breaker plate *d* are flanged, which bears the mandrel *e* and alloys the insertion of different dies. The gap between the funnel-shaped end of the cylinder and the tapered mandrel can be altered at random by intermediate rings of different thickness. The dosing plunger *g* transports the material to the cylinder *a* where it is compressed by the high-pressure plunger *i* and pressed through the shear gap and breaker plate to the die.

For the processing of moulding compounds to finished parts this system would make use of an already known method (injection transfer moulding according to Drabert & Sohne, Minden, West Germany) to remove the cured material from the cylinder. For reacting polymers the short dwell time of remixed materials would be especially advantageous, because fast reactions at high mass temperatures would be possible.

4.3 *Expandable materials*

Apart from the injection moulding of homogeneous polymers or plastic materials filled with solids, the processing of expandable materials has special advantages. On one hand, the foaming agent reduces the melt viscosity so that the mould can be filled by the expansion pressure of the foaming agent. Because of this, lightweight moulds and low-power clamping units can be used. On the other hand gas is cheaper than plastics. Some special-purpose machines are already on the market[14]. The technical development in this field has not yet ceased. The reintroduction of preplastification units with accumulators of a large volume, is of considerable interest.

In injection moulding of foamable thermoplastics there exists the disadvantage that the cycle time for thick-walled parts is longer the better the foam structure. Therefore, it seems more logical to use cross-linkable materials for foam moulding, as the heat must not be removed but is necessary for curing. Because of this, we try to add cross-linking agents to the melt just in front of the mould (Fig. 9). Economic advantages, because of considerable cycle time reductions using higher temperatures, must be added to the procedural advantages of processing only thermoplastic materials by the screw.

Besides the processing of thermoplastic melts there is the well-known method of casting monomers which polymerise in the mould. This procedure has been of almost no importance for the production of structural parts, apart from semifinished products of polymethyl methacrylate. There are attempts to improve this process by using a technique similar to die casting of metals. By this the weak point of air adhering to the mould walls being carried out could be overcome. Here, the low specific weight of plastics compared to metals is a disadvantage. However, there are still the difficulties of a high mould shrinkage of 10–15 per cent and of the formation of fins.

Therefore, it seems to be advantageous to add foaming agents to melts of this kind as is done for instance with polymethane. In this way, all monomer melts seem to have a promising future, which is already evident for polyurethane structural foam. In this case, the process is known in principle, but some problems still exist for its introduction into industrial practice. The reason for this is mainly that the process requires, because of its high investment, not only

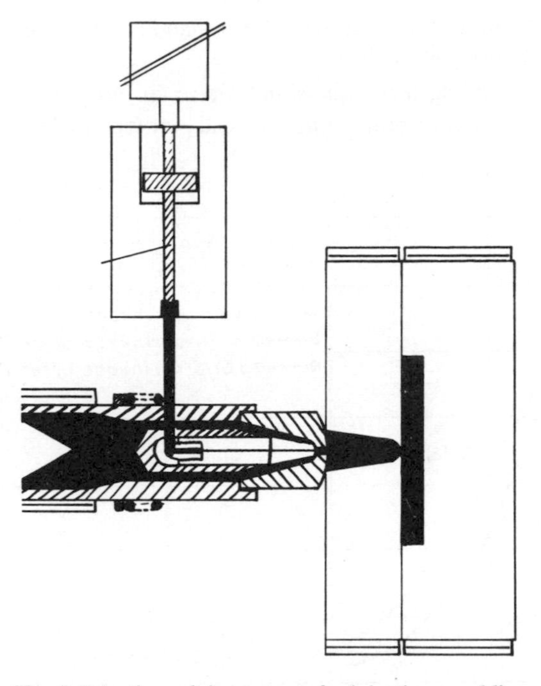

Fig. 9. Injection mixing process for injection moulding

large quantities of finished parts for the amortisation of the moulds, but also a fast shot sequence to obtain a high rate of utilisation of the rapidly conveying injection units.

4.4 *Other processing techniques*

Beside these methods there is the production of containers by sintering in rotating moulds, especially useful for powder processing and large parts which cannot be injection moulded. In the future, there will be attempts to operate these units more economically by processing other materials with them, for example monomers suitable for casting. A special advantage of this method is the possibility of producing several layers of different materials successively thus creating interesting material combinations in a finished part.

With regard to the manufacturing of finished parts, moulding techniques must be mentioned, especially compression moulding of semi-crystalline materials at temperatures near the crystallisation point. This technique allows a very small range of tolerance and shortest duty times of the moulds even for thick-walled parts, because there is only a minimum of heat required, which has to be removed after the process (Fig. 10). This method would be exceptionally interesting in the future if the base material–up to now expensive semifinished products–were considerably cheaper.

Fibre reinforced, wound, moulded or hand laminated parts of low-pressure resins still have with less than 4 per cent a small share in plastics processing. However, the method has its qualities and will gain importance by improvements being introduced. Regarding hand and wound laminates the possibility of laminating sandwich structures wet-onwet by using syntactic foam (that is without foaming pressure–Fig. 11) will make production cheaper because stiffness of structural parts can be improved at low cost[16]. Also, the processing of paste will lead to further rationalisation as the impregnation of hand laminates is transferred from manual to mechanised preparation (Fig. 12).

Finally, the filament winding technique should have an interesting future for big containers resulting from the introduction of programmed thread guiding or computer control.

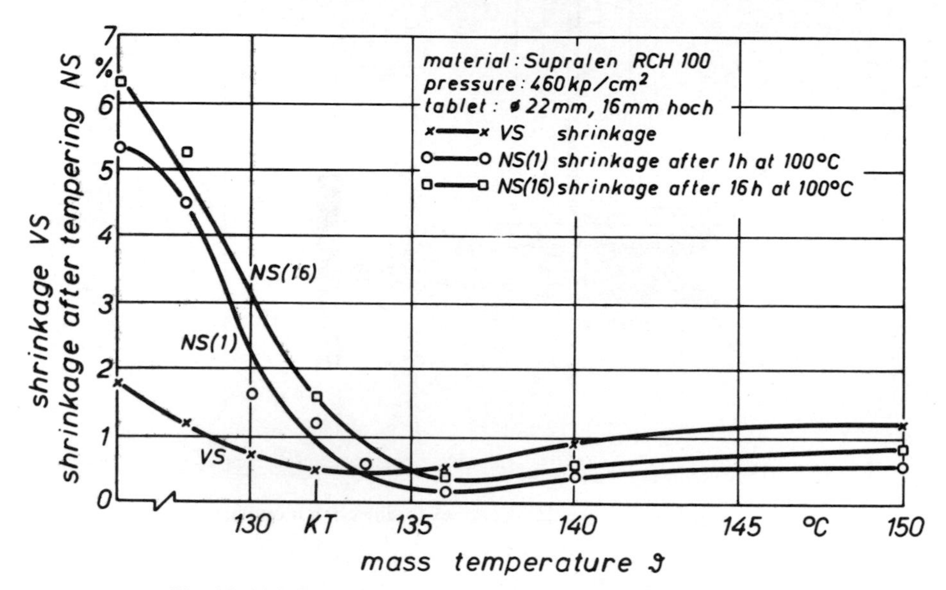

Fig. 10. Shrinkage of compression moulded parts *vs* temperature

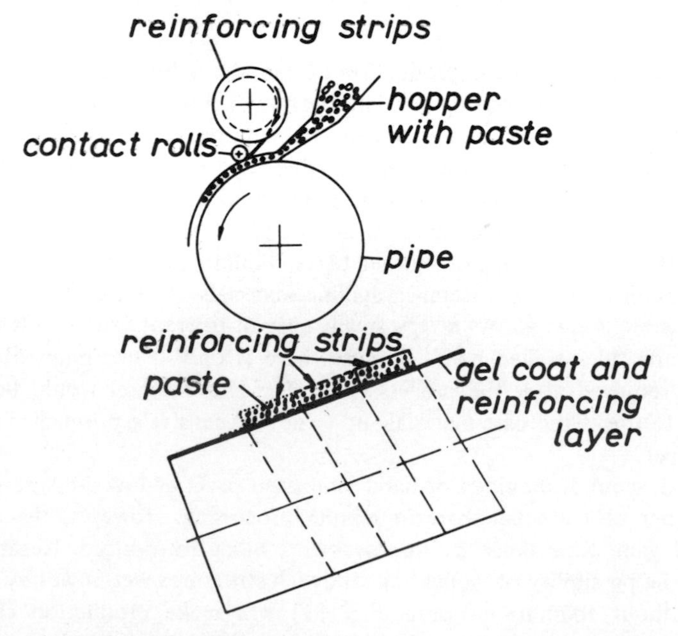

Fig. 11. Proposal for container production using a resin-foamed glass mixture

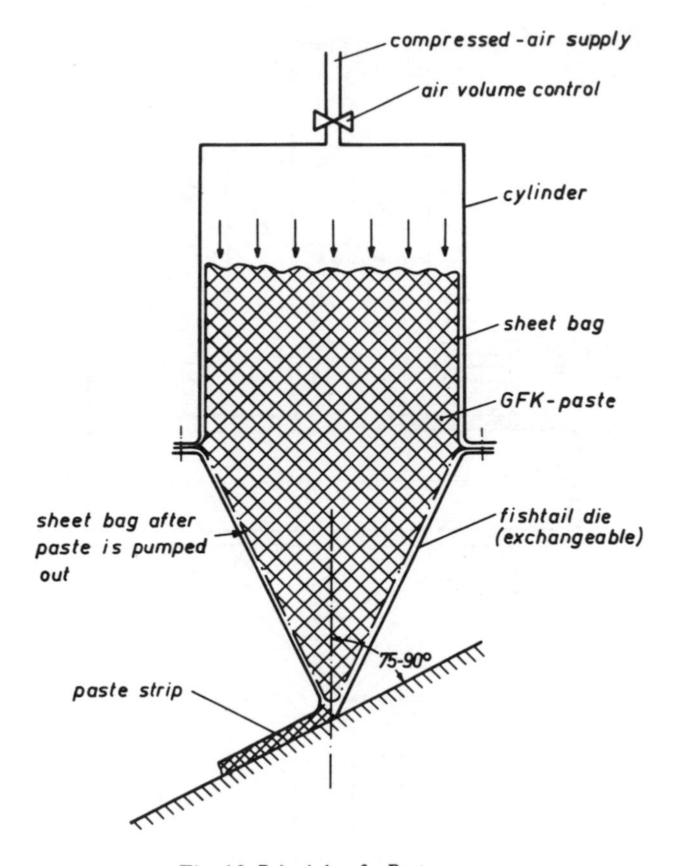

Fig. 12. Principle of a Paste press

4.5 *Extrusion with high pressure plastification*

High-pressure plastification can also be used successfully for extrusion. Not only can conventional single-screw machines be fed with melt from these units, but also completely new ways can be conceived.

With a radial configuration of a high-pressure plastification units for multiple feed of a mixer (Fig. 13), the units can be controlled easily in such a way that a constant output can be obtained. A particular advantage of this configuration is the possibility of producing tubes or hollow sections without marks from the wedge normally required to carry the mandrel. Apart from this application to extrusion the high-pressure plastification process is suitable for producing parisons for blow moulding and to feed calenders.

Also, the conventional extruder must change. A certain output can be obtained with extruders of high quality at low investment. This can already be seen in production where the competition is very hard, for example polyethylene films. Here the cheap working extruder without external heating was successful. Other fields will follow. Particular disadvantages of big machines for this method of operation are the long times required to reach stable operating conditions, that is thermal equilibrium, and the discharge of inhomogeneous melt at higher screw speeds. The first difficulty can be overcome by a cascade control[18], for example with an

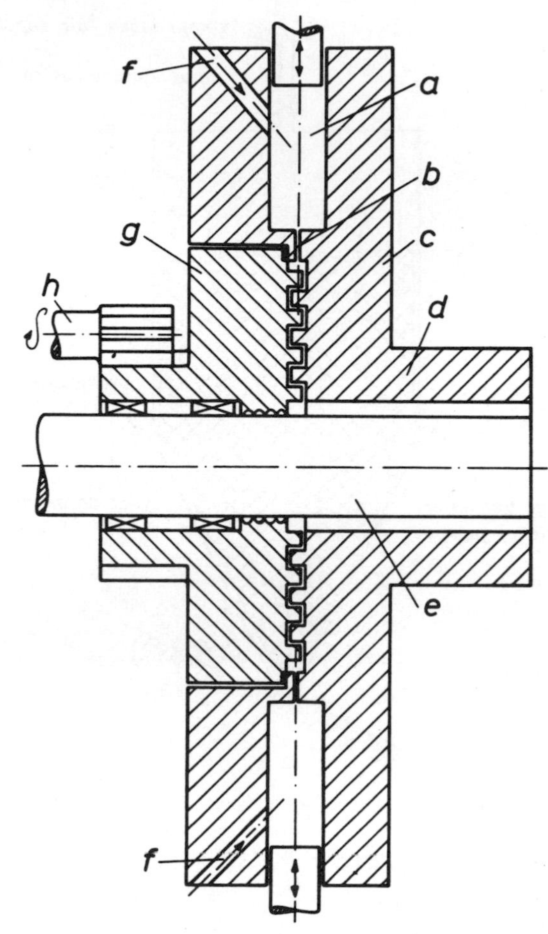

Fig. 13. High-pressure plastification units in radial configuration with disc mixer and fixed mandrel for (webless) markless extrusion of films, pipes and tubes (IKV-BASF-suggestion).

a high-pressure plastification unit
b shear gap (width 0·05 mm)
c stator of the mixer
d die for tube extrusion
e mandrel for tube extrusion
f material supply
g rotor of the mixer
h drive

ultrasonic sensor (Fig. 14) to measure the melt viscosity directly[19], and a special start-up control. The second problem of inhomogeneous melt is caused by the poor mixing ability of screws. Therefore, the screw must be altered to give genuine mixing as has been already shown successfully for big rubber extruders[12]. This means a complete change in the operation of the screw, because a good mixer which shall be made from the second part of the screw cannot convey any, but has to be fed by different mechanisms. For this purpose the simple concept of the solid pump is suitable. In the first part of the single screw a spindle nut is made by compression of the powder or granules which is locked against rotation by a grooved cylinder. A very efficient fluid tampering prevents the melting and shearing of the material in the grooves. This continuously formed spindle nut is conveyed in direction to the die, melts equally

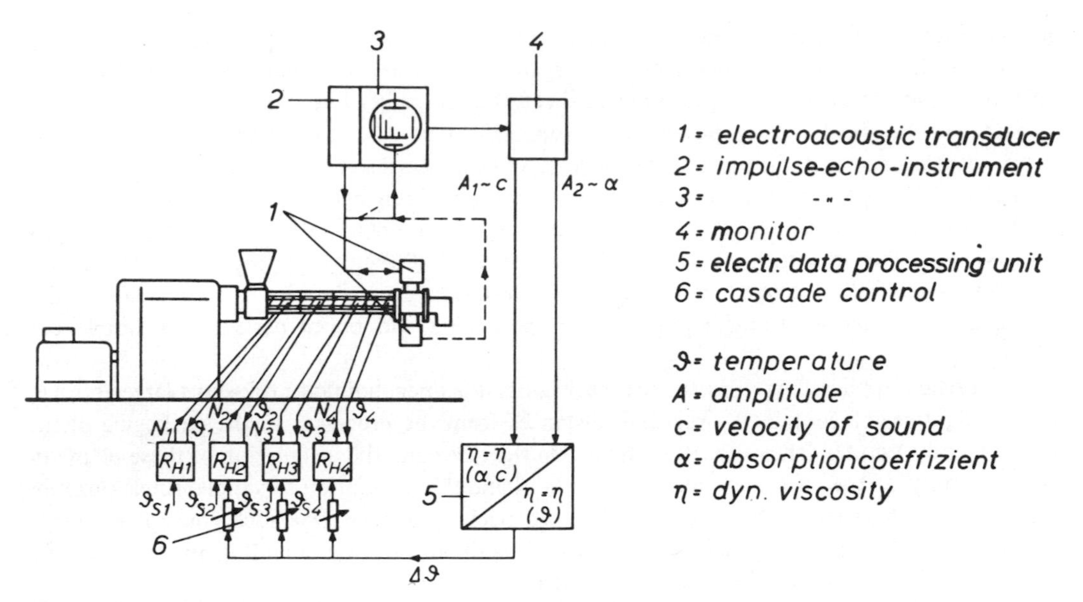

Fig. 14. Measuring the melt viscosity as controlled variable for extrusion by ultrasonic power (according to Pohlmann – IKV)

continuously off, and forces the melt through the following mixing zone. Of course, this method is exceptionally suitable for powder processing, because the high pressures which can be obtained in the powder bed just behind the hopper are sufficient to squeeze out trapped air.

Also auxiliary agents for processing, and other additives will increasingly be given directly to the powder and the melt respectively and mixed in the processing machine. To what extent the twin-screw machine will succeed beside the single-screw machine for processing depends on the success of the suggested improvements of the single screw.

Moreover in the future extrusion plants will be controlled by computers. Regarding the consecutive aggregates of extruders, cooling off will be the bottleneck, which limits off-take and thus production speed. Efforts to improve heat removal from the product, but also to put only the absolutely necessary amount of heat into the extruded material will continue. For example, we find sometimes very intricate cooling systems for pipes, profiles, and films, for instance with vacuum sections and spray-cooling.

4.6 *Blow moulding and calendaring*

The same problems arise in the field of blow moulding. Here, efforts are made to improve the extruder for powder processing–for example only single-screw machines for pvc. Machines with multiple heads or moulds also show here that heat removal is a problem. For big machines accumulators are predominant which will give little competitive chance to reciprocating screws. Similar arguments apply to injection blow moulding, which have proved until now to be successful only for wide-necked containers of materials such as polystyrene if the quantities were big enough.

Finally calendering must be taken into consideration, for the bulk of films and sheet is still produced with these high-quality machines. Present capacities of 100 metric tons/cm working width could be increased if a computer was used for self-adaptive control and if there were a greater number of calendar nips. The calendar's advantages of high production speeds for smaller batches and best qualities especially regarding the tolerance range, cannot be disputed. Calendaring will exist as long as pvc film and sheet of high quality is in demand.

Summarising the trends in plastics processing it can be said that there is a wide field for future developments. As in previous years (Fig. 15) the plastics machine-building industry can count on high rates of growth. Its share of the machine-building market will grow equally.

According to U.S. estimates[2] the projection of injection moulding machines will reach 6000 units/year in the seventies, in 1980 it is said to be 7000 units. In regard to extruders a production of 4000 units/year is expected in 1980. There will be also a big increase in the production of thermoforming equipment for packaging and bulky parts (for example car bodies).

For certain applications, for instance, packaging, it is uneconomic to transport large-volume, empty, lightweight containers over long distances from the processor to the packaging plant. Consequently two tendencies are developing. On the one hand the installation of these plants in the proximity of the customer, and on the other hand the integration of simple, single-purpose machines which can be incorporated easily into packaging lines. One such concept envisages that a fluid which is to be packed will be filled directly at the blow moulding machine into the just-moulded container, for faster colling of the plastic container.

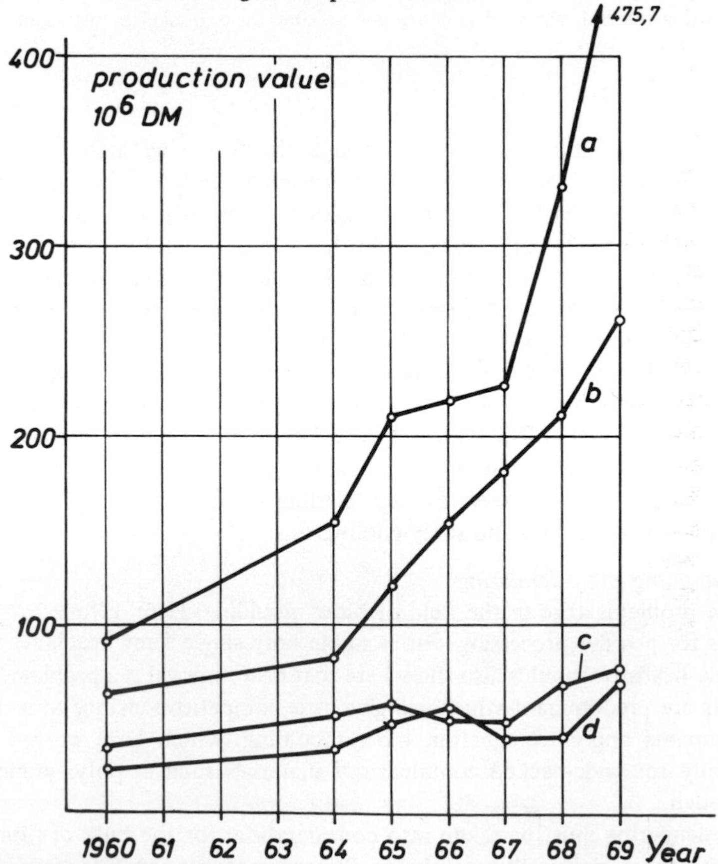

Fig. 15. Production values of rubber and plastics processing machines in West Germany.

a **injection moulding machines**
b **extruders**
c rolling-mills, calendars
d moulding presses

REFERENCES

1 HOUWINK, R., 'Das Zeitalter der Chemiewerkstoffe', *Kunststoffe*, Bd. 56 (1966), S. 597–598.
2 FRADOS, T., 'Plastics in the 1980s–a 15 year outlook', *Modern Plastics*, Bd. 45/15 (1968), S. 120.
3 N.N., 'Statistisches Handbuch 1970 des VDMA'.
4 MENGES, G. and DALHOFF, W., 'Die Entwicklung der Kunststoffe und ihre Hauptanwendungen: In: 1980 ist morgen', Droste Verlag, Dusseldorf, 1969.
5 TRIESCHMANN, H. G., 'Mineralolwirtschaft und Chemiewirtschaft', *BASF-Nachrichten*, Bd. 18 (1968) E1/E8.
6 GATH, R., In: Neue Verfahren, neue Produkte. Beiheft zu Nr. 40 von Der Volkswirt vom 3 10 1969, S. 19–24.
7 MENGES, G., In: Neue Verfahren, neue Produkte. Beiheft zu Nr. 40 von Der Volkswirt vom 3 10 1969, S. 14–19.
8 N.N., 'Gaps in Technology', 3rd Ministerial Meeting on Science of OECD Countries, 11 and 12, March 1968.
9 MENGES, G., 'Aspekte der Kunststoffverarbeitung und Aufgaben fur die Aachener Gemeinschaftsforschung', *Plastverarbeiter*, Bd. 21 (1970), S. 305–312.
10 MENGES, G. and HEGELE, R., 'Extrudieren pulverformiger Polyolefine auf Einschnecken-Extrudern', *Plastverarbeiter*, Bd. 21 (1970), S. X-1 – X-9.
11 BIELFELDT, F. B. and MORELL, H. J., 'Elektronisches Steuern und Programmieren von SpritzgieBmaschinen', *Kunststoffe*, Bd. 60 (1970), S. 373 bis 377.
12 LEHNEN, J. P. and MENGES, G., Dechema-Kolloquium am 7. 11. 1969, Chem.-Ing. Technik.
13 MENGES, G., DALHOFF, W. and MOHREN, P., 'Hochstdruckplastifizierung von Kunststoffen, ihre Bedeutung und Anwendung', *Kunststoffe*, Bd. 60 (1970). S. 85–88.
14 LETTNER, H., 'Fortschritte bei SpritzgieBverarbeitung treibmittelhaltige Thermoplaste', 1. Teil: *Kunststoffe, Bd. 60 (1970) S. 216–221, 2. Teil: Kunststoffe*, Bd. 60 (1970), S. 367–373.
15 RHEINFELD, D., to be published.
16 MENGES, G., KELLENTER, M. and KLEINHOLZ, R., 'Mit Schaumglasgranulat gefullte modellierfahige Pasten als Stutzkerne fur Sandwichkonstruktionen', Vortrag auf der 9. AVK-Jahrestagung, 1970, Freudenstadt.
17 MENGES, G., and KELLENTER, M., 'Mechanisierung der GFK-Fertigung durch Pastenverarbeitung', *Plastverarbeiter*, Bd. 21 (1970), S. XX-1 – XX-6.
18 MENGES, G. and MEISSNER, M., 'Erkenntnisse auf Grund regeltechnischer Untersuchungen an Extrudern', *Plastverarbeiter*, Bd. 21 (1970), S. X1-1 – X1-8.
19 POHLMANN, Diskussionsbeitrag zum 5. Kunststofftechnischen Kolloquium des IKV Aachen, Marz 1970.
20 SCHONBORN, H. H., 'Probleme der Abfallbeseitigung unter besonderer Berucksichtigung der Kunststoffe', *Schriftenreihe des Deutschen Rates fur Landpflege*, H. 13 (1970), S. 16–20.

NEW MATERIALS IN THE CAPACITOR INDUSTRY AND THE INFLUENCE OF A BUSINESS ENVIRONMENT

D. S. Campbell
Plessey Capacitor Division

SUMMARY

The capacitor industry in this country manufactures into a total UK market of approximately £30M and a European market of roughly four times this. Advances in the market arise from the need to increase performance, reduce size and reduce costs and this very much depends on the use of new materials – new plastics, new treatments of aluminium or tantalum, new encapsulation etc. This paper describes the business environment of capacitor manufacture and the various types of capacitors that are produced. The constraints in terms of finance for development are then examined and this is followed by a short discussion of the organisation that can be evolved to manage the allowed expenditure. Having set the total scene, the effects of materials advances are examined for two examples, namely the use of polypropylene sheet in capacitors and also the development of tantalum capacitors. The constraints of the business environment on these two developments are discussed.

1.0 INTRODUCTION

This paper is concerned with the influence of a business environment on new materials research and development in the capacitor industry. The capacitor industry in this country manufacture into a total United Kingdom market of approximately £30M per annum and a European market of roughly four times this. Advances in the market rise from the need to increase performance, reduce size and reduce cost, and this very much depends on the use of new materials – new plastics, new treatments for aluminium or tantalum, new encapsulation etc.

This paper is divided into five parts. First the present business environment of capacitors is described. Secondly the products, that is the different types of capacitors available, are briefly discussed. Thirdly the financial constraints imposed on the materials research and development are examined. Fourthly the organisation that can be built up to allow of materials research and development, given the financial constraints, is considered, and finally two examples of the way materials have developed in the total capacitor environment are discussed in some detail.

2.0 BUSINESS ENVIRONMENT

Capacitors have been developing ever since 1746 when Pietre von Musschenbrook accidentally discovered glass dielectric capacitors, (the Leyden Jar Capacitor). Many different dielectrics have been used since then and the electrodes have taken many different forms. All these forms tend to have slightly different properties and it is the optimisation of the properties, rather than any change in the fundamental behaviour of the capacitor, that has led to the advances that have occurred.

Nowadays capacitors are used for a wide variety of purposes and these uses tend to call for different properties. Table I gives a summary of the main applications. However, as with so many products the main emphasis in all cases is in making the capacitor more reliable, smaller and at the same time cheaper.

TABLE I: MAIN USES OF CAPACITORS

Market	Application
Entertainment	T.V. Radio
Industrial	Fluorescent lighting Cars Washing machines Vacuum cleaners Suppression Food mixers Power factor correction – industrial electrical installations including power stations. Energy storage Electrical motor starting and running High-voltage (voltage dividers, smoothers, surge suppression)
Professional	Electronics (filtering, smoothing, decoupling) Computers (power supplies, circuits) Telecommunications

The three main markets shown in Table I can be separated out from the business planning point of view and the information is shown in Table II.

In order to predict future trends it is necessary to examine the past history of these three markets using the statistical information that is available (for example[1,2]). The results for the entertainment market are shown in Fig. 1, where the number of pieces manufactured per annum is shown against the year, from 1950 to the present day[3]. It can be seen, particularly with regard to the monochrome television market, that it is subject to wild fluctuations from one year to the next, these fluctuations going in roughly a four-year cycle. It is apparent that manufacturers tend to over-produce during certain years and then to cut back heavily

TABLE II: BUSINESS PLANNING INFORMATION ON THE THREE MAIN CAPACITOR MARKETS

Market	U.K. expenditure (M£/annum)	Average growth (%)	Best growth (%)	
Entertainment	9	10	Colour T.V.	30
Industrial	11	5	Washing machines	10
Professional	10	15	Computers	20

when the over-production is realised. A similar situation applies, though on a smaller scale, to all the other items classified as entertainment business. It should be noted that colour television has, until 1970, shown a considerable rise although the first of the cut-backs is now with us even in this area.

For the other markets there are not such wild fluctuations. Fig. 2 shows the United Kingdom output of professional equipment plotted in terms of the total business per annum; figures are available only from around 1960 to the present time[3]. Here it can be seen that there is a fairly uniform rise both in computers and in telecommunication equipment.

An examination of the industrial market reveals a situation intermediate between that of the entertainment and professional markets.

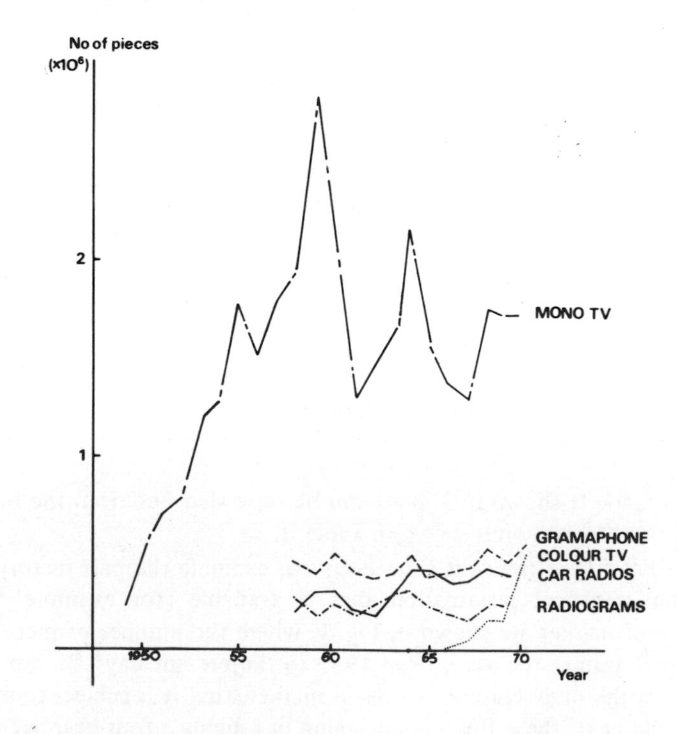

Fig. 1. U.K. output of entertainment equipment as a function of time.

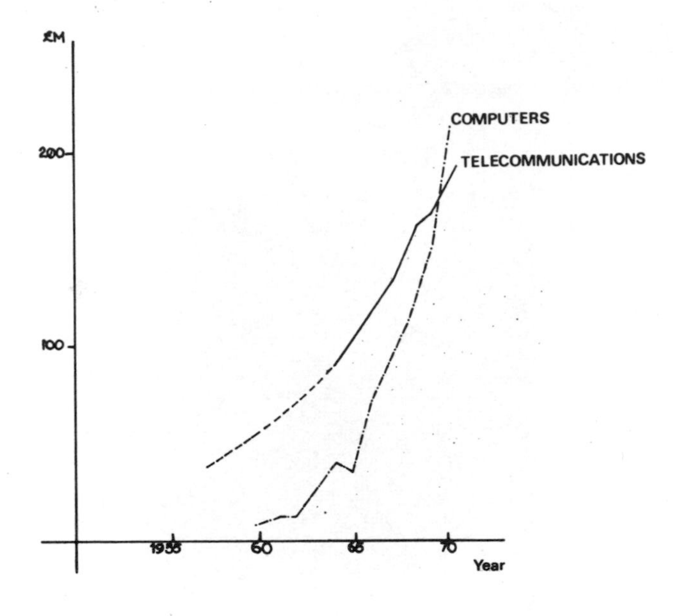

Fig. 2. U.K. output of professional electronic equipment as a function of time.

Because of the approximately equal size of each market a viable capacitor manufacturing business will try and have within it elements of all three markets. Attempts should be made to overcome the worst effects of market fluctuations by careful attention as to which capacitors to produce and also to stocking policy.

3.0 PRODUCTS

The basic construction of a capacitor consists of a metal electrode, a dielectric layer and a second metal electrode[4,5,6]. The dielectric can be of many different materials – air, ceramic, glass, evaporated silicon monoxide, paper, plastic, oxide films prepared by anodisation of metals, to name but a few. Capacitors based on the last three dielectrics are the only ones that are considered in this paper.

3.1 *Paper capacitors*[4]

These are made using very thin paper tissue down to 5 microns in thickness, and materials advance has given very strong papers for this application. Even so, it is usually necessary to wind the papers in pairs so that any small holes or flaws present in one tissue layer are most unlikely to occur opposite similar holes or flaws in the second layer – thus the dielectric strength of the total capacitor is increased. To give a high dielectric constant, the paper capacitor tissue is impregnated, usually with an organic oil, and the total capacitor is enclosed in a case which prevents seepage of this impregnating fluid. Figure 3 shows a typical example of a paper capacitor used for discharge applications at a repetition rate of two discharges per second.

Fig. 3. Typical paper dielectric discharge capacitor (repetition rate: 2 pulses/second).

3.2 *Plastic capacitors*[7]

Over the last ten years or so, very thin sheets of electrically useful plastic material have become available and these can be as thin as 2 μm. Capacitors made from plastics (that is polyester, polycarbonate, polystyrene, polypropylene) do not have to be impregnated in any fluid if they are going to be used at voltages less than 250V a.c. However, above this value it is advisable to impregnate if only to prevent the internal discharges that will otherwise occur in the air gap that will unavoidably exist between the metal electrode and the plastic sheet itself.

3.3 *Suppressors*[8]

Using either paper or plastic as the dielectric a third class of capacitor structures can be recognised; suppressors and filters. These units are basically capacitors which not only use the capacitance of the unit but also the inherent resistance and inductance that is found to be present[9]. In their construction it is often necessary to enhance the inductance by, for example, winding the capacitor on a core of magnetic material. The resistance, can also be enhanced by making the electrodes of an evaporated metal film, evaporated directly onto the plastic sheet, but deposited in such a pattern as to give a much higher resistance than that which would normally be obtained[10].

Fig. 4 shows some typical suppressor capacitors. Items *A* and *B* are car radio suppressors. Item *A* is manufactured using a paper dielectric and *B* is the equivalent capacitor using a much thinner plastic dielectric sheet. Items C and D are networks used in vacuum cleaner suppressors.

Such constructions can in fact save a lot of space. Fig. 5 shows an LC network prepared for use in sewing machine suppression[11]. The suppressor wound with plastic dielectric is shown as Item A and this replaces the four components shown on the board as Item B, namely two separate capacitors and two inductors. There is thus a considerable saving in space and there is also a saving in cost.

3.4 *Aluminium electrolytic capacitors*[12]

In aluminium electrolytic capacitors the dielectric is prepared by anodising a sheet of aluminium foil. This process can give a very thin oxide layer on the aluminium foil, down to 0·01 μm in thickness. As a result aluminium electrolytic capacitors tend to have very much higher capacitance per unit area and hence per unit volume than do paper and plastic type capacitors. However, because of the method of construction in which an electrolyte is necessary to give the capacitors the self-healing properties required and also so as to enable contact to be made between the dielectric deposited on the etched surface of the aluminium and the second aluminium electrode, these types of capacitors tend to have very much higher electrical losses than the equivalent paper and plastic types.

Fig. 6 shows some typical examples of aluminium electrolytic capacitors. These vary from Item 1, large smoothing capacitors for computer power supplies, which go up in size to around 200 000 μF down to Item 12, small plug-in electrolytic capacitors, used in television circuitry. These latter capacitors have capacitance values of around 8 μF.

3.5 *Tantalum electrolytic capacitors*[12]

Tantalum may be anodised in a similar manner to aluminium, and, the oxide that results is even more stable than that of aluminium. Therefore tantalum capacitors tend to be used for applications that require high stability, particularly in terms of shelf-life. The disadvantage of using tantalum is that the material is very much more expensive than aluminium, costing approximately the same as silver.

It is possible to anodise tantalum foil and thereby obtain capacitors equivalent in construction to those obtained by anodising aluminium foil. However, it is also found that sintered tantalum powder can be used in place of the foil. This sintered powder, which is in effect a tantalum sponge, can have a very large surface area, and as a result the total unit can have a high capacitance per unit volume. With this type of construction a wet electrolyte can be used to contact the oxide surface giving the 'wet' tantalum electrolytic capacitor. It has also been found possible to use a dry electrolyte, in effect a semiconducting oxide, to connect the oxide to the second electrode, giving the 'solid' tantalum electrolytic capacitor. In both the 'wet' and 'solid' types the second electrode is often made of silver.

Fig. 7 shows various types of capacitors that are made in tantalum. Items 1 through to 6 are of types using tantalum foil. Item 7 is a 'wet' tantalum electrolytic capacitor (that is using sintered tantalum as the basic material) and all the other items from 8 through to 22 are using sintered tantalum and a solid electrolyte the capacitor being encapsulated in various cases (metal tubes, Items 14 to 16, or plastic cases, Items 8 to 22 etc.).

4.0 FINANCIAL CONSTRAINTS

The amount of money that should be spent on research and development is a function of many different things. Firstly, one has to find a level of expenditure which is consistent with the amounts being spent in the industry as a whole, particularly with regard to one's

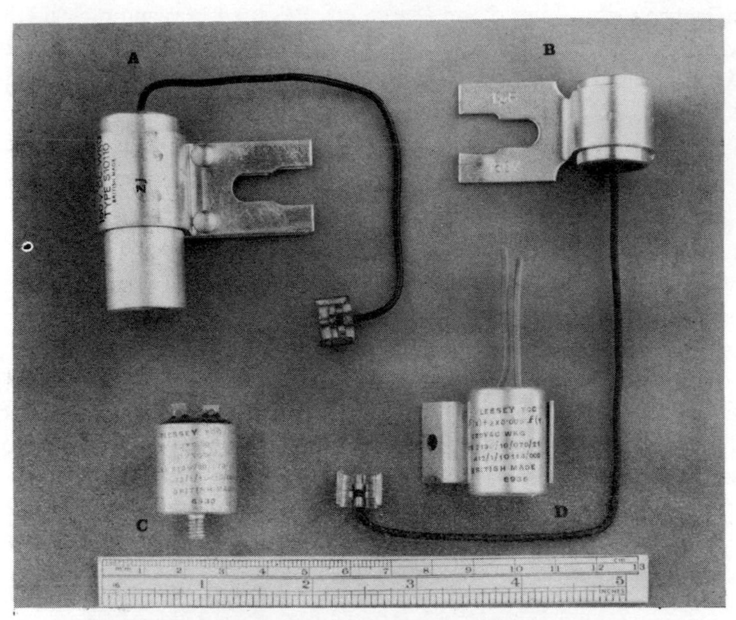

Fig. 4. Typical suppressors. *A* & *B* for car radio. *C* & *D* for vacuum cleaners.

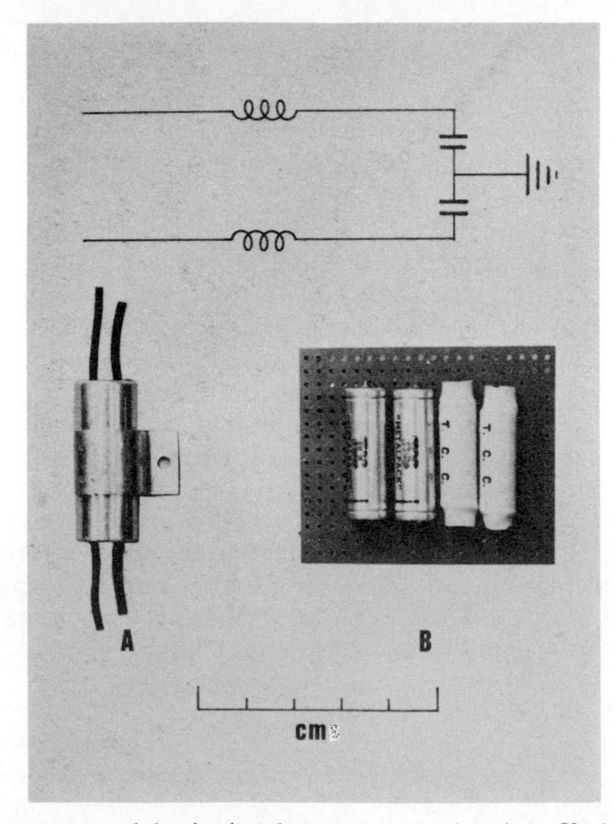

Fig. 5. Typical *LC* suppressor, and the circuit and components that it replaces. Used for sewing machines.

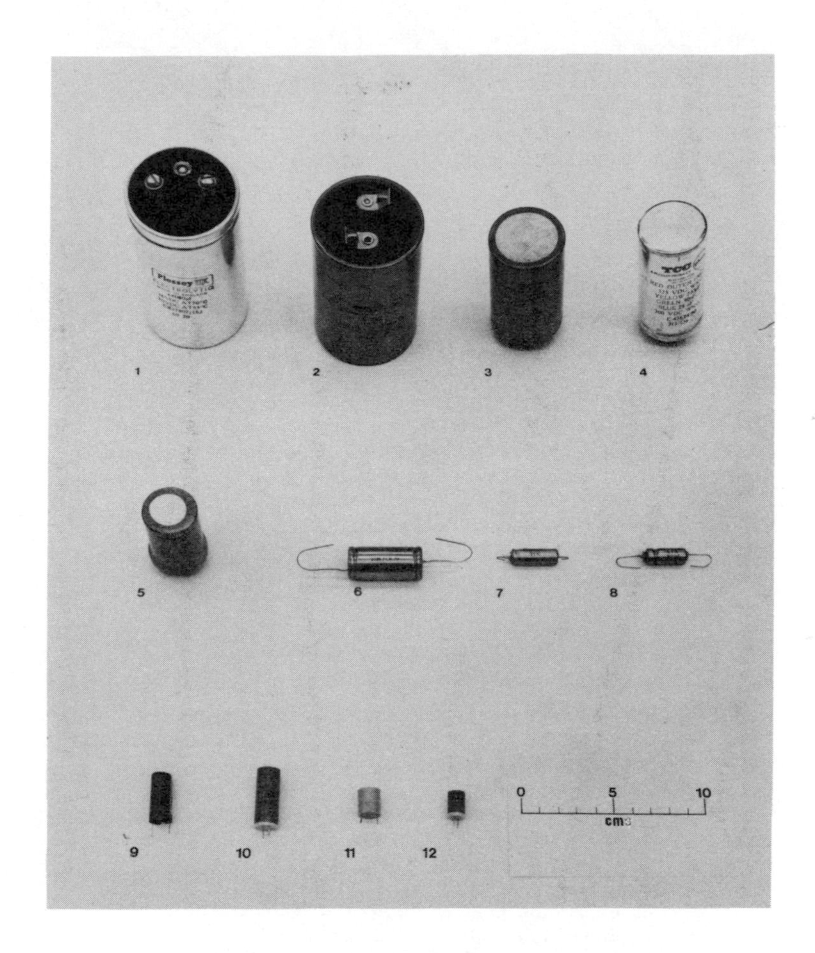

Fig. 6. Typical aluminium electrolytic capacitors.

1. 68 000 μF, 10V. Computer power supply
2. 1750 μF, 325V. Photoflash.
3. 250 μF + 500 μF + 50 μF, 300V. Triple – for television circuits.
4. 100 μF + 200 μF + 50 μF + 25 μF, 300V. Quadruple – for television circuits.
5. 20 000 μF, 2·5V. Low voltage.
6. 500 μF, 25V. Tubular.
7. 560 μF, 10V. Sub-miniature.
8. 220 μF, 5V. 125°C working. Extended temperature range.
9. 50 μF, 12V. Plastic case.
10. 4 μF, 350V. Plastic case, plug-in.
11. 1 μF, 250V. Plastic case, plug-in.
12. 2 μF, 63V. Plastic case, plug-in.

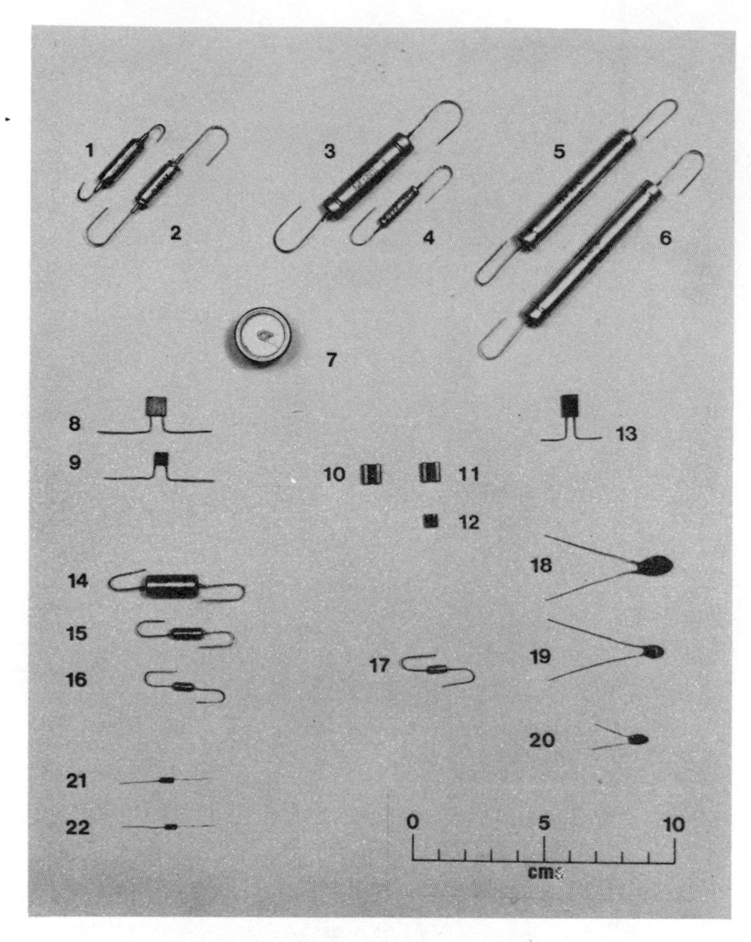

Fig. 7. Typical tantalum electrolytic capacitors.
1. Plain foil 16 μF, 6V. 85° C.
2. Plain foil, Reversible. 8 μF, 6V. 85° C.
3. Etched foil. 5 μF, 150V. 125° C.
4. Etched foil. 0·22 μF, 160V. 125° C.
5. Etched foil. 580 μ F, 10V. 125° C.
6. High-gain etched foil. 3000 μF, 10V. 125° C.
7. Wet electrolyte. 50 μF, 70V.
8. Plastic case, solid. 3·9 μF, 10V.
9. Plastic case, solid. 4 μF, 20V.
10. Flip-clip, solid. 6·8 μF, 20V.
11. Same type as 10 – reverse side.
12. Flip-chip, solid. 1 μF, 50V.
13. Transfer moulded plastic case, solid. 6·8 μF, 20V.
14. Metal case, glass seal, solid. 15 μF, 35V.
15. Metal case, glass seal, solid. 1·5 μF, 50V.
16. Metal case, glass seal, solid. 0·068 μF, 35V.
17. Metal case, resin seal, solid. 0·088 μF, 35V.
18. Tear drop, solid. 68 μF, 15V.
19. Tear drop, solid. 3·3 μF, 33V.
20. Tear drop, solid. 1 μF, 35V.
21. Hearing aid, solid. 4·7 μF, 6V.
22. Small hearing aid, solid. 3·3 μF, 6V.

competitors. Secondly, one must consider the influence of the business size and the planned growth rate of the business. A company may expand in a particular product either by the development of new or better products from within or by acquisition of other companies that are already making new products. In the latter case, the level of expenditure required on r. & d. can be minimal, consistent with retaining enough staff to advise on the technical aspects of any proposed merger and to help consolidate the product position after merger.

A third factor that is a useful guide in determining the amount of turnover spent on r. & d. is the obsolescence time of the product[14,15,15a]. Such a relationship between the percentage of turnover spent and obsolescence time is plotted in Fig. 8 and it can be seen that a business which manufactures products which have a very short life before they are replaced by something newer and possibly better may well require an expenditure of 20% or more of turnover on r. & d. to even keep the business standing still. On the other hand if the products have obsolescence times of 100 years or more very little expenditure, in terms of r. & d., will be required. If one applies this type of curve to the electronics industry as a whole the obsolescence lifetime of electronic equipment is generally taken as between 2 and 10 years. Thus the r. & d. expenditure necessary to maintain a constant level of progress should be somewhere between 10 and 15%.

A relationship such as that shown in Fig. 8 will only act as a guide to the expenditure required on the basis of a policy of planned growth by the development of new or better products from within the organisation. Any other business strategy will not require such large expenditure.

Table III taken from MacLeod[13], gives the average figures for several different industries. The percentage of turnover spent on r. & d. and design is given against the particular industry.

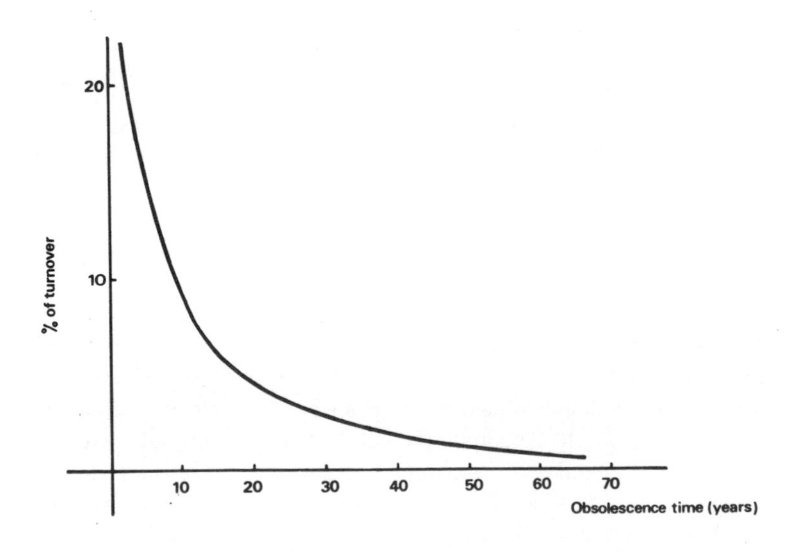

Fig. 8. % of turnover to be spent on r. & d. as a function of obsolescence time.

TABLE III: % OF TURNOVER SPENT ON R. & D.
FOR VARIOUS INDUSTRIES (1968)

	% turnover
Electronics	14
Instruments	6
Chemicals	5
Machinery	3
Vehicles	1

Such a relationship arises from the influence of all the factors discussed above. From these figures it can be seen that on average the electronics industry is spending r. & d. money at a rate consistent with planned growth by development from within.

When one applies such a concept to the capacitor industry one observes an obsolescence time of the order of 20 years, although there is a very wide spread on this figure (−10 to +30 years) dependent on the design or dielectric that one considers. From such a figure it would therefore seem that the amount of money spent on r. & d. should be around 5% of the turnover for growth from within the organisation.

5.0 ORGANISATION

In order to spend efficiently the money that is available it is necessary to have a reasonable understanding of the most satisfactory organisation of resources to achieve this purpose[10]. A major point that one must bear in mind in setting up such an organisation is that it is useless to concentrate all one's money on research as such. A breakdown of costing on an efficiently run developing business should give figures for expenditure of the following relative percentages in terms of the total expenditure:

- 5% should be spent on research
- 40% should be spent on developing products discovered during the research
- 55% should be spent on the industrial engineering necessary to get the developed product into production.

This split of funding is not always followed. It is only too easy to spend money on research and then not to be able to follow this up with the necessary expenditure on development and industrial engineering.

Fig. 9 shows a typical organisation diagram which defines various activities that are necessary to make efficient use of r. & d. expenditure. Various areas can be identified. Basic research as such, which is not necessarily oriented to any particular product, is probably best conducted in a university environment. When ideas have got to the stage of application to actual devices, then the work can be transferred into the applied research area, which is quite satisfactorily situated in a central laboratory belonging to a company, in a government research establishment or in a research association laboratory. When the product has got to a stage of proved feasibility then it should be transferred to a new product development laboratory which ideally is situated in the factory environment. In this environment it is possible to begin to get the industrial engineering aspects into perspective so that after some expenditure and time the new product can be brought to production.

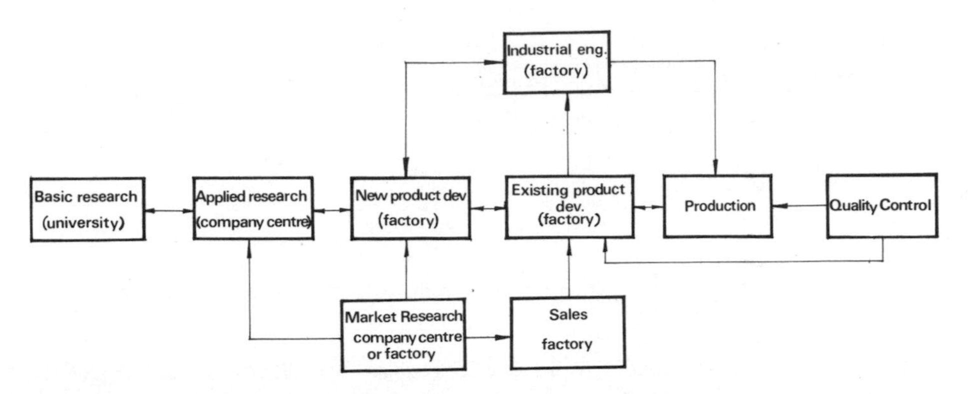

Fig. 9. Activity diagram showing the different types of r. & d. in relationship to production, market research, sales, industrial engineering and quality control.

Also included in Fig. 9 are other activities which bear on the r. & d. activity. Most important of these is that of market research which can be centred either at the factory or at the company centre. The importance of this activity cannot be over-emphasised. It is because of this activity that the applied research group and also the new product development group know the products to investigate. It can be seen that market research also influences sales, and it is necessary to distinguish between market research, which is looking into the future with regard to market changes and market trends, and sales as such, which is a factory-based operation selling the product which the factory produces.

Also included in the diagram is an activity shown as existing product development and trouble-shooting. Even in the best-run organisation things go wrong in the product on a day-to-day basis, either in terms of customer complaints or in terms of production troubles, and a technical group is necessary to solve the problems that arise in this manner. There is a relationship between this activity and new product development, sales, and industrial engineering, as shown in the diagram.

Finally, the function shown as Quality Control must be involved[18]. This is the activity that vets the output of the factory and passes it in relationship to the published specifications. Should trouble arise where the products of the factory do not meet the specifications then it will be necessary to refer the problem back to the existing product development group for solution. In the case of a new product which has come through basic research, applied research, new product development and industrial engineering, the quality control must agree that the new product can meet the specification to which it has been designed. Therefore quality control can be said to be the ultimate arbiter on the production of a new product.

In the organisation which is illustrated in Fig. 9, and using money available as discussed in Section 4.0, it is necessary for a decision system to be set up[19]. This is shown diagrammatically in Fig. 10 as a decision diagram for taking ideas to production. When the idea has been formulated it is necessary to reach a decision as to whether or not the idea is worthy of laboratory examination, and, to decide this, both market research and r. & d. expertise are required. The decision does not necessarily mean the expenditure of large amounts of money but rather that this project is going to be looked at instead of something else.

In this context a conscious decision must be made to conduct the research and development on a parallel or series base[17]. If only a limited amount of money is available it is often very

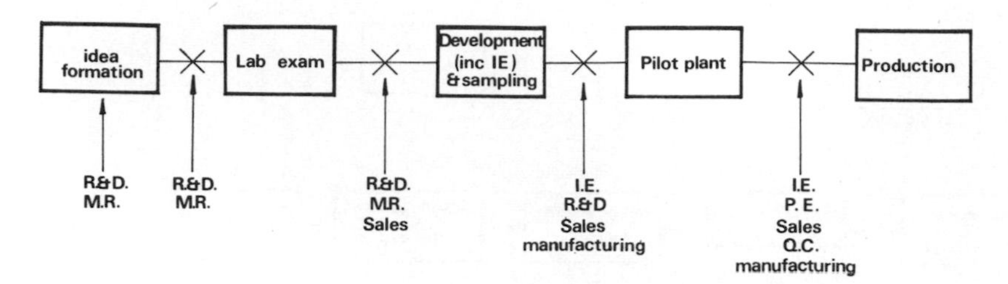

Fig. 10. Decision diagram for taking ideas to production (major decision points at *X*).

good sense to organise the expenditure on a series base as this will enable one project at a time to be brought to fruition and furthermore will enable market research to determine the validity of delayed items to a much greater degree. Helier[17] has in fact calculated that for certain cases there can be a 35% increase in profitability if projects are run on a series base. It does however mean that a certain ruthlessness and single-mindedness in the selection of projects is vitally necessary in the decision.

If the project passes laboratory examination then it is necessary to decide whether to start spending larger amounts of money on actual development, including a certain amount of industrial engineering. If development proves satisfactory and the market has shown a willingness to receive the product (samples have been supplied to the market and there has been adequate come-back), then the decision has to be taken as to whether or not to start spending what can amount to considerable amounts of money on a pilot plant which will manufacture the product in a similar way to the final production manufacture. Finally, a decision point is given with regard to the taking of the product to full-scale production and at this stage it can be seen that the skills of industrial engineering, production engineering, sales, quality control and manufacturing as such are all required before such a decision can be made.

Such a diagram indicates only four major decision points and, of course, in actual practice these decision points may be introduced at many different stages.

The total problem then is to set up an organisation that uses available finance and personnel[20] in the most efficient way. Of course it must never be forgotton that the personnel aspect is a very important one, because without a motivated staff the rate of development will be slow.

6.0 EXAMPLES

In order to see how the business environment that has now been discussed influences materials research and development in the capacitor industry two examples will now be examined.

6.1 POLYPROPYLENE

Figure 11 shows a materials development diagram for the polypropylene sheet used as the dielectric in plastic capacitors. The initial work on this, the basic research, started in the middle 1950s, when it was found that it was possible to make a satisfactory solid out of polypropylene. This work led to applied research on solids and also plastic sheets. The plastic sheets were relatively thick and were not intended for electrical use.

It can be seen from the diagram that this work led to the development, and thence to the

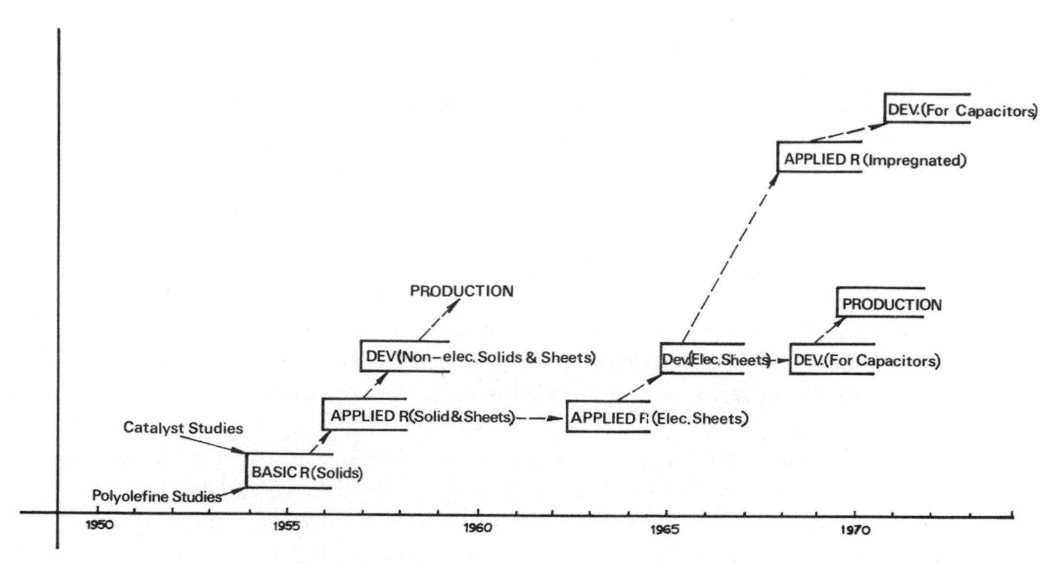

Fig. 11. Materials development diagram for electrical grade polypropylene.

production, of solids and sheets of a non-electrical nature. The market situation with regard to polypropylene meant that the activities of the manufacturers would be centred on applications other than electrical ones: the electrical applications of polypropylene were not seen to be more than 10% of the total market. It therefore was necessary to wait until around 1963 to see applied research activity starting up on the applications of polypropylene sheet to electrical uses.

To obtain sheets that were thin enough, 12 μm or less, and that were completly free from pin-holes, raised problems other than those that had been examined in the 1950s. However, this work was carried on, and eventually satisfactory electrical sheets were developed using extrusion systems.

Because of the availability of this sheet it was possible to start development work on the use of this material for capacitors. The problems involved are those of metallisation and the end-connections that have to be applied. These have now been more or less solved, so that production of capacitors using polypropylene sheet has now started up at different factories across the world.

The thinnest sheets that are used at the present time are 6 μm in thickness. The properties of such a capacitor are listed in Table IV.

TABLE IV: PROPERTIES OF 6 μm POLYPROPYLENE DIELECTRIC CAPACITORS

ϵ	=	2·25
Dielectric loss	$\leqslant$	0·01%
Capacitance/unit volume	=	0·5 μF/cm^3
Working temperature	=	70° C
Working voltage	=	250V a.c.

This however is not the end of the story, as the type of capacitor that has been developed has a limitation in voltage application for reasons which have been mentioned. It is therefore necessary for work to continue on the use of polypropylene with impregnants, and applied research has been in hand since 1968[22,23]. The requirement is for an impregnant that will not affect the polypropylene so that the capacitors cannot be used at high voltages and at high frequencies.

A subsidiary requirement that is coming in at the moment is for an impregnant that is not a pollution hazard – a particularly strong requirement in the United States. One of the organic impregnants used in paper capacitors, a material known as Aroclor, is toxic and stable and it is expected that very shortly there will be a ban on the use of this material in America. It is therefore necessary to find an adequate substitute impregnant for paper capacitors as well.

Development of impregnated polypropylene for capacitor application is now starting up at various centres.

Thus, from the basic work on solids in the middle 1950s, has arisen a whole range of activities both in terms of non-electrical and electrical requirements. The driving force from the point of view of capacitors has been the need to reduce the size and reduce the cost and increase the reliability of capacitors. Fig. 12 shows the result of this type of activity as applied to ballast capacitors used for fluorescent lighting. The capacitor on the right is a paper capacitor, 8·4 μF in size, using impregnated Aroclor paper. A similar capacitor developed out of polypropylene is shown on the left. This has the same voltage and capacitance rating but is half the size and one third the weight, and furthermore cannot leak impregnant.

Fig. 12. Paper dielectric fluorescent lighting capacitor (right) and equivalent polypropylene dielectric fluorescent lighting capacitor (left).

6.2 *Tantalum electrolytic capacitors*[12]

The development diagram for tantalum electrolytic capacitors is shown in Fig. 13. Two basic-research starting-points can be seen: first, basic research on sintering of tantalum, which was started in 1935; and second, basic research on semiconducting oxides, started around 1950.

From the basic work on sintering, which led to applied research on sintering, the 'wet' electrolyte, anodised tantalum, capacitor was developed. This came into production in 1947. Work on sintering has in fact continued, leading to increases in the surface area of the powder, and recently very-high-gain powders have become available that have enabled capacitors with higher capacitance/unit volume to be prepared[24]. The older types of sintered tantalum capacitor gave around 2000 μF per gram, but nowadays powders permit 7000 μF per gram.

Basic research on semiconducting oxides started in the 1950s, and it was soon realised that this could be applied to tantalum electrolytic capacitors, thereby eliminating the wet electrolyte that had previously been used. As the wet electrolyte had often been concentrated sulphuric acid there was obviously a considerable incentive in replacing it with a dry contact material. Such a development gave the 'solid' tantalum capacitor. This use of a 'dry electrolyte' in its turn led to the development of plastic encapsulation which would not have been possible in the wet electrolyte system. Nowadays production has started on plastic capacitors prepared using dry electrolyte systems and these plastic-cased capacitors are finding application in computer circuitry.

In parallel with this work has been work on the development of etched tantalum foil capacitors. As in the case of aluminium it is possible to etch tantalum so as to increase its surface area compared with a plain foil, and initial work, starting in 1950, led to gains in surface area of roughly 10 to 1, for 20V anodising. This in turn enabled tantalum foil capacitors to be brought into production around 1957. Work on etching has continued ever since, and recently, at the Plessey laboratories in Bathgate, it has been found that surface gains of around 50 for 20V anodising are possible by the use of suitable surface treatments[25]. This widens considerably, the application of tantalum foil in capacitors as it is thereby possible to use this material in a much less expensive manner.

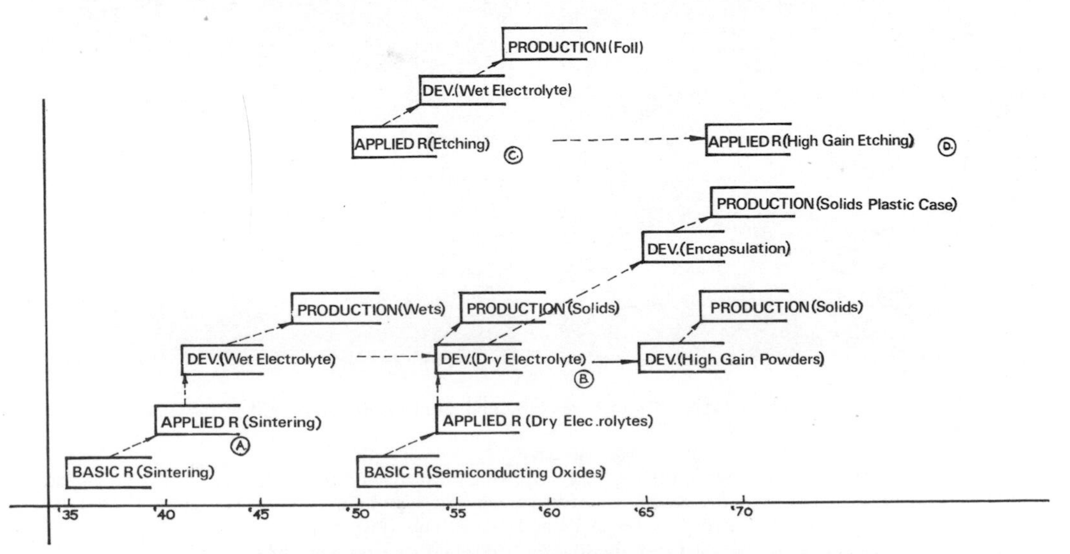

Fig. 13. Materials development diagram for tantalum electrolytic capacitors.

Electron micrographs illustrating materials development as presented in Fig. 13 are shown in Fig. 14. *A* shows a sintered compact exemplifying the applied research on sintering. It can be seen that a completely porous sponge of tantalum results from the sintering. *B* shows a section through a sintered tantalum capacitor using a 'dry electrolyte'. At this magnification three different regions can be distinguished; first the central tantalum core, secondly a skin of oxide which has been grown smoothly onto the tantalum by an electrolytic process, and finally a rough coating of semiconducting electrode material, in this case, manganese dioxide. *C* and *D*

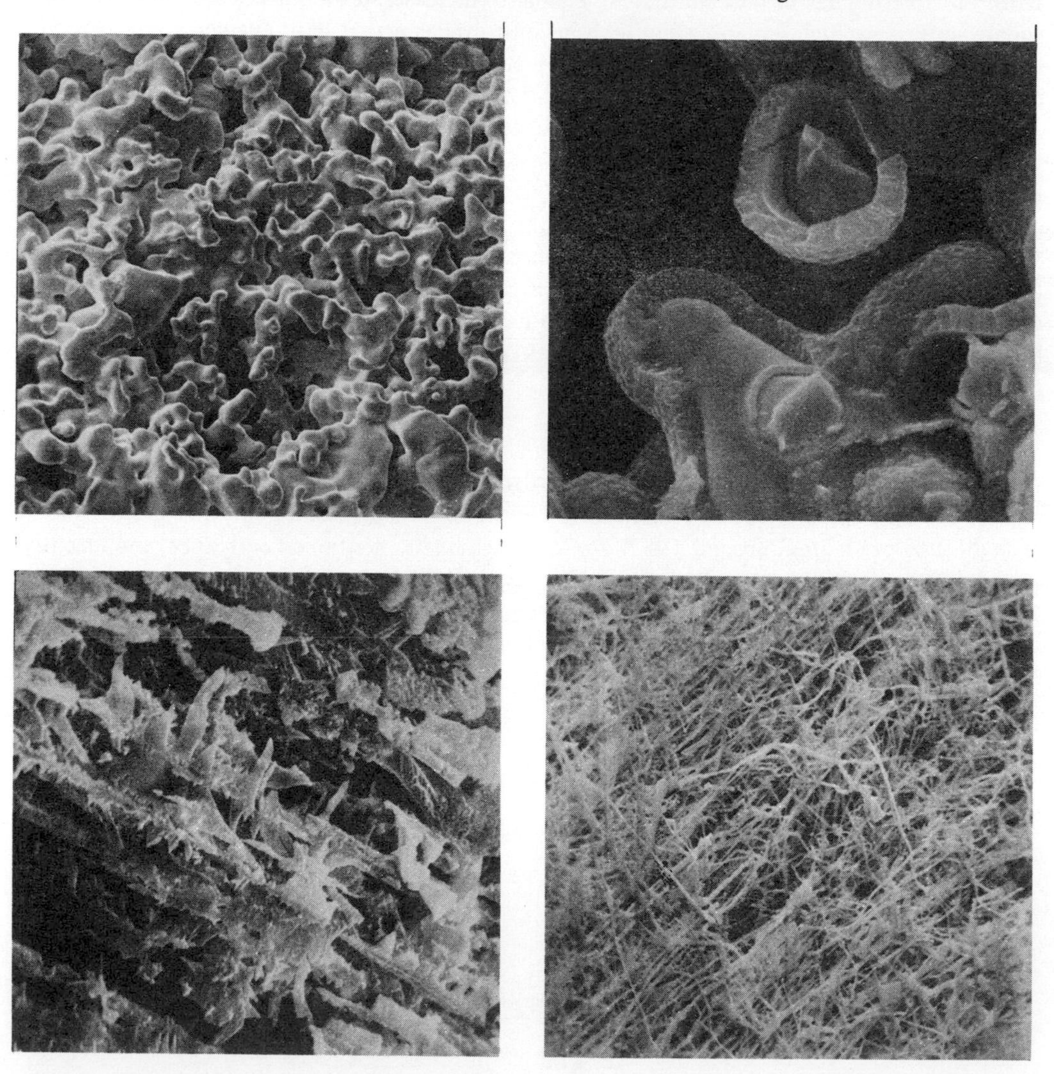

Fig. 14. Scanning electron micrographs of structures obtained from the development of tantalum electrolytic capacitors.
A. Sintered tantalum compact. Length of edge 100 μm.
B. Sintered, anodised and pyrolysed tantalum body, showing both the anodic oxide layer and the manganese oxide dry electrolyte acting as the second electrode. Length of edge 10 μm.
C. Etched tantalum foil. Surface gain 10× for 20V anodising voltage (90°C).
D. High-gain etched tantalum foil. Surface gain 50× for 20V anodising voltage (90°C).

show the result of etching tantalum foil in various ways. *C* was obtained using an etch which gives a surface gain of roughly five times and which was developed in the early 1950s. *D* shows the result of etching with the new processes developed recently. Here a very large number of very thin channels have been obtained, giving a surface gain of approximately 50.

It can therefore be seen that, following basic work on sintering and on semiconductor oxides, the capacitor industry has developed a number of different products. The rate of progress may appear to be relatively low but is consistent with expenditure in relationship to an obsolescence time of roughly 20 years.

7.0 CONCLUSIONS

This paper has attempted to show the effect of a business environment on materials development associated with the capacitor industry.

It has been shown that various constraints exist that limit the finance available for r. & d., and this in its turn limits the rate of progress.

It has also been noted that it is not possible to set up a viable r. & d. effort without the backing of an efficient market research organisation. Without such an organisation it is not possible to decide rationally on the projects to be encouraged and therefore it is not possible to develop products that keep the company in the forefront of the business activity.

Finally, it should be noted that if one did not spend the amount of money indicated by one's examination of the obsolescence rate and the related factors, one could be storing up trouble for the future. This is because, in the environment which we find ourselves in at the moment in the U.K., the industry is under considerable pressure from developments both in the United States and in Japan. Thus the lack of r. & d. expenditure may be a short-term gain on company profits but could ultimately result in collapse in the face of overseas competitors.

8.0 ACKNOWLEDGEMENTS

The author would like to acknowledge the discussions he has had with his colleagues, particularly Dr P. J. Harrop, Business Planning Manager at the Plessey Co. Ltd, Capacitor Division, Bathgate, West Lothian, and Mr K. Fearnside, Technical Director of the Components Group of the Plessey Co. He would also like to thank the Plessey Co. Ltd for permission to publish this paper.

REFERENCES

1 'Annual Abstracts of Statistics', H.M.S.O.
2 'Statistics of Science and Technology', H.M.S.O.
3 HARROP, P. J., (1971), Private communication from Plessey Co. Ltd., Bathgate.
4 DUMMER, G. W. A., (1964), 'Fixed Capacitors', Pitman, London.
5 MULLIN, W. F., (1967), 'A.B.C. of Capacitors', Foulsham, England.
6 HALL, M. M. J., (1970), *Engineering*, Dec., p. 655.
7 HABERMEL, P. D., (1970), *Radio & Elec. Eng.*, **40**, p. 259.
8 HUSAIN, I. M. A. and HARROP, P. J., (1970), *Component Technology*, **4**, p. 29.
9 HUSAIN, I. M. A. and HANSEN, F., (1969), British Patent No. 29605/69, 'Improvements relating to Electric Filters (L.C. Filter).
10 HUSAIN, I. M. A., (1970), British Patent No. 53374/70, 'Improvements in or relating to capacitors (R.C. Module).
11 HARROP, P. J. and CAMPBELL, D. S., (1970), *Component Technology,* **4**, p. 19.
12 CAMPBELL, D. S., (1971), *Rad. & Elec. Eng.,* **41**, p. 17.
13 MCLEOD, T. S., (1969), 'Management of Research, Development and Design in Industry', p. 89, Gower Press, London.
14 HART, A., (1963), 'Symposium on Productivity in Research'.
15 WHITE, E. A. D., (1971), 'The effective use of research and development funds in the UK Electronic Industry', Industrial Research Budgeting Project, Electrical Engineering Department, Imperial College, London.
15(a) DUCKWORTH, W. E., (1966), *New Scientist*, Volume 32, No. 524, p. 564.
16 MONTEITH, G. S., (1969), 'r. & d. Administration', Iliffe, London.
17 HELLYER, F. G., (1967), *New Scientist,* **33**, p. 222.
18 GIRLING, D. S., (1970), *Rad. & Elec. Eng.,* **40**, p. 173.
19 NICHOLLS, S., (1969), *Work Study and Management Services*, Oct. p. 667.
20 BURNS, T. and STALKER, G. M., (1961), 'The Management of Innovation', p. 178, Tavistock Pub., London.
21 GOLDIE, W., 91969), 'Metallic Coating of Plastics', **2**, p. 375, Electrochem., Publications, England.
22 KRASUCKI, Z. and CHURCH, H. F., (1969), E.R.A. Tech. Pub. No. 5260.
23 COIT, J. P., DESVAUX, M. P. E. and KRASUCKI, Z., (1969), E.R.A. Tech. Pub. No. 69.79.
24 Norton Company, Metals Division, Newton, Massachusetts, USA, (1970), Sales Technical literature on SGQ, SGP, SGN and SGQR tantalum powders.
25 HARRISON, D. K., (1970), British Patent No. 10536/70, 'Contamination etching of Tantalum'.

SESSION IV

Celebration Day Papers

Chairman: Mr A. L. Stuchbury, O.B.E.
President of the Institution of Production Engineers

RAISING PRODUCTIVITY

Phil. P. Love
Glacier Metal Company Ltd

SUMMARY

This paper deals with two areas in which, it is suggested, the use of models should be considerably developed as a means of raising productivity. In the field of industrial relations an indication is given of the already existing use of 'models'. It is suggested that further applications, particularly with 'dynamic models', would be of value, provided that the models used were realistic. The second area of potential application of models is in the management accounts field. The author claims that insufficient attention has been paid to the presentation of management accounting data, to the extent that wrong decisions or no decisions have been taken as a result of the presentation of data in an unrealistic manner.

1.0 THE ROLE OF FINANCIAL AND SOCIAL MODELS

Although a great deal has been written about productivity and about ways and means of increasing productivity, there are two factors which are very much to the forefront at this time and which clearly require further study. These factors are *industrial relations* and *management accounts*.

There will be little question as to the relevance of industrial relations to productivity, but in case there is some doubt about the relevance of management accounts, a short note is required. The managers of manufacturing companies depend first upon what they see and hear directly of what is going on around them and, secondly, upon information collated for them by others. Management accounts constitute an important part of the information on which managers make decisions about productivity. Hence management accounts are relevant to the subject of productivity.

It can also be said that a great deal has been written about these two subjects, namely, industrial relations and management accounts. However, the use of models as a means for understanding these factors has had little attention, despite the fact that in many other fields of activity models are extensively used and have been of immeasurable value.

A word about models. A model is a representation of reality and may be used to study those aspects of reality which have corresponding analogues in the model. Models can be classified as static or dynamic, as physical or abstract, and can be further sub-classified.

If the model is dynamic then, to the extent to which there are analogues of salient factors in the real situation, the model can be used to predict the outcome of change.

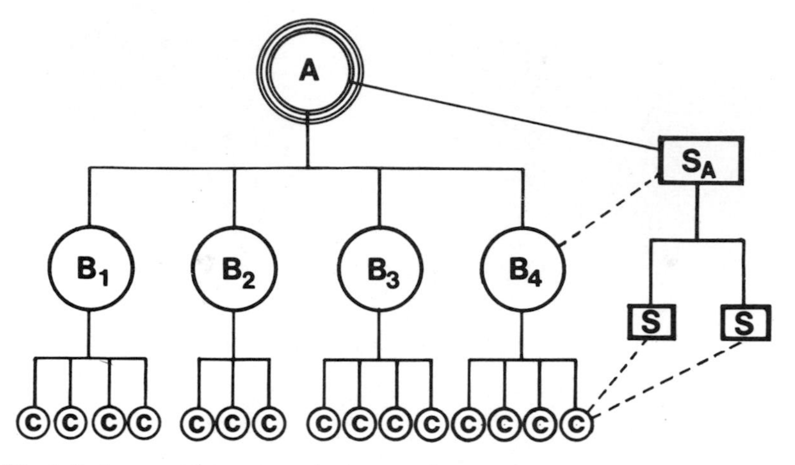

Fig. 1. Basic organisation chart. *A, B* and *C* are executive roles. *S* denotes a staff role.

First, industrial relations: models are not new in this field. The most usual model is a static one, and is often called an *organisation chart*. An organisation chart is a model of the relationships between various members of employees in a company. Figure 1 is an example of a fairly simple organisation chart. Each role in the company is represented either by a circle or a rectangle. The fact that there is a relationship in the organisation between these roles is modelled by joining up the roles on the chart with lines. The level of each role in the company is shown analogously by the position on the paper in which each role is shown: the higher up in the organisation, the higher up on the paper. Various kinds of role can be modelled by various shapes. Thus, in Fig. 1 roles denoted by a circle could be so-called *executive roles* and roles denoted by a square or rectangle could be called *staff roles*. Various natures of relationships between roles can be indicated by various kinds of lines which join up the roles. Thus, a solid line can indicate a direct manager-subordinate relationship and a dotted line can indicate a staff relationship.

This is a static model which can be displayed and which enables everyone in the company to know what their manifest role is. Thus, if some newcomer comes into the company at level *C* it is not necessary to go through the whole process of introducing him to the chief executive or to the senior managers B for him to know where the role he occupies stands in relation to the rest of the organisation. Moreover, in each circle, square or rectangle, there can be a little note which briefly describes the nature of work of the role. The model can be used by *A* to see whether or not all the kinds of work which he wishes to be carried out in the company are appropriately manned without his having to go round every individual in the company to find out, and so forth. Indeed, although this is only a static model it is, nonetheless, a useful one: and it is a very inexpensive one.

Models of this kind can be very greatly augmented, not merely to show that there is, for example, a manager-subordinate relationship between, say, two roles, or a colleague relationship between other roles, but it can also be used to indicate the operating routines in the company.

Routines very frequently involve collateral activity so that, say, the occupier of role *B*3, although he may have collateral relations with the occupier of *B*2, cannot be instructed by *B*2, and communication between them is governed by the routines laid down by *A* or his staff officer. In order not to make the model too complex, these routines are usually displayed on other charts. Production flow charts are yet another example of a static model very much concerned with productivity, and can range from the activities of every individual stated in very general terms to the activities of a series of individuals in highly specific terms, related, for example, to one particular part being produced in the company.

It will be appreciated, therefore, that without the use of models, even static models, it would be very difficult indeed to run a modern factory. It is, however, rarely realised that models are being used.

2.0 INDUSTRIAL RELATIONS

In what way, however, can models help us with industrial relations?

So far reference has been made only to abstract static models in human organisation. In the field of science or engineering, physical and dynamic models can be much more dramatic and effective in the solution of complex problems. But, before considering physical dynamic problems in the field of industrial relations, it is necessary to repeat the definition of a model as a representation of reality which may be used to study those aspects which have analogues in the model.

If a model either has inadequate analogues, or if what are considered to be analogues do not have properties analogous to the aspects of reality which they represent, then it is possible to be led astray by the model. At the risk of being tautologous one must state that it is therefore paramount, when one is making real decisions based upon the behaviour of a model, that the salient factors of reality have analogues in the model and that these analogues are, in fact, analogues: that is to say, have properties which model the properties of the real factors.

Take a very simple numerical example. If one wished to find out how long it would take to reduce the diameter of a 6-inch-long bar of steel from 3 inches to 2·9 inches in a lathe whose spindle speed was 150 revolutions per minute, the depth of cut was 0·05 inches and the tool feed was 0·01 inches per revolution, a simple calculation could indicate a cutting time of 4 minutes. And the same time would be derived if the diameter of the bore were 4 inches and had to be reduced to 3·9 inches, and so on. Extrapolation however, would be limited by a number of other factors not shown in this numerical model. For example, the power of the lathe would be finite, and there would be a limit to the speed at which the tool would continue to cut. These are but two out of many limitations. Hence, wrong conclusions will be reached if based on the behaviour of this model beyond the limits of reality.

In industrial relations there is recourse to a number of models. Unfortunately, however, these models are based on an Aristotelian approach to industrial relations, and although the more obvious factors have their corresponding analogues in the model their properties often do not correspond and, indeed, very frequently a models is derived from a *modus operandi* which, although it appears to work well in other situations, has little or no bearing on many of the problems of industrial relationships which we have to face.

Industrial relationship problems are often cited as problems between managers and unions. This is frequently a gross oversimplification as the words tend to result in an inference that only two parties are involved and, certainly, problems of industrial relations are very often spoken of as if only two parties were involved.

Assume for a moment that we are involved with two parties. It is commonly supposed that, since the civil courts are set up to deal with disputes between two parties, then it is only right that, when two parties are in conflict, the conflict should be resolved by a third party known as an arbiter. If we examine the real situation, however, we start with two parties interacting with one another in the broadest terms. That is to say, the behaviour of one party is felt by the other party who, in turn, adjusts his behaviour, that is to say, reacts. This reaction may cause the first party to modify his behaviour in such a way as to reduce the effect of that behaviour on the other party, or it may have no effect at all, or it may even cause the first party to behave more aggressively, and the interaction will escalate until either one of the two parties retires or they come to blows.

Fortunately, people in general are reasonable. Sooner or later they will begin to communicate verbally. Because of common misconception they may decide that the problem between them is too tricky and that they should agree upon an arbiter, but an examination of the results of arbitration indicates that, in reality, whatever an arbiter may decide, one or even both parties will feel aggrieved. Frequently the two parties come together and reach a different solution more acceptable to them than that defined by the arbiter. In short, they make a bargain, and no arbitrary judgment is regarded as a bargain.

What kind of model have we got here? The situation is that there are two parties who interact with one another and who wish to optimise the position. That is, if the interaction is mutually helpful but at some shared cost, then they want the best result, or, if the interaction is mutually interfering, then they wish to arrive at minimum total interference with maximum total freedom. Without a model they can carry on interacting as indicated, but, with a model, we can endeavour to demonstrate the effects of their prospective behaviour and, in the end, draw up protocol by which they can continue to exist.

What factors are represented in this model? First, the two parties themselves are represented (the model may indeed contain the actual parties involved). Secondly, the relationship between the parties is modelled in verbal terms, so that each party is able to picture the developments which might, in reality, occur, and in turn display to the other party in verbal terms his reaction to those developments. That is to say the iterative nature of the real interaction is present in the verbal model. Thus, verbal communication is the painless and inexpensive model of what otherwise would be behavioural communication. Thirdly, the parties themselves would, in reality, have to come finally to some *modus operandi,* and this condition is built into the model. The important thing to note is that there are no third parties: the introduction of a third party such as an arbiter, destroys the model.

Problems in industrial relations, however, are rarely problems involving only two parties. Hence, if the model which is set up to examine the situation and to deal with the problem assumes that in fact there are only two parties, such a model is almost bound to lead to a dead-end. It is unfortunate that one so often hears managers say that the problem in their factory is between the unions and that they can do little or nothing about it. If for no other reason they pre-empt themselves by speaking of the unions as if they were one party on one side of the table while management was another party on the other side when, the reality of the situation is that there are several parties. Themselves as managers, union A, union B, union C and, say, union D, and that if a situation has developed which is intolerable to any one of these parties, and it may be intolerable to only one of the unions, then the manager has an industrial relations problem on his hands, and, as he is paid to manage the outfit, he must then become actively concerned in the problem.

If, however, the only model concept which he has of the situation is restricted to a two-party model, then he will feel himself to be powerless, and as frequently happens, abdicate, hold his hands up in the air and say: 'This is an inter-union matter and I can do nothing about it'. On the other hand, if he does take recourse to his two-party model, what he will do is see the particular second party that he considers to be causing the problem and he will get involved in the two-party verbal model. He may arrive at a protocol acceptable to him and this particular second party. When the details of the protocol are made public, however, some other parties, which the manager tends to group under one party, will find the protocol unacceptable, and the manager will find himself involved in a second two-party verbal model and will come out with a second protocol. Thus, he will find himself involved in a series of two-party-type verbal models, and, inevitably, having to adopt a highly defensive attitude because if he gives way to one of his second parties he will have to give way in turn to every one of the others.

Is this not a familiar situation? Is this not a microcosm of what is happening throughout the country today? Is this not a result of the fact that the bilateral discussions which take place seriatim are not rigorous models of reality? The procedures have implied that for a long time it has been assumed that a two-party model is adequate to deal with a multi-party situation. Thus the fact is that the constitutional models which are brought into operation do not contain some of the essential analogues. A two-party model cannot be used to explore a situation which contains more than two parties, and, because of the application of the two-party model to solve the problems of more than two-party systems, the parties involved have to resort to behavioural communication which is manifested by petulance and resort to power.

If we are going to avoid these very serious behaviour patterns which interfere with productivity then we shall have to design our constitutional models to represent, not *some* of the salient factors in the real situation, but *all* of the salient factors in the real situation. This means that the model will be a lot more complex. The proposals which are being put forward by people of considerable standing are nearly all Aristotelian, that is to say, they are the outcome of cerebration rather than of analysis of the salient factors and their properties in the real situation. Consequently, ideas about models of situations having some vague resemblance to the current industrial situation get transferred without any test of relevance.

The essential problem in industrial relations today is that of differential entitlement to the proceeds from industrial operations. One of the salient factors in the situation is that, although what an individual may feel to be fair pay for the work he does is itself a subjective assessment, it has been shown that, in aggregate, these statements constitute a multi-model function, and that people are able consistently to state what they should be paid for the work they do in terms of the average pay rates for occupations somewhat above and also somewhat below their own. For example, various categories of local government employees have very clear ideas as to what would be fair relative rates of pay. They have sufficient contact with each other in their respective occupations to make judgments which, although subjective taken individually, can be taken as a group by virtue of their consistency, and cannot then be regarded as other than objective. There are good grounds for assuming that, given a total available amount for distribution as wages and salaries, they could reach agreement on differentials and, if this is so, it would be possible to determine equitable differentials without resort to a series of two-party bargains.

This proposal has been put forward by Lord Brown and it is worth a brief examination to see if it passes the test of a requisite model.

Each year the goods produced in the country as a whole have a total determinable value, which is what the market is prepared to pay for these goods. The goods, in fact, are the outcome of production processes applied to raw materials. Some of the raw materials have to be brought into the country from outside, some of the evergy required for the production processes has to be brought in from outside, all the rest is indigenous. The net proceeds of the country's productive activity is the difference between what the market will pay for the products, and what we have to pay for materials and energy imported and it is this net sum which has to be equitably distributed.

This net sum can be established periodically, say quarterly, or every six months, or even every year. Any mechanism which, in effect, distributes in numerical terms a greater amount than the amount of the net proceeds will merely result in a reduction in the value of the currency used. Thus, in a model the object of which is to arrive at an equitable distribution of the proceeds, the total of the proceeds can be directly stated in numerical terms: both numbers and currency have the same operational properties.

There are large numbers of interested parties; indeed, every adult member of the population is an interested party and, indeed, every adult is interested in more than one sense. For example, a person who is an employee in a company making furniture will be interested not only in ensuring that he gets a fair payment for the work he does in relation to the general standards of payment which prevail, but that the environments, first that in which he lives and second that in which he works, are better than tolerable. As regards the environment in which he lives, he will be prepared to agree that taxes should be levied in a variety of ways, not only to maintain that environment but continually to improve it. Likewise he will agree that, so far as his work environment is concerned, some of the proceeds should be used to improve conditions, to improve machinery, to keep pace with international developments etc., and that the rest be then distributed by way of pay for work done.

There is a constitutional means by which he can be represented in determining what proportion of the net industrial proceeds should be spent on national environment. That means is Parliament, with government setting the policy and the government departments doing the necessary staff work. It would be possible, period by period, probably annually, for Parliament in a budget to state in numerical terms the amount of the proceeds which would be left after meeting the environment maintenance and improved expenditure for distribution as between local industrial environment and pay. Lord Brown has proposed the setting up of a National Council for Regulating Differential Wages, on which would be represented all people properly interested in the distribution of the remainder and the differentials would be decided by unanimous or near-unanimous vote.

Does this model contain the requisite number of analogues? In reality people in employment are grouped (some of the groups are organised) by occupation, and the owners of capital (not infrequently also themselves employees) have formed a limited number of institutions by which they are represented. Although the total number of representative bodies is limited, this number would, nevertheless, be very large. But there are good examples of means by which representation can be effected which would only involve a relatively small number of people. For example, the Cabinet can make decisions on behalf of Parliament and which Parliament can reject. The CBI and the TUC are other examples in which a group of groups can be effectively represented with the appropriate checks. Thus it would be possible for a body such as NCRD reasonably to represent a vast variety of differential interests.

What is the operational analogue here? In reality it is the respective powers of the interested parties which will carry the day if they decide that the situation is intolerable. Their power can take the form of withdrawal of cooperation so that, in fact, they can, by withdrawing their co-operation, negate any proposals made as we are at present constituted. In other words, they can veto any proposal. This power of veto must have its analogue in the model. Hence, the need for unanimous agreement. Without this operational analogue built into the constitutional model, a requisite element would be missing and the model would be valueless.

Let us test an alternative model which has recently been submitted, namely, that Parliament should determine the amount available for differential distribution but that there should be introduced an arbiter or tribunal or other body to which each interested party would make representations as to the relative merits of his claim for a share in the proceeds and that the arbiter or tribunal or judicial body should make the decisions.

Does such an arrangement model reality? Up to the point at which Parliament would determine the amount available for distribution by way of differential entitlement there is no incompatibility, but if the arbiter or tribunal makes a decision which is not acceptable to one or other of the power groups, then in reality the power group can veto the decision and the model is thereby rendered ineffective. The proponents of this model, which introduces a factor which is not really in the situation, will then say: 'But we shall have a law which gives authority to the arbiter or tribunal'. The weakness of this, however, is that the people who are interested in a fair distribution of differential entitlements comprise the majority that elects Parliament, which would sooner or later have to repeal any such law: unless, of course, Parliament itself ceased to be a democratic institution.

Indeed, in the last analysis, the persistent application of unreal models is a very dangerous practice by which we may find our hard-won democratic institutions in a Procrustean bed.

3.0 MANAGEMENT ACCOUNTS

In *The Daily Telegraph* of 18 April 1967 the Institute of Chartered Accountants in England and Wales advertised a short summary of the annual statement of the President, Sir Henry Benson, to members of the Institute. In this statement he said, 'the basic role of the practising accountant is that of auditor and it is his work in this field which probably makes the greatest impact on the public. In recent years improved techniques have anabled his work to be more penetrating in character and of more constructive help to management'.

This section of the paper is not concerned with the actual financial auditing function of accountants as required by the Companies Act, the Income Tax Act, and other legislated requirements. Rather, your attention is sought to consider the second part of the accountants' work as referred to by Sir Henry Benson, namely, the improvement of techniques to enable the accountant to be more penetrating in character and of more constructive help to management.

Accounts are numerical models. Conventionally a 'balance sheet' is a static financial model of a company at a particular point of time and a 'profit and loss account' is a static financial model of changes which have taken place during a period of time (usually one year). Conventional accounts are models presented to shareholders in accordance with legal requirements.

In such accounts the analogues are numerical and the operations are arithmetical. In order to carry out these operations, each of every physical asset and transaction is subject to one or other of a number of factors by which the quantity of each asset and transaction is converted into monetary terms. Numerically money is a precise quantity. If you have 37½p in your pocket then you have precisely that. If you think you have about 35p in your pocket, give or take 3p, an accountant tends to reject such a statement because, as he says, you can turn your pocket out and check precisely what you do have. Because the accountant is a genus of auditor, any statement involving possible error is anathema to him and the training which most accountants undergo inhibits their understanding of the concepts of mathematical statistics. They are, for example, quite prepared to add a really subjective valuation of imponderable assets, say, amounting to £200,000, to a measured quantity of assets, say, valued at £1,528,242, and express the total assets as the sum of these two quantities, namely, £1,728,242. In this latter figure the subjectiveness of the assessment of the £200,000 is lost and what appears is a figure which is clothed with a degree of precision quite unjustified as a model of the reality.

In the preparation of accounts for shareholders the best that the accountant can do at present is to round off figures to, say, the nearest thousand pounds or, if the company is immense—a General Motors—to the nearest million dollars, but even then quantities are expressed in terms of four or five significant figures when it could be shown in most cases that three significant figures would be more than adequate.

Then we come to management accounts, that is to say, accounts which are prepared to enable managers to make better decisions about increasing productivity. A great deal of development is required to make the accounting models really worthwhile. Managers in the field of engineering or other applied science are subconsciously suspicious of many of the accounting data presented to them and only too frequently do they discover when it is too late that something has gone wrong. In the light of events, for example, it has been stated that, had due account been taken of the data presented to the Rolls-Royce managers by their accountants, then Rolls-Royce would not be in trouble today. But there is an article in *The Sunday Times* dated 28 March entitled 'Cabinet scrapped Rolls on wrong RB211 figures', and the article could only have been possible as a result of the inadequacy of the accounting data as a reasonably accurate and unobscure model of the situation. If it is possible for such high-level professional accountants to become confused over such an important issue how can the professional engineering manager be expected to make operational sense out of the morass of the so-called management accounts which are presented to him?

I have made a study of the management accounts, as distinct from the statutory accounts, of over a score of substantial companies, with a view to determining whether or not the data can be relied upon for making management decisions. I have found that there was a significant number of instances where the data presented made such an obscure model of reality that it resulted in decisions being taken where they were not required and failure to take decisions where they were required.

As indicated earlier in the paper, a model must contain analogues representing the significant aspects of reality and that the properties and interrelationships of the analogues must themselves correspond to reality. A model in which the only operations are arithmetic is not adequate because the processes of reality have statistical properties, some of which are quite simple statistical properties, others more complex.

A manager's job can be summarised as *deciding what has to be done to meet the policy of the company and then issuing instructions in pursuit thereof.* He can perceive some of the results of his decisions by direct physical observation, but modern substantial businesses are so complex that he has to rely in the main on numerical models and to use only a relatively small part of his time by way of physical sampling. It is thus more than ever important that the models are fully representative of the reality, and many of managers' failures today are due to inadequacies of the numerical models presented to them.

The plans which a manager has to make have to be capable of being stated in numerical terms, that is, in inputs and outputs which are measured in units compatible with the respective properties and by means of factors converted into monetary terms. A plan is rarely stationery; it usually incorporates trends. These can be conveniently budgeted, making use of interest tables so that the feasibility of these trends can be assessed.

In executing a plan a manager relies in the main upon the numerical models presented to him by his various specialists. The manager does not have time to explore every aspect of the company in depth. He relies on the accountant to prepare numerical financial models representing the use and outcome of all the resources. The management accountant is anxious to present *all* the data, with the result that most managers get much more information than they can use. But worse still, the real situation often has significant properties which are not represented in the financial model. One of these properties is the statistical nature of many of the measurements which are converted into monetary terms, so the possible errors in assessment are not manifest.

In general, there are two classes of variation in any multifactored system, namely, decreed variation and casual variation.

Decreed variation is imposed in the situation as a result of legislation or custom or season, though there are many other sources. One of the dominant decreed variations in the operation of a factory is the variation in the number of working days in the calendar month or in the accounting period. Decreed variations of this sort can be predicted precisely and regression analysis of the main use of company resources should be carried out to determine the effect of such obvious decreed variations as the variation in the number of working days.

Casual variation cannot be predicted but measures can be taken to limit its effects. Indeed, perhaps the main task of a manager is to notice the existence of casual variations, to determine the causes and to make on-going decisions to reduce, if not to eliminate, the effects and get back to the plan. The manager constitutes an essential part of the system by which input is modified to minimise deviation of output from plan.

In mathematical statistics a function called the 'standard deviation' is used to measure the effect of variations in the data. ('Standard deviation' is not standard in the sense of being arbitrary, and it is a pity that the word 'standard' has been used. But the term has now been used for so long in mathematical statistics that I am afraid we are stuck with it.) It is very important to eliminate the effects of decreed variations in an accounting model. The manager can then avoid taking actions on what appear to be deviations from target but are really the result of decreed variations, and should more readily notice the effects of casual variations. A sample of about twenty management accounts of companies shows the effects of decreed variations as, on average, about twice those of casual variations, and in one case the effect of decreed variation was nearly four times that of the casual variations. It must be clear that failure to eliminate decreed variations from management accounts render these accounts much more difficult to use.

Now the manager does not need to look at everything. He ought to be presented with the sufficient and necessary control data. Minimally, his attention should be drawn to those categories of data:

- Where the ratio of the standard deviation (calculated after eliminating the effects of decreed variations) to the decayed average exceeds the value which the manager decides is a tolerable limit. (Where the resource is directly measured this ratio should generally not exceed 0·05, that is to say, a control deficiency of 5% would be reasonably tolerable. Where the data refer to a difference, for example profit as the difference between income and expense, then a control deficiency index of 10 to 20% is not intolerable.)

Where the change which has taken place in any particular quantity (for example material usage, overtime, order intake, delivered sales, stocks, debtors) exceeds the standard deviation.

By highlighting the figures selected on these two particular criteria the manager is very unlikely to waste time attending to things which in reality do not need attention and, on the other hand, is unlikely to fail to consider things which do need attention.

It might be said that this is management by exception. So it is, except that display of the exception by making use of mathematically statistical processes is much more in accord with the properties of the situation than is the case when display is by means of simple differences.

Finally, accounting techniques can be applied to the dissemination of policies by means of regulators. One of the problems which faces a manager is how he can issue operational instructions to his subordinates which still leave them the discretion to make decisions in matters where they are best able to make good judgments. For example, one casual variation may be an epidemic of sickness which will force a subordinate manager to exceed an overtime rule and which causes him to seek a concession from his manager. The granting of a concession weakens not only the particular instruction to which it refers, but also weakens the force of all instructions coming from that manager. Again, overtime may be due to faulty programming or to customers' scheduling. If, however, the senior manager applies a regulator to the effect that every manhour of overtime will carry a charge of £x against the particular subordinate's operating statement, then that subordinate has a measure of the degree to which overtime may or may not be used. He has a measure of the policy on overtime and to make judgments as distinct from receiving some hard and fast limit to which he can use overtime.

The use of regulators in production factories is not explicitly recognised but they exist all the same. An increase in the cost of a particular material will influence those who can affect the extent to which that material is used, causing them to redirect their operations to minimise the expense arising from the increase. If this additional expense is appreciated to be greater than some other expensc to which they are directing themselves, then redirection is almost automatic. Why not, therefore, make use of regulators as a means of control to effect company policies, that is to say, why not apply regulators from within as well as accepting what amount to regulators applied from without?

To do all these things requires the accounting data to constitute a model of reality, that is, to incorporate analogues of all the significant factors in the real situation. There is a great field here for research in management accounting technology and in which the Chartered Institutes of Accountants have been guilty of professional dereliction.

4.0 CONCLUSIONS

In this paper I have attempted to draw the attention of production managers to the use of models in two very important fields of their activity, namely, the field of industrial relations and the field of management accounts. There is absolutely no reason why the concept of models, which has proved itself to be of immeasurable value in many of the activities of mankind, cannot be applied effectively to these two fields in particular.

APPENDIX: A NOTE ON 'MODELS'

The term 'model' can be applied to any device which is used as an aid to examining potential realities without being involved in setting up the full-scale real situation.

If one is proposing to build a large office block, or a cathedral, or a large ocean-going liner, one of the simplest and most elegant ways of conceiving the potential reality is to build a scale model. In these particular models many of the essential features of the potential reality are incorporated in the model and there are two characteristics, namely, that there are consistent analogues in the model representing the main features of the potential reality, and the scale and juxtaposition of the main features correspond respectively with the scantlings and space relationships of the potential reality.

An architect's model is a physical static model, but models can be classified in three more categories, viz. physical dynamic, abstract static and abstract dynamic. In more specific terms a model represents a system and it must contain analogues of all the significant aspects of reality.

These analogues must have, or be attributed, properties which correspond with the properties of the features which they represent, and the model must also incorporate relationships which correspond with the relationships existing between the features of reality.

Dynamic models are generally more useful than static models because, not only can they represent a situation at a specific point in time, but, by changing the intensity of analogues representing the aspects of reality, they can represent what will happen if the corresponding changes are applied in reality.

If the model is a physical dynamic model the application is dramatically evident, as for example in a wind tunnel.

But abstract dynamic models are even more dramatic when they are understood. For example, through a series of abstract models was evolved the first set-up to extract the fuller implications of the Michelson-Morley experiments, the relationship between mass and time. This, together with further experiments and observations, all involving model concepts, made atomic energy available to us now. Perhaps less exciting but so far of greater benefit is the outcome of Maxwell's electromagnetic theory, which uses abstract dynamic mathematical models.

But models are designed by men, and if the device which purports to be a model does *not* contain analogues of *all* the essential features of the reality, and if the operations by which the model is made to work do not correspond to the operations of the reality, then such a device can be misleading.

However, it is not necessary for the model to have absolutely precise analogues and operations to be useful. For example:

$$T = 2\pi\sqrt{(l/g)}$$

can fairly represent the periodicity of a pendulum, and it is possible by means of this mathematical model to find out the effect of lengthening the pendulum. But this is not a precise model and has an error normally negligible for small angles of swing. There is a more precise mathematical model, involving rather ponderous elliptic functions, which is more precise but even then ignores damping effects. In turn, damping effects can be incorporated in the model. All the models are useful. It is important to recognise the limitations of each, and it is only too easily possible to assume that reality has some of the properties which are built into the model but which are not present in reality.

The paper refers to two models. The first, physical dynamic (relating to industrial relations) and the second a numerical abstract with a combination of static and dynamic elements.

MANUFACTURING PROCESS RESEARCH WITHIN THE COMPANY

W. J. Arrol
Joseph Lucas Ltd, Research Centre

I wish to deal with the organisation of manufacturing process research in a large company which relies heavily on its production engineering for its commercial success. In particular, I must deal with the components section of the automotive industry with which I am familiar, although the principles involved apply to other industries as well.

The profession of production engineering has been a development over the last fifty years which is of the greatest importance. We are going to continue to rely on it, but the people who are doing the production engineering and the manufacturing process research may in future be somewhat different from those who have done it in the past.

Before dealing with the organisation of manufacturing process research I should like to refer to a matter already mentioned at this conference and to put forward an opposing view. Neither in production engineering nor in product engineering do we suffer from lack of ideas – we suffer from far too many ideas – very good ones and patentable ones. However, of the ideas we have on the product side, probably not more than one or two per cent ever finish up in a product; probably not more than one in five thousand finishes up as a product in its own right. These are depressing figures, but they demonstrate that a director of research, or a production engineer, or a design engineer, has to use a good deal of judgment in spending his company's money on those ideas available to him which are most likely to succeed.

For improvements in manufacturing processes we have in the past depended on the traditional approach, which relies heavily on ad hoc methods. Ideas have come perhaps from the shop floor, perhaps from production engineers, but from people who on the whole have not had the opportunity of treating their work in the way in which a research man would treat his. It is natural that such production and methods engineers should improve a process step by step more or less by feel rather than by analysis. Ad hoc development is a perfectly legitimate way of making progress in manufacturing processes. In my opinion it will always to some extent be a necessary way. Its working medium consists of tools which are expensive and which frequently get broken. The method also requires a good deal of material which finishes up as scrap. It is a legitimate way of making progress, but it is slow and costly.

In the last few years there have grown up in a number of the larger manufacturing companies which have research centres, and also in universities and research associations, what might be called a somewhat more classical research approach to production engineering. This involves the amassing of a large amount of information, developing from this some form of hypothesis, testing the hypothesis and, if necessary, rejecting it and going back to the beginning. It also involves mathematical modelling where this is appropriate, and, where such modelling is inappropriate, it may very well involve analogue techniques. An example where mathematical analysis certainly is just not worth doing – where it fails from sheer complexity and where analogue techniques are useful – lies in the design of tools for high-pressure diecasting. Here the problem is the study of the flow of materials and the flow of heat in a shape so complex as simply not to be worth dealing with mathematically, but which can to a large extent be studied by analogue methods.

The company with just about the highest reputation in the world for its production engineering is the Delco Remy division of General Motors. This has still a heavy dependence on people who retain the traditional cut-and-try approach. But of course Delco also has available to it the very large effort in manufacturing process research carried out at the General Motors Technical Center not so very far away. If this company chooses still to make use of both approaches there is clearly no case for those of us who can use sophisticated techniques to be contemptuous of those who do not use them. We are in a transient state of affairs, where people who profess the scientific approach are doing things in parallel with people who are unable to adopt it. In my own company, over the last few years, I think it is fair to say that we have had a number of cases where both forms of approach to the improvement of manufacturing processes have been applied at the same time. As a result of this I have the feeling that, if we are going to make the maximum progress, we must not allow antagonism to grow up between the traditionalists and the scientists. There will be a dichotomy of approach for many years to come and thus the organisation which we must develop should enable these two to carry on separately on the same problems.

I should like now to give you a few examples with which I hope to point out as we go along roughly how such dual approaches can occur. The first is a case taken from our own experience for the development of a cold and warm extrusion sequence for producing a roller clutch sleeve of great complexity. It is a five-roller clutch and the inside shape of the sleeve has on it five cams and five buttresses, and it has obviously got to be produced as cheaply as possible. In this particular case it was felt on all sides that the right way to produce would in the long run be by extrusion. It was realised both by the production engineers on the shop floor and in the research centre that there were two different ways of going about developing the sequence. One was the traditional way and the other would be by analogue techniques using a lead – 10% antimony alloy with mild steel tools. The behaviour of this alloy under light loads can be extrapolated using a scaling factor to the loads which would be achieved in real life with extrusion steels and using tools made of the proper material.

It was decided in this case to have a friendly race. One size of sleeve was adopted by the factory responsible for making the final product and intent on developing the sequence by traditional methods. Another size of what was essentially the same design was taken by the Group Research Centre to be attacked by analogue techniques. It so happened in this case that the analogue method won in time and very much so in expense. It does not follow from this one case that it always would win, but here the two approaches were applied simultaneously, with enthusiasm on both sides, and without any acrimony. Such competition ensured that there would be no resentment on the shop floor when either sequence would finally be put into production.

Figures 1 and 2 illustrate the sequence. The slug is cropped straight from bar and dumped to a squat form. The hole is drilled, and that is the only piece of machining in the whole component. It could be avoided, but in this case drilling is cheaper than extrusion. After

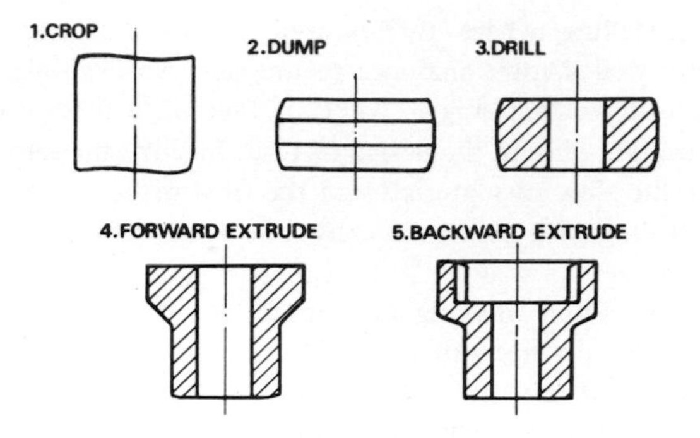

Fig. 1. Schematic extrusion sequence for deep cam sleeve.

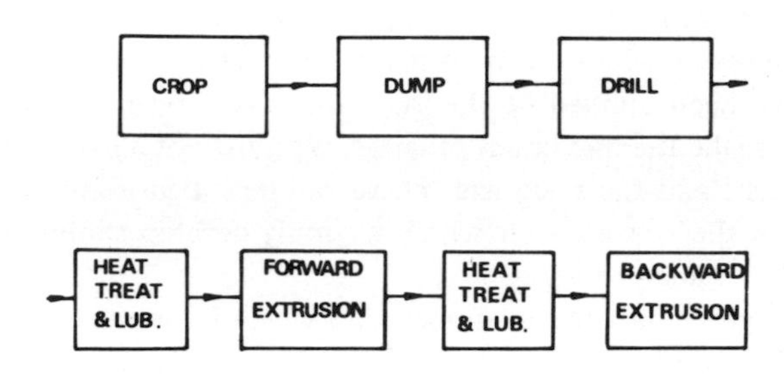

Fig. 2. Extrusion process sequence for deep cam sleeve.

annealing and lubrication there is a stage of forward extrusion, and, after lubrication again, final backward cold extrusion, giving the very complex cam form.

We have problems in such an entirely different field as gas carburising procedure, which is not only a matter of engineering but quite largely of chemistry as well. Rather sadly, one frequently finds that furnace manufacturers do not have access to the shop floor where their furnaces are being used in full production, so the components manufacturer has to give to the furnace manufacturer quite a lot of help. The user thus takes part in the development of equipment suitable for his own purposes. This is true not merely of furnaces but of a wide variety of other equipment. In gas carburising carbon passes from the gas phase to the component, and the chemical nature of the atmosphere changes at the same time as the composition of the product. This poses quite complex chemical control problems.

In this case, the Group Research Centre set out to try to develop an automatic sequence for the control of the gas atmosphere of the furnace and of the temperature in such a manner as to produce an acceptable product in the shortest possible time and with minimum reliance on the judgement of the operator. This problem would no doubt have been soluble by ad hoc means but the time taken to develop a programme would have been intolerable.

I should like to give you just one more example of what may in the future be an even more logical sequence of process developments than has yet been mentioned. This particular one is the idea of the Petroforge, with which many of you will be familiar.

Experiments at PERA showed that if a piece of mild steel is struck by a tool travelling at a speed of the order of 70 ft/s the fracture is entirely different in mode from that occurring if the tool is operated more slowly. At PERA a weight was dropped from an appropriate height. At Birmingham University, in Professor Tobias' department, a gasoline-powered press was designed, capable of delivering individually controlled blows. Burning hydrocarbons is certainly one of the cheapest and most convenient ways of releasing stored energy. The machine was successful in that it demonstrated fine blanking without the necessity of special tooling and the extremely accurate cropping of quite heavy bar.

At this stage it was clear that if fine blanking or bar cropping or possible powder compacting were to be involved on a production scale, some company with facilities for, and experience in, large-scale production ought to have a look at the equipment. Accordingly, my company bought a Petroforge from the University and put it into the Group Research Centre, where the next phase of development was carried out in collaboration with the Ministry of Technology. This was to evaluate the device and to provide a Petroforge to a design more nearly like that which would have to be made by a machine tool manufacturer for use in full production. This phase of the development took about seven man-years of work, and on completion the new design was passed to the Ministry for transfer to a machine tool manufacturer.

The company had thus been able to evaluate the Petroforge as a machine tool, for the company's own information, and to carry out a stage of development based on its production experience before passing on the machine for final design to a machine tool manufacturer who alone would be capable of making a saleable product.

The overall question may now be asked: 'What organisation ought a company to establish in order to optimise its efforts on manufacturing process research'? The experience of my own company and of those in similar engineering fields does not suggest that there is any one ideal organisation, but one can put down certain principles of general applicability.

Firstly, alongside the company's normal process development facility, a research facility needs to be established, fully equipped to carry out process research on full-size production machines in an environment away from the factory. It should have available to it services including generous model-shop facilities, computer services, instrumentation services and the like.

Secondly, the research facility must be so organised, and its relationship with the production engineering function of the company be so close, that it has access to such factories as are concerned with the manufacturing process under investigation. In the long run any innovation has to be tested on actual components in production, and this can be done most economically on the shop floor. However an important point of discipline arises. When any member of a research centre is working in a factory he must come under the day-to-day discipline of the factory manager. He must work factory hours and must become known as an equal of the production and methods engineers employed in the factory, and not as a privileged outsider.

The third organisational principle is in the careers policy of the company. People who enter the research department of a company should not stay there too long but after a few years should be retrained for some other form of management. There is no part of a group research centre where this is more important than in manufacturing process research. The danger is that a man approaching manufacturing processes from the scientific point of view for too long will run the risk of ultimate isolation from the shop floor at the same time as he becomes out of date scientifically. The best state of affairs exists where the production director of a subsidiary company looks on the research centre as a natural recruiting ground for senior methods and production engineers.

We have, in manufacturing process development in this country, passed from complete reliance on step-by-step improvement to the application of the scientific method in some but not all cases. The future will undoubtedly see this trend continuing, but never, I suggest, to completion.

SESSION V

Production Research

Chairman: Professor D. S. Ross
Rolls-Royce Professor of Production Engineering,
University of Strathclyde

TRENDS IN MACHINE TOOL DEVELOPMENT AND APPLICATION

C. F. Carter, Jr.
Cincinnati Milacron Inc., Ohio.

SUMMARY

This paper deals with three areas of machine tool technology: machine utilization, machine structure, and the process of metal removal.

The utilization of machines is introduced by summarizing the statistics available concerning the operations on the average workpiece as it travels through a shop. Comment is also made on the effort spent on different kinds of workpieces. An analysis of these data leads to conclusions about the importance of non-metal cutting time during the use of a machine tool. The importance of emerging control techniques in attacking this problem is stressed.

The continued importance of structural improvements is explained in terms of the requirement for machines to utilize new cutting tool materials. These materials generally allow higher cutting speeds and demand greater rigidity than is presently available. Techniques which will afford better than a trial-and-error response to these demands are discussed.

Finally, developments which influence the cutting process itself are discussed. The relative importance of different degrees of sophistication of adaptive control is discussed. From this, predictions can be made concerning the most fruitful areas to apply technology.

As an example of a process undergoing considerable technological change, grinding is discussed on the influence of high wheel speeds.

1.0 INTRODUCTION

Machine tools have traditionally been improved and changed in response to outside developments, and the improvements have related to the machine's use as an independent element in the manufacturing process. Increases in rigidity and horsepower have been the result of improved cutting tool materials. Improvements in accuracy are made in response to the demand that parts made on machine tools be assembled with less skill, and that these parts perform better in their assigned function. Pressures which bring about these kinds of change in machine tools will not abate, and changes in response to them will continue to be made. However, the pressure to reduce product cost will focus attention on machine tool time utilization, and it is change in the time utilization of machine tools which will do more for reducing product cost in the next decade than improvements in structure, accuracy, and cutting capability.

The literature on metalcutting economics usually divides the time for a workpiece to be produced into such segments as load, unload, cutting, positioning, tool change, etc. Most authors then concern themselves with methods to minimize product cost (or maximize profit) by optimizing the cutting time. As a result, there is very little literature which sensitizes us to the total time utilization of machine tools and the *relative* importance of cutting time.

2.0 PART GEOMETRY

In order to consider the relative importance of cutting time and machine tool time utilization to the overall manufacturing system, let us consider some numbers in order to gain perspective. First, consider part geometry. Table 1 indicates the relative amount of shop effort on three broad categories of parts.

TABLE 1

Part geometry	% shop effort
Rotational	47
Box-like	26
Plate- and beam-like	27

This, of course, represents a composite for all metalworking and may not be representative of a particular shop. Incidentally, the numbers presented in this paper generally refer to batch-type production situations and do not include the influence of mass production techniques where dedicated transfer lines are used. Table 1 would tell us that a given amount of effort to reduce costs on parts of rotation will have considerably more influence on total manufacturing costs compared to similar efforts on other types of parts. The gains to be made here are still significant in spite of the fact that the efficiency of producing parts of rotation already exceeds the efficiency of producing other types of geometry. This is attested to by the fact that 70% of the incidence of parts is rotational, even though only 47% of shop effort is expended in producing them.

The statistics represent a composite of our own studies and the work of others[1,2,3]. The term 'effort' is used to convey the cost or man-hour content of an operation, and is considered more appropriate than the numbers (incidence) of part or machines.

3.0 MACHINE TYPE

Another set of numbers which may be helpful in providing a perspective for the manufacturing operation is the relationship of machine type with respect to shop effort. This is shown in Table 2.

This gives us some insight concerning the relative importance of different processes, as well as machine types.

4.0 WORKPIECE TIME IN SHOP

As a further insight, let us consider the life of the average workpiece in the average shop, Fig. 1. Our studies would indicate that about 5% of the life of an average workpiece is spent on the

TABLE 2

Machine type	% of shop effort
Lathe	40
Drill	9
Mill	7
Boring mill	10
Cylindrical grinder	6
Surface grinder	6
Gear cutter	9
Other	13

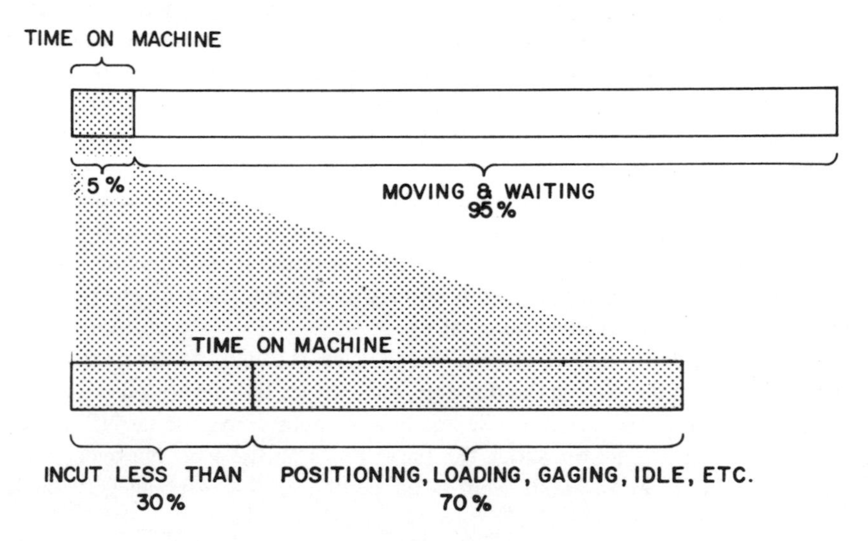

Fig. 1. Life of the average workpiece in the average shop.

machine tool. The rest of the time is spent in being moved or waiting in line. This fact should immediately tell us that in order to reduce the time a workpiece is in the shop (and hence inventory), we must work on the time between machining operations, and not just on speeding up the machining operation itself. If we look at the 5% of the time which the workpiece spends at the machine, we can further break that down into what is going on at the machine.

Detail studies have been made which break down the time utilization of a machine into such factors as cutting, setup, loading and unloading, gauging, and idle. These studies indicate that the cutting time is seldom more than one-half the time the machine is used, and is usually about 30% of the time the machine is being used. A clear-cut definition of cutting time is seldom given; however, it can be assumed that parts of the operation, such as the return stroke on a surface grinder, or return for an additional pass on a lathe, or indexing and tool change on a tool-changing machine, are considered part of cutting time when generalized numbers are given. It is therefore necessary from a metal-removal point of view to consider another term which would more precisely describe the influence of the cutting operation itself. This term might be called 'time in cut.' It then becomes evident that improvements made specifically to the material removal process in a given situation will have to be factored by about 30% in order to measure the impact on total machine tool operation.

The bigger savings that can be achieved in such areas as part handling, machine configuration to combine processes, and part flow through the shop is becoming widely recognized, and machine tool developments are largely concerned with these points. The point to be made here is that even though the material removal process remains at the heart of the manufacturing system, and its control is essential to product quality and cost, improvements in what goes on between the tool and the work (where parts are already being made successfully) may not bring about savings as great as equivalent improvements made at some other point in the system.

5.0 DIRECT NUMERICAL CONTROL

One development which will have great impact on the utilization of machine tools, as that utilization relates to the time a workpiece is in the shop, is direct numerical control. In order to develop a perspective toward DNC, we need to start with numerical control. It introduced two things, new methods for the control of machine slides using servo-mechanism principles and a new method of getting motion signals to the machine in the form of punched paper tape. It took us some time to lose our preoccupation with the mechanics of how the machine is controlled through the use of NC, and recognize the importance of the added management control brought about by a new way of communicating with a machine. We are just now becoming well aware of the problems of transposing information from a drawing to a form that can be utilized by a machine. And just now becoming aware that the efficient transposition of this information is the key to effective use of NC.

If we stand at a machine in a DNC system and look up through the organization of people from the operator to the supervisor to the manager, etc., we see that information is handled in a different way. The machine is still controlled as it was under conventional NC; however, the paper tape is gone and information about the part comes to the machine over wires from a central data source. Information about what is happening at the machine goes back to the central point over wires.

This new way of communicating with a machine is the key characteristic of DNC, which opens up all of the advantages of DNC. In fact, one should think of a DNC system as a communication and information network more than a control system. The supporting elements of a typical DNC system, as shown in Fig. 2, attest to this concept. Two points are worthy of note here. One is that what we call peripheral devices play an important role in this system. DNC relies on more than a central computer, it relies on cathode-ray tubes, line printers, disc memory, data links, etc. Without the reality of these devices, DNC could exist only as a concept. The second point is that, to the extent that the system handles information which does not control actual machine slide motion, it is not dependent on numerical control. It can be considered as part of a general management information system used for the real-time reporting and control of all manufacturing operations. It is this broader aspect of the control which will give us detailed knowledge on the progress of a part through the manufacturing process and allow us to attack the problem of machine time-utilization and workpiece time-in-process.

The advantages of DNC which can be most easily measured will be those which receive the

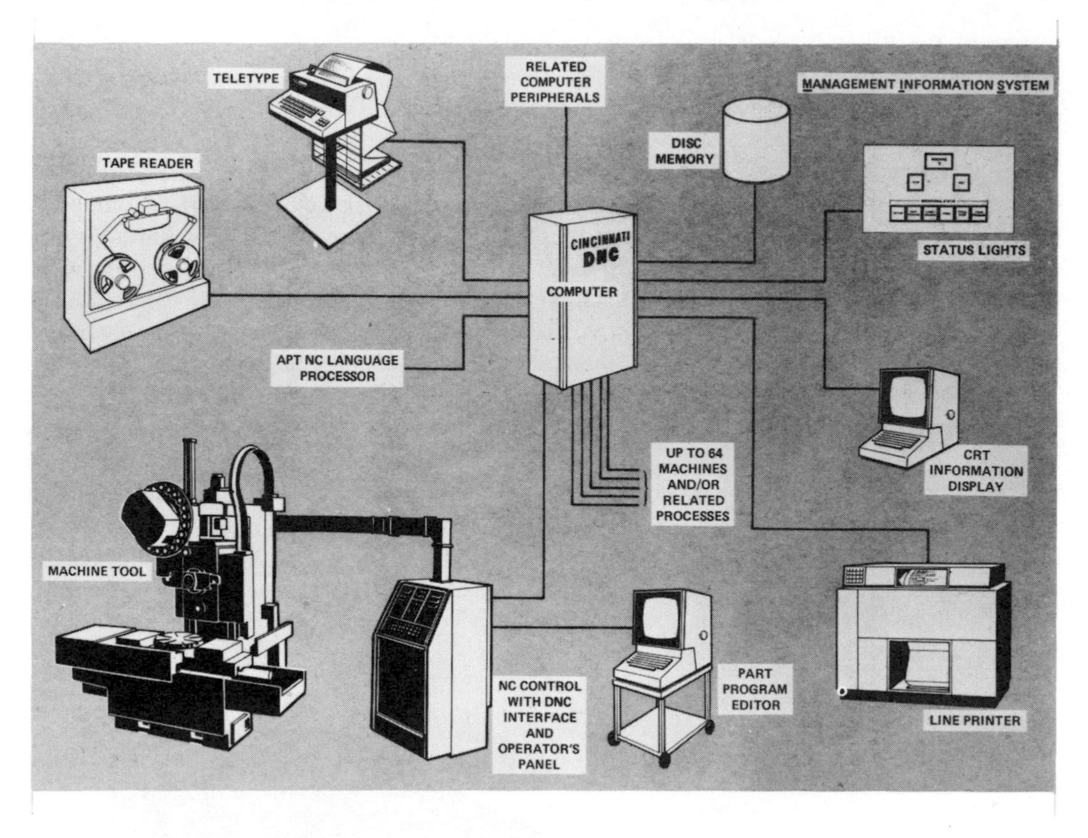

Fig. 2. The support elements of a typical direct numerical control system.

predominant, initial attention. For instance, the time it takes to prepare a tape which will successfully make a part, and make it under optimum machining conditions, is something which we can easily measure. It is usually the case that when a tape is tried on the machine, changes are required in order to make a successful part. This is the case in spite of the fact that the geometry data may have been previously checked on some type of plotting device. Actual running of the tape reveals such problems as errors in speed and feed selection, interference with workholding fixtures, need for programmed stops to clear chips, etc. The mechanics of getting these changes on the tape can require from four hours up to a matter of days. Valuable productive time is lost while the tape is changed, and due to the awkward nature of the procedure for modifying the tape, many changes which would improve productivity are not made once a successful part is being produced.

Under DNC control, when a part is being run for the first time and errors are discovered or clearances should be checked in advance, machine position data and other control information may be displayed on the cathode-ray tube at the machine (Fig. 2). A keyboard is provided from which changes in the data may be made and the new condition tried. Since the procedure for changing data is simple and instantaneous, even minor modifications which will improve the operation will be made. Once changes have been proven, they can automatically become part of the permanent program in the disc file.

Potentially, the area of savings on which DNC will have the greatest impact is that which relates to the flow of a part through the shop. This is because of the information network which DNC represents, and the fact that it may be used as a basis for a management information system. Information such as part number and process, cycle time, number of parts in queue, tool life, etc. will be readily available on a real time basis – note peripheral equipment in Fig. 2. Admittedly, the value of this information is a direct function of how management uses it; and therefore, it is difficult to arrive at specific dollar justifications which relate to the cost of gathering the information. We recognize also that the nature of the information which management needs to improve its effectiveness is not clearly known now and is difficult to predict. However, there is a growing realization that the general problem of workpiece in-shop time must be attacked by a system approach and that more real time data will be needed to solve the problem successfully. DNC will therefore be central to the solution.

The subject of DNC would not be complete without some mention of the fact that computerized control techniques make possible the introduction of complete multi-station manufacturing systems. These are systems which utilize numerically controlled machines tied together by automatic parts-handling equipment. In this way, a variety of parts may be manufactured with complete control of not only the machining cycles but the flow of parts between operations. Fig. 3 shows a typical layout of such a system which would include automatic part-handling conveyors, loading stations, switching stations, and gauging stations.

In order to gain a mental image of how machine configuration can affect manufacturing

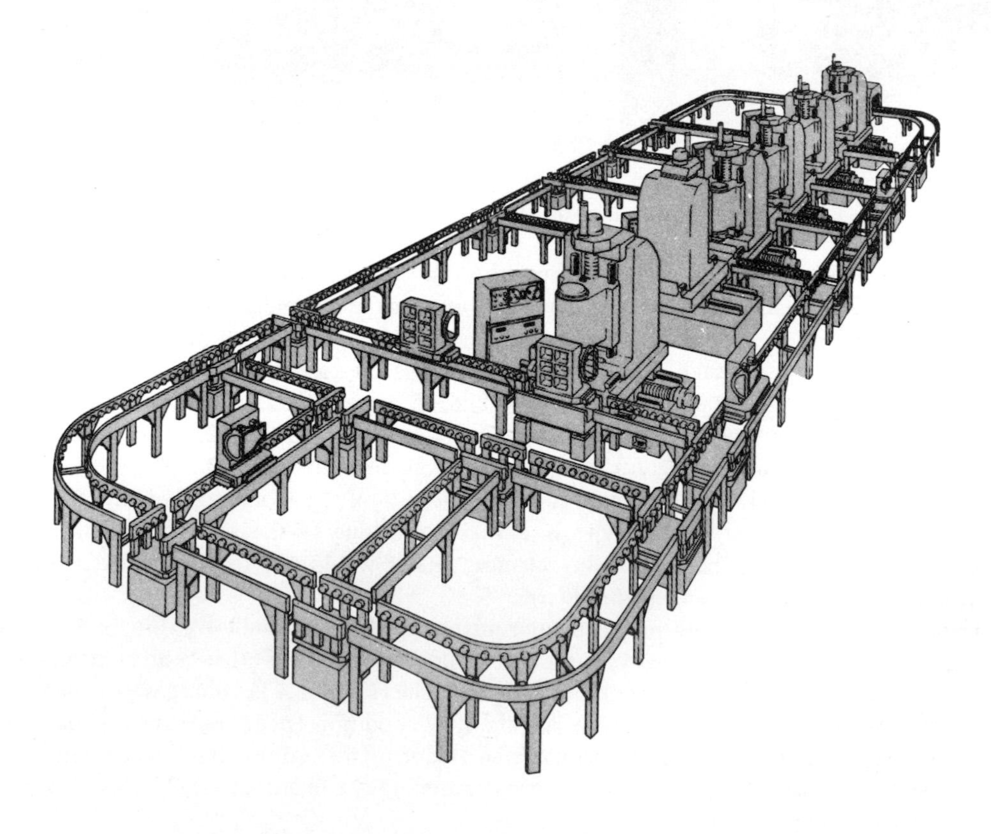

Fig. 3. A multi-station manufacturing system.

cost, we should refer to Fig. 4. This is a highly generalized family of curves which shows the area with respect to lot size in which certain types of machine tools have proven most effective. Since multi-station manufacturing systems using complete computer control are a relatively new concept, the length of its line on the figure is obviously not precisely known; however, it represents an important development in machine utilization in the general area indicated[8].

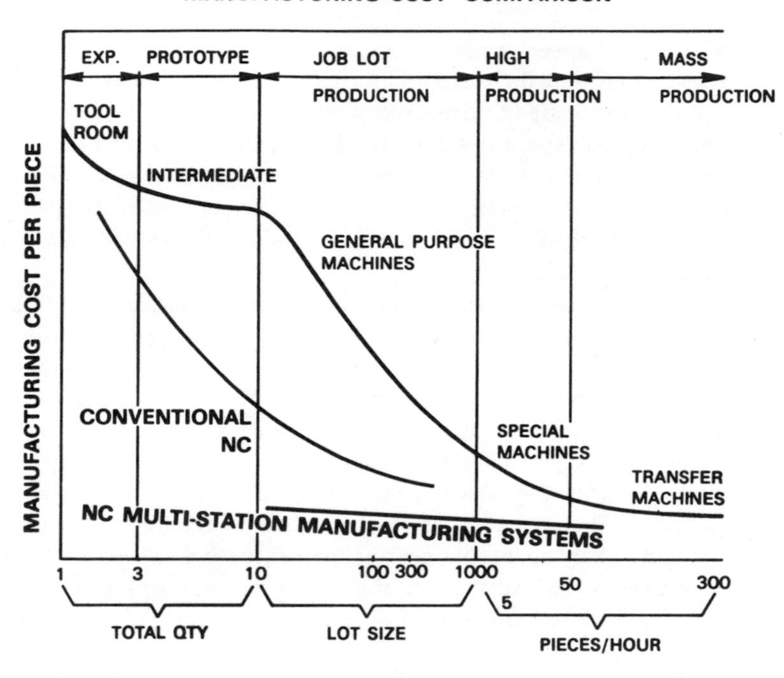

Fig. 4. Cost comparison of a numerical control multi-station manufacturing system with an ordinary manufacturing system and a conventional numerical control system.

6.0 MACHINE TOOL STRUCTURES

Although methods of controlling and monitoring machines will have the most significant influence on total shop productivity in the next decade, important developments will also occur which will improve the productivity of individual machines. Machine tool builders will be under increased pressure to provide means for quicker part loading and reduced setup time. However, there is one area in which the machine tool builder must respond to the results of considerable research effort carried on in the decade of the 1960s. That is the improvement in cutting tool materials. The increased cut speed capability of new tool materials brings with it the demand for increased horsepower and reduced susceptibility to chatter.

Machine tool chatter is a phenomenon about which machine operators and designers have been aware for many years. Most operators have procedures which relate to such things as cutter geometry, cut speed or clamping forces of tooling, which allow them to improve the conditions of chatter. However, it was not until recent years that a body of scientific knowledge was accumulated with respect to the phenomenon of chatter. We now have a good understanding of the relationship between the natural frequencies of various machine elements, their mode shapes, and the nature of chatter. We are beginning to develop some ability to predict the conditions under which chatter will occur, and make design decisions which will allow greater productivity without chatter. Instrumentation and design techniques used to study and reduce chatter, which have gained a foothold in the 1960s, will become commonplace in the 1970s.

There is a direct relation between the use of new cutting tool materials, which provide for significant increases in speed, and the susceptibility to chatter or vibration of the machine. If we consider the significant increase in cutting speed on conventional work materials, which can be attained by going from high speed steels to carbide to ceramics, we must remember that the materials in that order are more likely to chip under conditions of vibration. For this reason, the machine tool builder must respond to advances in cutting tool materials by providing equipment which has higher rigidity, and, more specifically with relation to the problem of vibration – higher dynamic rigidity.

Instrumentation is now available which allows us to excite the machine tool through various frequencies of vibration. A typical setup is shown in Fig. 5. The amplitude of the structure in response to this forced vibration is plotted on an *x-y* recorder. A typical plot is shown in Fig. 6. Troublesome frequencies can be easily spotted from this plot. By moving the vibration pickup to different points on the machine, we may obtain the relative amplitude of different points of the machine structure, and from that draw a diagram of the mode shape of the machine in response to various frequencies. Fig. 7 is a typical result. Note that the right view shows the point of force application. These data tell us which frequency (usually the one with the lowest dynamic rigidity) is most likely to be the chatter frequency, and also the optimum point at which to mount a damper or add stiffness to the structure, since we know the mode shape. Static rigidity must, of course, be maintained in order to achieve the required flatness or other

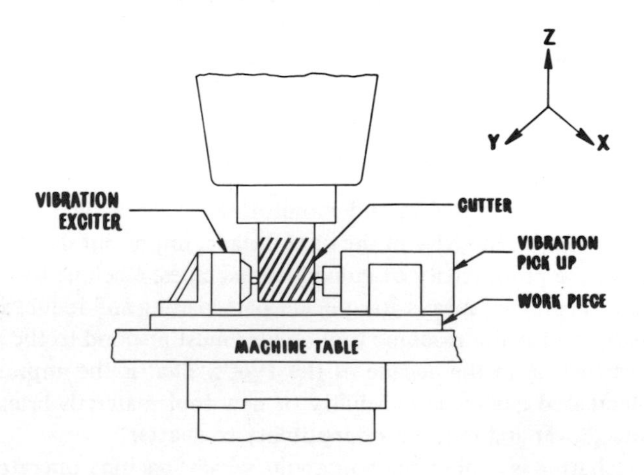

Fig. 5. A typical setup for measuring the stiffness of the principal axis.

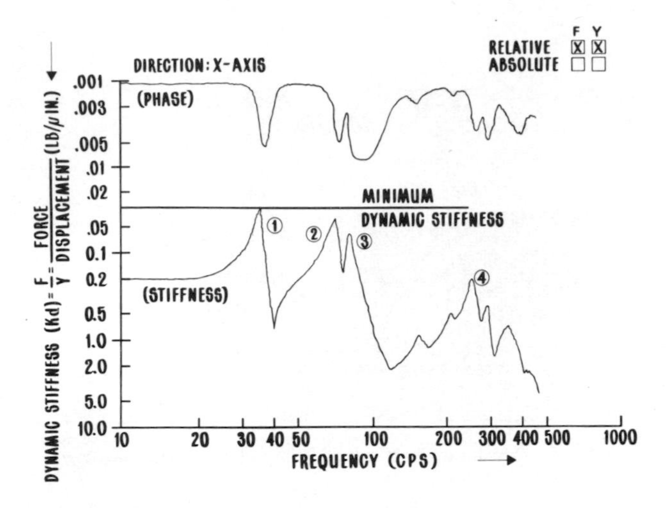

Fig. 6. A typical response plot from a stiffness measuring test.

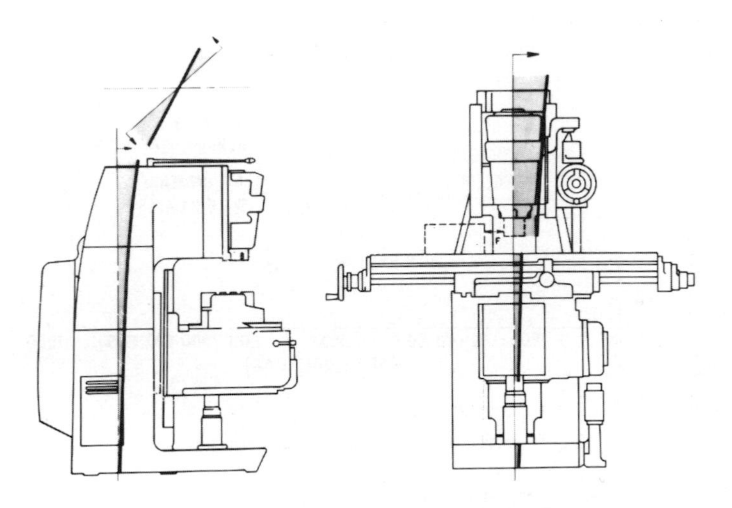

Fig. 7. Dynamic deflection plot of a machine.

geometrical accuracy of the workpiece. However, when static rigidity is adequate, a damper may be added to one of the machine elements in order to decrease the susceptibility to chatter. Dampers, Fig. 8, usually take the form of a mass coupled to the machine structure through an energy-absorbing plastic. A damper is most effective when tuned to increase the dynamic rigidity of one frequency; however, it can be effective in increasing the dynamic rigidity of frequencies close to the same value. The influence of a damper on dynamic rigidity is shown in Fig. 9.

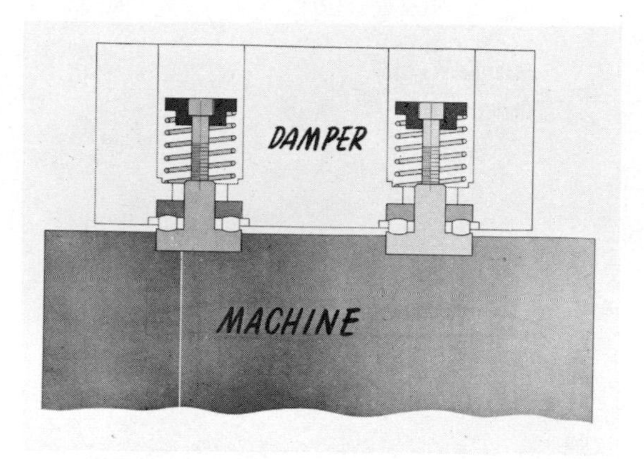

Fig. 8. Development of tuned damped vibration absorber.

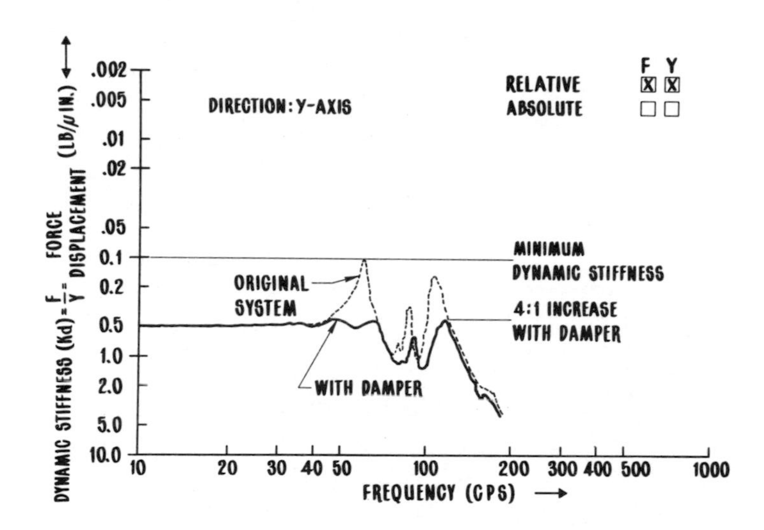

Fig. 9. Dynamic rigidity response with damper.

7.0 THE PROCESS

Machine tool builders are also improving the productivity of their equipment by developments concerned with what is going on between the tool and the work. Two such developments will be discussed here: one is the broad area of adaptive control, the other is the advance being made in the field of grinding by going to higher wheel speeds.

7.1 *Adaptive control*

Since adaptive control is such a broad term, it may be well to review the characteristics of some systems which may be referred to as adaptive control. This can be done with the aid of some simple block diagrams of typical control systems. In Fig. 10 we see a simple feedback control system. In this case, the command may be a feed rate of slide velocity. The object of

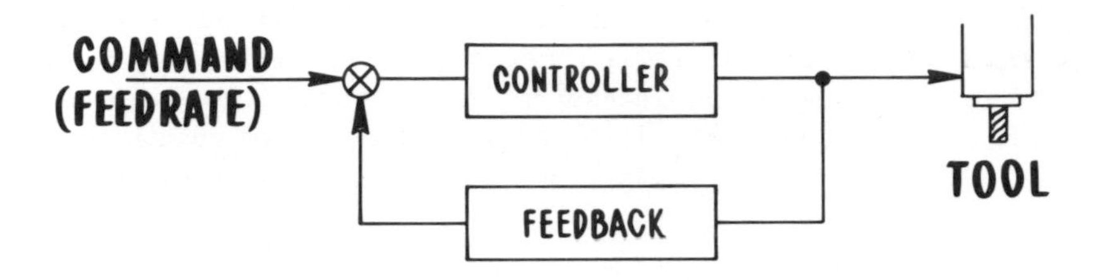

Fig. 10. Simple feedback control.

the control is to make the machine tool slide move at the command velocity regardless of outside conditions. The feedback element would sense the slide velocity and compare it with the commanded velocity or feed rate. In this case, the control is independent of and unresponsive to any process variables.

In Fig. 11, we see that another loop has been aded to the control. In this case, some process variable is being measured, and the initial command is being altered, depending on the condition of the process variable. This is called 'constraint control,' since some limiting value, or constraint, is usually selected for the process variable, which may be torque, deflection (force), horsepower. This is the simplest form of adaptive control, and with proper sensing of the process variables, which will be used as constraints, it can be very effective.

In order to talk about more sophisticated forms of adaptive control, we must introduce a new term called 'performance index.' This is the relationship of all process variables which leads

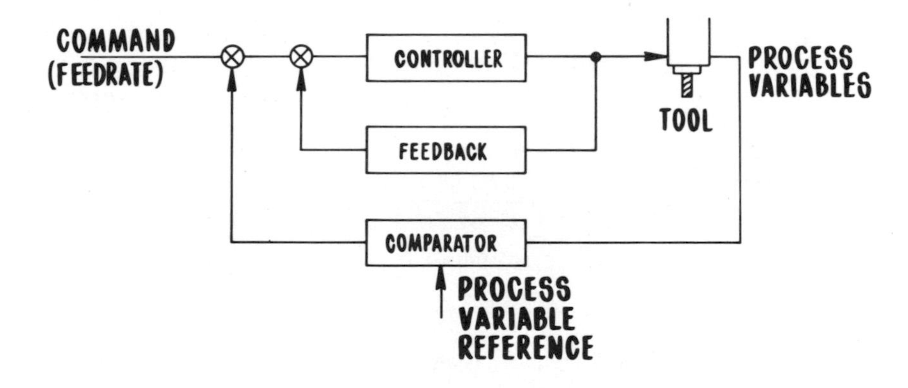

Fig. 11. Adaptive constraint control.

to the best or optimum performance. It is usually expressed in terms of cost or production rate, but it could be expressed in terms of accuracy or surface finish, if these represented the most demanding or important outputs of the process. Much has been written about the relationship between cost, production rate, and tool life; however, the mathematical models found in the literature are seldom used in actual practice. They now take on a new significance because these models and the general philosophy on which they are based are important in deriving the performance index for an adaptive control system.

Fig. 12 indicates the block diagram of a system in which a performance index is used. Here the process variables are measured as in constraint control. However, the system's reaction to the process variables now depends on a performance index rather than simple limits for the variables. In this case, the performance index is calculated off-line and a strategy or reaction to the variables has been predetermined – hence the term 'programmed adaptive control.' One system in use today falls into this category [4]. In this case, the performance index is cost, based

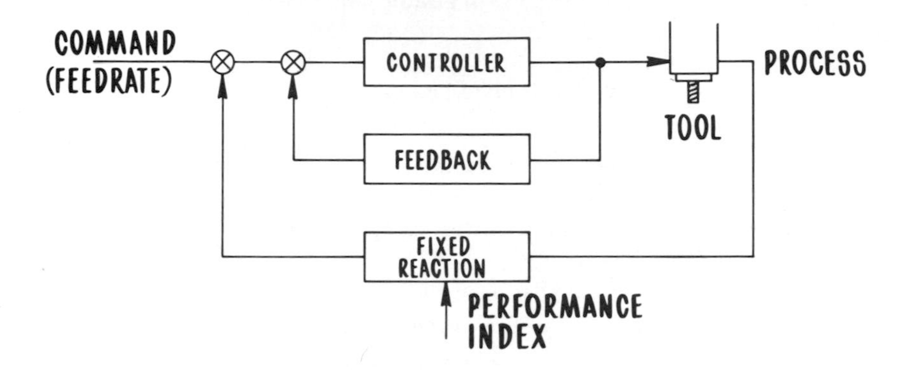

Fig. 12. Programmed adaptive control.

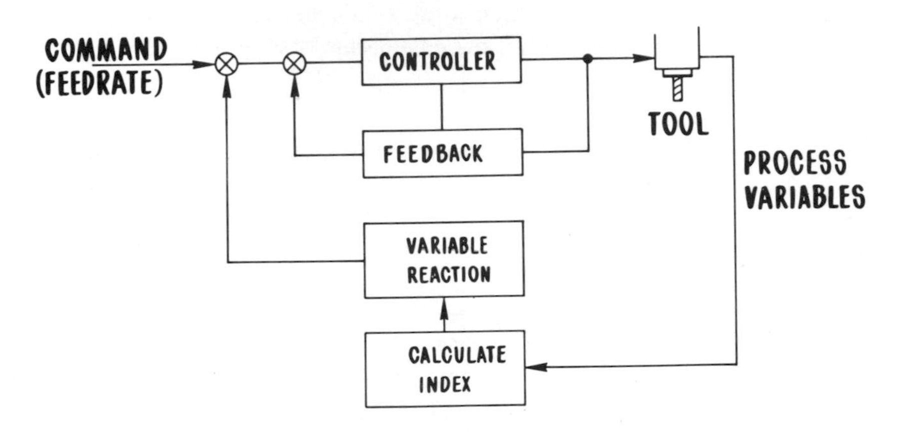

Fig. 13. Optimal adaptive control.

on tool life. This calculation is made off-line and the direct performance index for the system is actually tool life. A fixed strategy is responsive to the process variables of torque and deflection which act to drive the system to achieve the desired tool life.

Fig. 13 represents the more sophisticated form of adaptive control. In this case, the mathematical model of the performance index is set into the control and the controller continually calculates the index. The reaction of the system to the calculated index is to attempt to maximize (or minimize) it. Much has been written about these techniques to maximize or optimize the performance index, and most written discussion of adaptive control centres around selecting a mathematical 'hill climbing' technique. However, the real problem is to be able to measure enough process variables on an instantaneous basis in order to satisfy the mathematical model of the performance index. In no case has this been done to date, and all adaptive systems running on machine tools presently are either of the programmed adaptive type or constraint adaptive type. In fact, the central problem in the development of adaptive control is the reliable measuring of process variables, and the strategies of control assume a role of minor importance in comparison to the selection, cost, and reliability of process sensors.

Fig. 14 indicates the influence of adaptive control on some common process variables. Under conditions where variable width or depth of cut will be encountered, a programmer must select

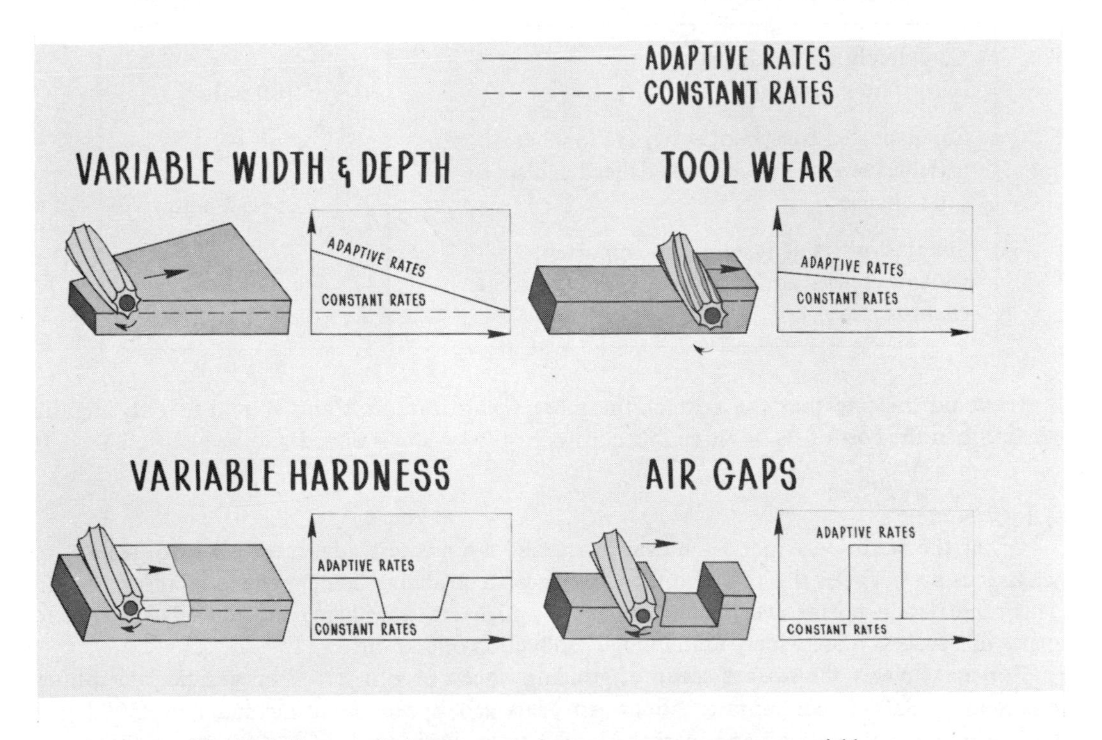

Fig. 14. Influence of adaptive control on some common process variables.

machining conditions so that the part will be in tolerance under the worst conditions. As the diagram indicates, the adaptive rate will slow down to the most conservative value only at the most critical point. As the tool wears, deflections increase, and depth of cut must be chosen to allow for machining under worst tool wear conditions. With adaptive control the system senses the effect of tool wear and slows down only as wear increases. Increased deflection brought about by increased hardness can also be sensed by the system and proper tolerance maintained in spite of variable hardness. Considerable improvements in productivity can be gained by speeding up over air gaps which the programmer may overlook unless the tape is completely optimized manually.

It can be readily seen that the use of adaptive control will greatly reduce the requirement for accurate selection of speeds and feeds during the planning stage. However, the need for accurate machinability data will continue to exist, since they must be used for the proper selection of cutting tools, and tooling of the machine tool, and broad feed and speed ranges.

In order to judge the desirability of various degrees of sophistication of adaptive control systems, we have made theoretical and laboratory studies. Table 3 shows the results of these studies.

TABLE 3

Type of adaptive control	Productivity gains
Completely optimizing system (wear rate sensor used)	100 units
Sophisticated constraint system (a number of variables sensed and controlled) feed and speed rates change	95–97 units
Simple constraint system (few important variables sensed and controlled) feed only changes	85–88 units

It would indicate that the cost of the more sophisticated systems should be only slightly greater than the cost of the more simple constraint-type systems in order to be justified.

7.2 *Grinding*

Of all the traditional metal removal processes, the greatest advancements with respect to what goes on between the tool and the workpiece are probably being made in grinding. Significant increases in horsepower and wheel speed which are now being introduced will tend to make this process more widely used in high production operations.

For many years the surface speed of grinding wheels used in precision grinding operations was held to 6500 ft per minute. About ten years ago, speeds were increased to 8500 ft per minute on some operations and recently, applications with speeds of 12 000 and 16 000 ft per minute have been introduced. It is not clear why wheel speeds remained at 6500 surface ft per minute for such a long period of time, since it is now evident that doubling or tripling these speeds does not produce excessive or catastrophic wheel wear. In contrast, cutting speeds in traditional operations, such as turning or milling, could not be doubled without undue cutter wear. These operations are usually run at speeds close to the limit of economic tool life. In grinding, factors other than the wear characteristics of the abrasive grain have tended to keep cutting speeds down. These factors include unbalance forces, machine rigidity, safety, and coolant application. These factors are now being worked on with what might be considered a systems approach, and the result is significant in gains in productivity for grinding.[5]

Due to improvements in our knowledge of the grinding process, [6] and data available on the effects of high wheel speeds, it is possible to arrive at a simplified explanation of why high wheel speed produces improved results. The explanation is made with the aid of Fig. 15. This set of curves illustrates the general relationship of some important grinding parameters. Normal force, F_n, displays a linear relationship with respect to wheel speed, V_{wh}, under conditions of constant metal removal. Surface finish also bears a linear relationship with respect to F_n, and

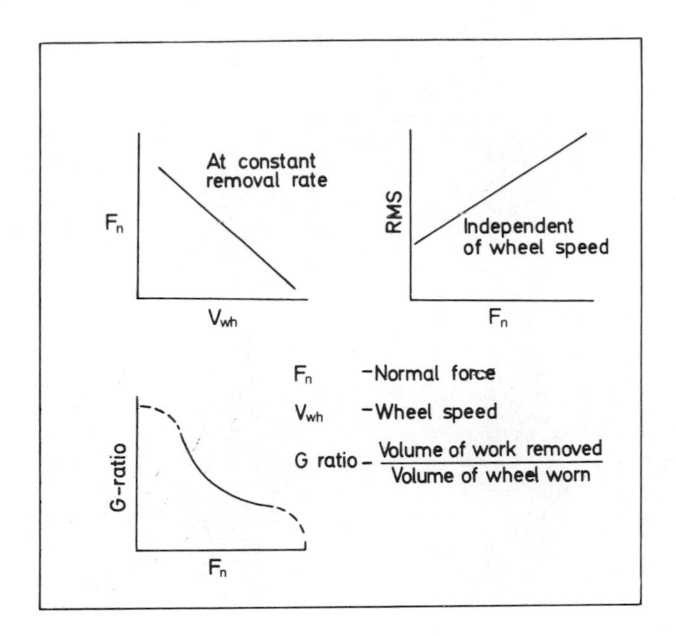

Fig. 15. Relationship of some grinding variables.

this relationship is independent of wheel speed. The third curve in Fig. 15 indicates that wheel wear ratio G bears a more complex relationship with respect to F_n, but generally, G increases as F_n becomes smaller. In addition, available data indicate that surface temperatures in grinding are reduced by low values of F_n and increased by increasing V_{wh}.

Since high wheel speeds reduce normal force, it follows that better surface finishes should be produced, and such is the case. Improved G ratios should be possible and this also has been reported. Usually, when wheel speed is increased, metal removal rate (feed rate) is also increased so that normal force is brought back to some value similar to that which had been attained under traditional conditions. When this is done, the effect of G ratio cannot be predicted because of the increased influence of surface temperature combined with the higher normal force. In general, surface finish is actually a function of final normal force during the last one or two revolutions of the workpiece in the case of cylindrical grinding, rather than the steady state force reached during most of the grind. Since high wheel speeds allow the induced normal force to reach much lower values during the transient conditions of infeed stopping and backoff, better surface finish should be observed under almost all conditions of high wheel speed grinding.

More rigid grinding machines have allowed the use of higher force intensities (normal force per unit width of wheel in contact with the work) and brought about greater values of metal removal rate per unit width. This factor, coupled with the use of high wheel speeds, increased the overall effectiveness of the grinding process to the point where it is competitive with the traditional processes of turning and milling in many applications.[7]

A typical candidate part for high stock removal rate in grinding is the sun gear shown in Fig. 16. This part is a hot forging made from 8620 steel. The maximum stock removal is $\frac{3}{8}$ in on diameter at the step, and nominal removal would be 0·100 in on diameter for most of the part.

The length of grind is 6 in and the grinding time is 20 s. Peak horsepower is 120 at 12 000 ft per minute wheel surface speed.

Fig. 16. A typical sample for high stock removal in grinding.

Considerations to be met in order for grinding to be more productive than turning are:

(1) Product of diameter $\times$ length $\geqslant 4$

(2) Less than $\frac{1}{4}$ in to be machined from diameter

(3) Tolerances $\pm$ 0·003 in or looser

(4) Surface finish 30μin and up

(5) Mass produced or large batch operations

(6) Special problems (for example, interrupted cuts) which might be more restrictive on turning than on grinding

In cylindrical grinding operations (internal and external), these improvements will be limited to high production applications until rapid and economical means of changing wheel shape and/or changing wheels are found. Future work will be aimed at improving the versatility of the process even more than improving its productivity.

REFERENCES

1 KOLOC, J., 'Problems Related to Automation of Small Lot Machining,' presented at *Annual Meeting of CIRP*, Cincinnati, Ohio, Sept. 1963.

2 KOLOC, J., 'A Contribution to the Manufacturing System Concept in Production Engineering Research,' presented at *Annual Meeting of CIRP*, Liege, Belgium, Sept. 1965.

3 GALLAND, 'Entwicklung einer werkstuckbeschreibenden Systemordnung zur Kostensenkung in der Kleinserienund Einzelfertigung,' Technischen Hochschule Aachen, *Doktor-Ingenieurs genehmigte Dissertation*, Feb. 25, 1964.

4 MATHIAS, R. A., 'Adaptive Control of the Milling Process,' *IEEE National Machine Tools Industry Conference*, Cleveland, Ohio, Oct. 1967.

5 Prof. Dr. -Ing. Dr. h. c. OPITZ, H., and Dr. -Ing. GUHRING, K., 'High Speed Grinding,' Technische Hochschule Aachen, *17th CIRP General Assembly*, Ann Arbor, Michigan, Sept. 1967.

6 HAHN, R. S., 'Variables in Grinding – Their Effect on Stock Removal, Wheel Wear, Surface Integrity, Grinding Chatter,' Paper No MR69-562, *ASTME*, Dearborn, Michigan, 1969.

7 MATSON, C. B., 'High Efficiency Centertype Grinding,' Paper No MR70-551, *Society of Manufacturing Engineers*, Dearborn, Michigan, 1970.

8 PERRY, C. B., 'Variable-Mission Manufacturing System,' *International Conference on Product Development and Manufacturing Technology*, University of Strathclyde, Glasgow, Sept. 1969.

SUPERPLASTICITY: A CONTRIBUTION TO INNOVATION IN FORMING

F. Jovane
Universita' di Bari, Italy

SUMMARY

Superplasticity – a combination of low strength, viscous behaviour, large capacity for stretching – and its main aspects are reviewed. Types of superplasticity, superplastic alloys, thermomechanical processing to produce superplasticity in various alloy systems, flow properties, equations for superplastic flow, mechanisms of deformation, properties after deformation, are reported. The industrial potential and applications of superplasticity to forming are then considered. The conditions for superplasticity to be industrially acceptable, and such forming processes as pressure and vacuum sheet forming, die-less wire-drawing, forging, coining and extrusion, are reviewed on the basis of the available literature and data. Advantages (such as reduced number of stages to obtain a part, reduced working loads, feasibility of intricate shapes with fine details) and disadvantages (such as strain-rate low compared with that of ordinary alloys and the need for high temperatures in some cases) are discussed for each process and the requirements for economic and technological assessment are pointed out.

1.0 INTRODUCTION

Superplasticity is viscous behaviour as exhibited by certain metals and alloys at a temperature above half the melting point in degrees absolute, within typical strain-rate ranges. When being deformed superplastically these materials have a highly sensitive flow stress and high ductility. Elongations up to 2000% have been obtained. They flow with the 'fluid-like characteristics' of hot polymers and glasses. Figure 1 shows tension test results of as cast and superplastic Pb-Sn eutectic alloy. Elongations are respectively 40% and 1650%.

The first observations which may be related to superplasticity were made by Rosenhain[1] and Sauveur[2] in the twenties. Later Hargreaves[3], Jenkins[4] and Pearson[5] found superplastic behaviour in tin-lead and cadmium-zinc eutectic alloys. In 1945 Bochvar, Presniakov and co-workers carried out in Russia an extensive research programme on superplasticity in non-ferrous eutectic and eutectoid alloys. Their work was reviewed by Underwood[6]. While Lozinsky and Simeonova[7] introduced the word *superplasticity* to the western scientific literature in 1959, it was only the study, by Backofen and co-workers at M.I.T., of fundamental aspects of superplasticity and its potential[8,9,10,11] that raised the interest of the scientific community in the phenomenon.

Since then research on superplasticity has been continuously increasing and an impressive body of knowledge has been developed. Surveys have been carried out by Jovane[12], Weiss and Kot[62], Chaudhari[13], Backofen *et al.*[11], Sherby[14], Weld[15] and Davis *et al.*[16]. Recently a very

Fig. 1. Tension test results of as-cast and superplastic Pb-Sn eutectic alloy – elongations respectively 40% and 1650% – Jovane, Shabaik, Thomsen[66].

comprehensive review of superplasticity, with particular reference to deformation mechanisms and superplastic alloys and metals available, has been made by R. Johnson[17].

The viscous behaviour and high ductility of superplastic alloys, and the good service properties of some of them, disclose a promising field for innovation in forming. Research work on forming in superplastic condition has been done and much more is in progress. Most of this is classified or confidential because of its industrial potential. The exploratory work is concerned with the development of new processes and the assessment of already established ones. The aim is to determine whether in the future, hopefully not too far into the future, it will be possible to satisfy the desire to reduce production costs or raise quality and the need for new products.

The purpose of this paper is to use the available data and results to analyse, from the production point of view, the contribution of superplasticity to forming processes. Therefore the main properties and characteristics of superplastic alloys, which are relevant to forming are reported. The framework within which any superplastic forming process should be assessed is given. The work done on superplastic forming is reported and discussed.

2.0 TYPES OF SUPERPLASTICITY

The various cases of superplasticity can be classified into groups[12]. Their number may be reduced to two[17], namely the groups related respectively to:

- Application of special environmental conditions
- Existence in the material of a special microstructural condition

2.1 *Environmental superplasticity*

Phase transformation plasticity is the most important case falling in this group. It occurs when cycling metal under low load, several times, through a high-temperature phase transformation. Oelschlagel and Weiss[18] obtained elongation of more than 500% in steel by cycling more than two hundred times, under low load, through the $\alpha \rightleftarrows \gamma$ transformation temperature.

Phase transformation plasticity for obtaining high elongations under low stress is of some interest, but, at present, its applications seem to be confined to a limited region of metalworking.

2.2 *Structural superplasticity*

The great majority of cases of superplasticity fall into this group. The following conditions must be satisfied for structurally superplastic behaviour to occur:

- The alloy must have a very fine grain[8]. The grains must be fairly equiaxed and their size within 1–10 μm. In ordinary metals the grain size is within 10–1000 μm. The fine grain requirement is a necessary but not sufficient condition. If grain boundaries are not appropriate, fine-grained materials may be brittle even at high temperature.

- Grain growth during deformation must be as low as possible. To obtain fine and stable grain size, using appropriate thermomechanical processing (see later), a two-phase (microduplex) structure must be produced. In this case one phase will inhibit the growth of the other. The two phases should have similar strength and ductility[17], but different composition to favour stability.
- Deformation should take place at constant temperature, above 0·4 T_m the melting point in degrees absolute and within appropriate strain rate ranges.
- The strain-rate sensitivity index, m (see later), should be higher than 0·3 (see Backofen *et al.*[8]). In hot working of normal metals $m \cong 0{\cdot}1$.

This paper will mainly be devoted to structural superplasticity.

3.0 SUPERPLASTIC ALLOYS

A review of superplastic alloys has recently been done by R. Johnson[17]. In Table 1 most of the alloys exhibiting superplasticity are reported[17].

The range of alloys available is already wide, but, by appropriate thermomechanical processing (see section 4.0), a larger number of alloys will eventually be rendered superplastic.

For the purpose of discussing the exploitation of superplasticity, the available alloys may be divided into three groups[19]:

- Low-melting-point alloys (T_m in the range 100–300°C), for example Pb-Sn. They have creep problems and they cannot be used in structural applications.
- High-melting-point alloys, for example titanium alloys and stainless steels. They are superplastic in the range 500–1000° C, that is the conventional hot-working range. They do not have creep problems at room temperature. Their high forming temperature, as will be discussed later, may give technical problems.
- Medium-melting-point alloys. Their forming temperature is not too high. Creep problems at room temperature are negligible. The best known alloy of this group is the Zn-22Al eutectoid alloy, commercially known as Prestal[20] in Britain.

4.0 THERMOMECHANICAL PROCESSING TO PRODUCE SUPERPLASTICITY

The fine equiaxed and stable grain structure necessary for superplastic behaviour to occur, may be obtained by:

- Hot working[12,19] (extrusion, rolling, forging, etc.) a eutectoid or as-cast eutectic alloy (for example Pb-Sn). In fact, by breaking the lamellar structure an intimate mixture of the two phases is obtained. The latter have an equiaxed fine-grain structure. The amount of work affects superplastic properties[9,21,22].
- Hot working an alloy in a region of the phase diagram where two phases are present. Recrystallisation and precipitation take place during deformation and cooling: a duplex microstructure is obtained. This process has been applied to several nickel-chromium-iron alloys[23,24] . These may also be obtained by heating to the single-phase region, quenching, cold-working and finally, heating to the two-phase region. During the last operation recrystallisation and precipitation take place, thus producing a duplex microstructure.
- Exploiting such solid-state reactions as spinodal reactions in quenched Al-Zn alloys[25], or precipitation when crossing the β transus in titanium alloys[25]. In the latter case defor-

mation must proceed while precipitation is taking place.

- Using powder metallurgy[9,26–28], pack-rolling alternate sheets of two metals[9], electroplating alternate layers of two metals[29]. Powder metallurgy has the advantage that alloys of any composition can be produced whereas conventional alloys are limited by the form of the phase diagram[19]. It could also be possible to use 4-5-6-component phase systems instead of a two-phase system: this would be impossible with conventional methods[19]. Much research work is now under way to find thermomechanical processes for producing new superplastic alloys and to reduce the production cost of superplastic alloys already available.

5.0 FLOW PROPERTIES

Flow and service properties are very important for, respectively forming and structural applications of superplastic alloys. This section deals with flow properties. Flow stress, high ductility and mechanisms of deformation will be considered.

5.1 *Flow stress*

The flow stress of superplastic materials depends on three main parameters: strain-rate, temperature, grain size.

5.1.1 *Effects of strain-rate*

Strain hardening in superplastic materials is negligible. A yield limit cannot be detected. The physical properties can be represented by a stress, σ, strain-rate, $\dot{\epsilon}$, curve. Typical (σ, $\dot{\epsilon}$) curves for three superplastic alloys[10,21,25] are shown in Fig. 2. For one of them the curve of the as-cast alloy is also given[21].

The dependence of flow stress on strain-rate may be evaluated, as proposed by Backofen[8], in terms of m, the strain-rate sensitivity index, from the relationship:

$$\sigma = K\ \dot{\epsilon}^m \tag{1}$$

where σ = true stress, $\dot{\epsilon}$ = natural strain-rate, K and m are functions of temperature, grain size, $\dot{\epsilon}$ and, in some cases, of strain. This, may be due to grain size variation during deformation.

The (m, $\dot{\epsilon}$) curves obtained from (σ, $\dot{\epsilon}$) curves, as suggested by Backofen *et al.*[8], are shown in Fig. 2.

High elongations are associated with high m (see par. 5.2.1). If m = 1 the material behaves as a Newtonian fluid and, it can be demonstrated, there is no formation of necks during elongation. Maximum m in superplastic materials is 0·9. Minimum m is 0·3 (see Backofen *et al.*[8]).

For ordinary metals, when hot-worked, m = 0·1, as in the case of as-cast Pb-Sn alloy (see Fig. 2). If the condition m = 0·3 is plotted in Fig. 2, three regions can be seen for each (σ, $\dot{\epsilon}$) curve. Only the central one corresponds to the typical strain-rate range where superplastic behaviour occurs. This shows clearly that the titanium alloy is superplastic at low strain-rate while Al-Zn is superplastic at relatively high strain rate. From Fig. 2 it can be seen that there may be a two-order-of-magnitude decrease in stress at the low ends of (σ, $\dot{\epsilon}$) curves. The low value of flow stress may be as small as a tenth of the flow stress in the same material non-superplastically deformed under the same condition, as in the case of Pb-Sn eutectic alloy.

The only extensive data available on flow properties are for tension: a few data have been obtained for compression[21,31,32]).

Torsion tests have been carried out by Ghosh and Duncan[33]. A bulge test, developed by Jovane and Naso[34,35], has been used to determine the flow properties of a Sn-Pb superplastic

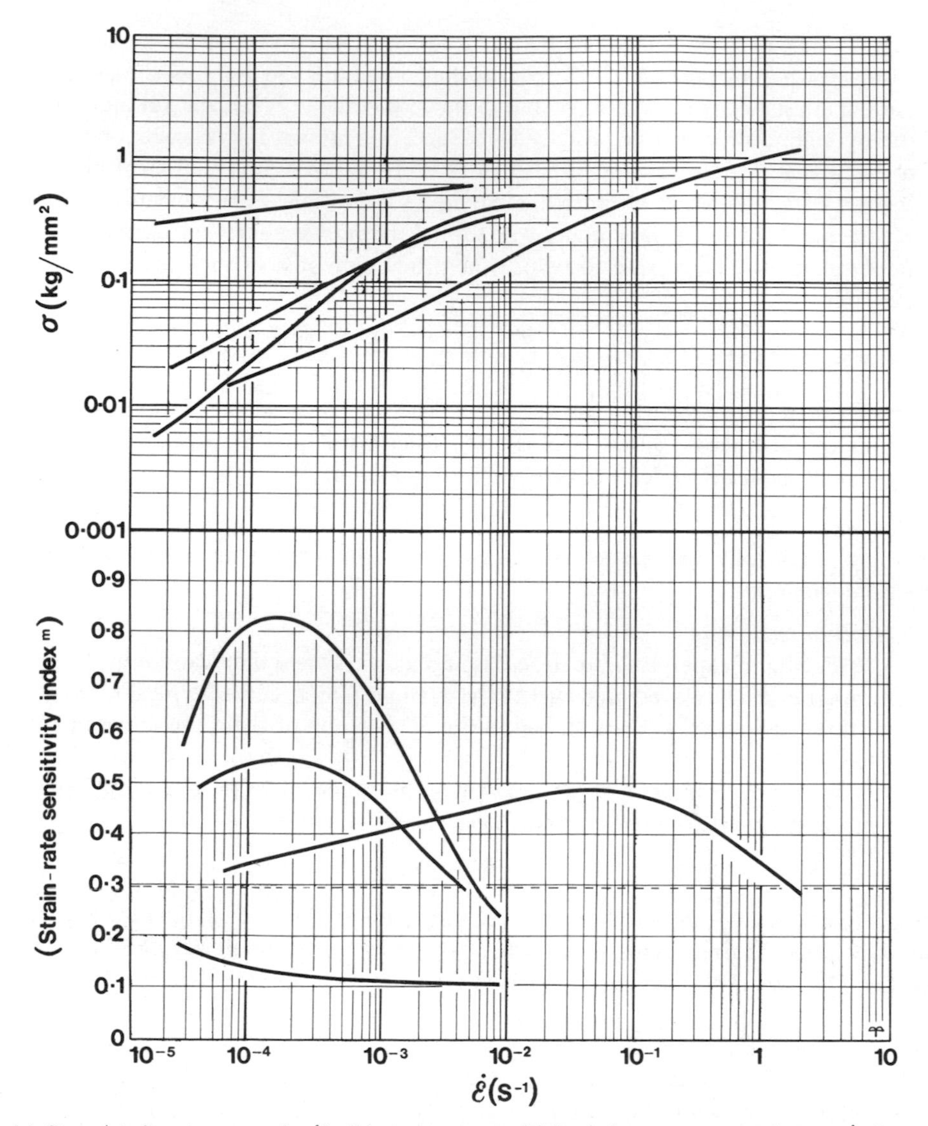

Fig. 2. (a) Stress/strain-rate curves (σ, $\dot{\epsilon}$); (b) strain-rate sensitivity index m versus strain-rate, $\dot{\epsilon}$, for various superplastic alloys.

alloy in biaxial tension, that is in a condition similar to that associated with pressure forming. Curves thus obtained are very close to tensile ($\sigma, \dot{\epsilon}$) curves.

5.1.2 *Effect of temperature*

Flow stress in superplastic alloys at any strain-rate is strongly dependent on temperature, more than in conventional metals. As previously said, K and m from eqn 1, are temperature-dependent.

5.1.3 *Effect of grain size*

Flow stress at any strain-rate and temperature depends on grain size. The quantitative relationship between flow stress and grain size has not yet been clearly defined, but it may be a direct proportionality of flow stress to a power of the grain size, that is the smaller the grain size the lower the flow stress.

5.1.4 *Equations of superplastic flow*

Equations of superplastic flow are necessary for studying the mechanics of forming processes and evaluating forming loads. Some equations have been proposed but they may be not easy to handle, and no agreement has been reached on them. Therefore the empirical eqn 1 has been proposed by Backofen *et al.*[8] to describe $(\sigma, \dot{\epsilon})$ curves.

Equation 1 is accurate enough in those strain-rate ranges where m is fairly constant. Rapid variation of m with $\dot{\epsilon}$, as in the case of the titanium alloy of Fig. 2, results in poor accuracy from eqn 1.

Jovane[36] has found that an empirical relation:

$$\sigma = A + B \ln \dot{\epsilon} \quad (2)$$

for some alloys describes well the $(\sigma, \dot{\epsilon})$ curves over a wide strain-rate range. A and B are material parameters and may be easily determined from a semi-logarithmic representation of $(\sigma, \dot{\epsilon})$ curves.

Padmanabhan and Davies[37] have developed a method based on multidimensional regressional analysis which can be used numerically to interpret superplastic data. It could be used in connection with the calculation of loads and power required in forming superplastically.

5.2 *Ductility*

One of the most striking and best known features of superplastic alloys is the high elongation, as shown in Fig. 1. But ductility is also present in compression and torsion.

5.2.1 *Ductility in tension*

High elongations (see Table 1) have been detected in superplastic alloys when tested at constant crosshead speed or under constant load, as in the case of phase-transformation superplasticity. Pressure-forming tests have shown that high deformation may also be reached in biaxial tension. A relationship exists between the strain-rate sensitivity index, m, and stretching capacity[9,25,30,38]. High elongations are associated with high m. If $m = 1$, the material during elongation behaves as a Newtonian fluid and forms no necks, or other localised deformations, as does an ordinary metal in tension.

As m is a function of temperature, grain size and $\dot{\epsilon}$, elongations are greatly dependent on the same parameters. Elongations are also dependent on surface conditions as shown by Jovane[39], Hart[40], Avery and Stuart[41] and Morrison[22,30]. Maximum elongations obtainable are very sensitive to the initial surface discontinuities in the specimen. Finally maximum elongation attainable is sensitive to the initial geometry of the specimen, as shown by Morrison[22,30].

5.2.2 *Ductility in compression*

The data available are limited[21,26–28]. Freche, Waters and Ashbrook[26] have shown that small cylinders of nickel-base alloy, produced by extrusion of pre-alloyed powders, could be hot-pressed by 70% without fracturing. The same alloys, non-superplastic, would fracture under little deformation. Similar results have been reported by Reichman, Castledine and Smythe[27,28], who hot-pressed small cylinders of In-100 alloy, produced by extrusion of pre-alloyed powder.

5.3. *Mechanisms of deformation*

High elongations and strain-rate sensitivity are not the only remarkable features of superplasticity. Examination of the microstructure of a superplastic alloy before and after an elongation of as much as 1000% shows that the structure has not changed, and, in general, there are no voids. This is quite different from ordinary metals, where the grains, after a much smaller deformation, are stretched in the direction of the applied stress. The mechanism of superplastic flow must therefore be different from that of plastic flow.

Several mechanisms have been proposed to account for superplastic flow, but so far there is no agreement. It can only be said that boundary sliding is widely accepted as playing an important part in superplastic deformation. The accommodation processes necessary to maintain compatibility between grains may be crystallographic slip, stress-induced vacancy migration (Nabarro-Hering or Coble creep) recrystallisation, etc., in a combination which is typical for each alloy, test condition etc. A review of mechanisms of deformation is reported in R. Johnson's paper[17].

6.0 PROPERTIES AFTER DEFORMATION

High- and medium-melting-point alloys, as pointed out in section 3.0, may be used for structural applications.

Room-temperature properties of high-melting-point alloys are generally good. The ultra-fine two-phase structure, as in the case of stainless steels, results in high strength, toughness, and good fatigue resistance[23,42,43]. As-formed materials may show poor creep properties if service temperature is increased.

Most of the mechanical properties of the medium-melting-point alloys are good at room temperature. They may have poor creep resistance. The creep problem may be solved by heat-treating[26,27,28,42,43] the formed material to coarsen the fine structure necessary for superplasticity. In the case of the stainless steels[42,43] a treatment has been devised which changes the grain structure so much that the formed part may be put in service at a temperature higher than the temperature at which it was superplastically formed.

In some cases forming and post-treatment have been combined to reduce costs[44]. An alternative to post-treatment is the method of Naziri and Pearce[45]. They added copper to the Al-Zn eutectoid alloy, thus improving creep properties without affecting significantly the flow stress and *m* above 150°C.

7.0 EXPLOITATION OF SUPERPLASTICITY

The aspects of superplasticity already examined are very appealing for exploitation in forming processes*. But manufacturability is not the only main concern in choosing an alloy for a given application. Serviceability and cost must also be taken into account.

The degree to which superplasticity will be industrially accepted depends on the following factors:

- Availability of economic superplastic alloys of technological interest
- Economic thermomechanical processing for producing superplastic alloys

* Such alloys as stainless steels may be used at high temperature while in superplastic condition: grain-boundary embrittlement is reduced by neutron irradiation[47].

- A new approach in design to fully exploit the possibilities offered by forming in the superplastic condition
- Introduction of new forming processes and use of already-established processes so that superplastic properties may be exploited economically at their best
- Improvement of some forming properties
- Good service properties of parts formed from superplastic alloys.

Any economic and technological assessment of forming in the superplastic condition must be done within the framework of the above factors.

Succeeding sections will be devoted to superplastic forming. Here it may be worth adding that the areas related to the factors mentioned are enjoying great attention by r. & d. people in research institutions and industry, and, as this paper shows, interesting results have already been obtained.

8.0 SUPERPLASTICITY AND FORMING

If it is to turn towards superplastic forming industry should be shown that it can thereby produce at lower costs, or manufacture parts that otherwise could not be obtained, or attain higher quality in production.

The processes in which exploitation of the viscous behaviour and high ductility of superplastic alloys have been or are more likely to be tried are:

- Sheet-forming techniques already in use for plastics. If superplastic alloys are treated[19] as 'high-modulus, high-strength, heat-treatment-sensitive, thermoplastic polymers' the whole range of forming techniques for plastics may be transferred to metal forming. Fundamental methods of thermoplastic sheet forming[49] are indicated in Fig. 3.

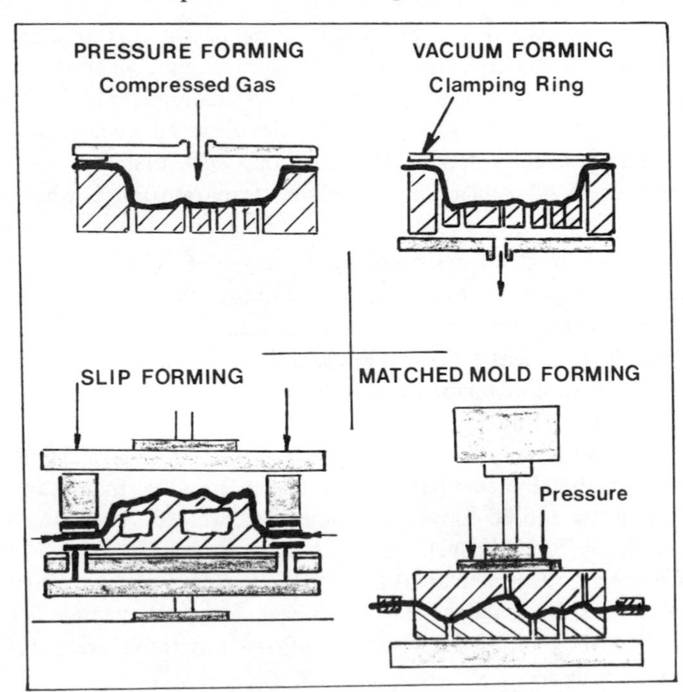

Fig. 3. Fundamental methods of thermoplastic sheet forming[49].

- New forming processes. Innovation is necessary to fully exploit superplastic properties. An interesting example of new processes is die-less drawing based only on simple tension.
- Forging and coining. Forming may take place under low and/or constant load. Excellent die filling may be obtained.
- Rolling and extrusion. As the flow stress of superplastic alloys is strain-rate-sensitive, forming loads may be lowered by reducing working speed. Complex cross-sections may be obtained.

In succeeding sections the use of the above processes for superplastic alloys will be reviewed and discussed on the basis of the data and literature available. Original results by the author are reported on extrusion in superplastic condition.

8.1 *Pressure and Vacuum Sheet Forming*

Of the typical processes developed for plastics and which may be used for superplastic alloys, pressure and vacuum sheet forming have been more extensively studied and used than the others. This review will be mainly concerned with them. A great deal of research work is being done in this field, but a large part of it is classified. Most of the published results will be reported here.

8.1.1 *Review of literature*

The first demonstration that superplastic alloys can be pressure-formed was given by Backofen, Turner and Avery[8]. Using an Al-Zn eutectoid alloy they blew a dome higher than a hemisphere.

Fields[50], using the same alloy, vacuum-formed a rectangular box with very fine details. He also proved that high draw ratios could be achieved.

One of the problems to overcome in pressure forming may be that of keeping the strain-rate in the material within the superplastic range. An approximate analysis has been made by Jovane[51,52] of the viscous deformation of a thin circular diaphragm clamped at the periphery and subjected to one-sided hydrostatic pressure. He obtained equations relating pressure, stress, strain-rate, time and deformation. The equations show that forming pressure could be varied so that the process takes place in a prefixed strain-rate range, where *m* may be satisfactory high.

The analysis has been used on pressure-bulged discs of 6 Al-4V titanium alloy while in superplastic condition[53]. Figure 4 shows some typical results obtained. The dome on the right is higher than a hemisphere. The forming time was 8 minutes.

Reference to the development work being carried out in industry is made in a review on superplasticity by Weld[15] and in a paper by Backofen[44]. A typical example of a part formed by I.B.M. developers is shown in Fig. 5.

Hundy[54] has dealt with the problem of exploitation of superplasticity by the motor industry. He has shown that the cost pattern is different from that in conventional sheet metal forming. Superplasticity would be convenient if complex welded assemblies could be replaced by a single superplastically formed panel.

North[55], employing processes similar to those at present in use when either vacuum-forming or blow-moulding plastics, has obtained parts in Prestal. A typical example, a refrigerator door, is shown in Fig. 6. The alloy offers the advantage of lower assembly costs allied with complex shapes and lower tooling costs.

Fig. 4. Typical 6 Al-4V titanium alloy membranes, pressure-bulged in superplastic condition. From left: H = (ratio of membrane height to die radius) = 0·45, 0·70, 1·33; Jovane[53].

Fig. 5. Trepanned rod, sintered iron mould and goblet blow-moulded in superplastic condition. *Courtesy of I.B.M.*

Neal[56] has reported that a furnishing panel on Concorde may be made in Prestal, thus replacing thermosetting plastics. Such a panel, still at development stage, is shown in Fig. 7.

Al-Naib and Duncan[57] have carried out experiments in which tubes and sheets of superplastic tin-lead eutectic alloys were formed using various pressure-forming techniques. These include, among others, free forming of a tube, forming a tube in a die and production of a bottle.

The superplastic deformation of circular diaphragms subjected to one-sided hydrostatic pressure has been investigated by Cornfield and R. Johnson[58], both theoretically and experimentally. Special attention was given to the thickness variation in bulged shapes. A stainless steel cup with corrugated base, pressure-formed by them, is shown in Fig. 8. Hollow parts in a titanium alloy and Al-Zn were also produced.

A similar problem has been studied by Holt[59]. Two cases were dealt with: (1) the bulging of a flat circular sheet clamped at its perimeter, and (2) the bulging of a sheet into 90° V grooves. The latter process is related to the problem of moulding a sheet to a die contour in detail. Bulge profile and sheet thickness distribution are predicted as a function of the variables of pressure, geometry, time, K and m.

Fig. 6. Refrigerator door inner panel, formed in superplastic condition. North[55].

Thomsen, Holt and Backofen[60] have studied the forming of superplastic Al-Zn eutectoid sheets in bulging dies. They separated the process into two components: stretching, to establish general shape, and another component, more related to coining, to generate surface detail. The results of the two studies combined allow prediction of the outcome of practical forming. Hollow-ware with fine details was obtained from flat blanks and tubular preforms. In one case 'plug-assist forming', from polymer processing, was adopted.

8.1.2 *Analysis and discussion*

The main results of the theoretical and experimental work done may be summarised as follows.

Pressure-forming of high- and medium-melting-point alloys in different shapes with fine details is feasible. Large panels replacing complex assemblies can be obtained, thus reducing die and assembly costs. Deformations in bulging higher than in conventional forming can be reached. Thinning over radius is reduced. The cost of machinery is low as the forces involved are small. Instead of two matching dies, only one is required. Depending on the material and other matters, such as geometry, the forming time goes from a few minutes (titanium alloys) down to less than a minute (Al-Zn eutectoid alloy).

For some shapes analysis enables one to control the forming process so that it takes place in a prefixed strain-rate range, compute the working pressure and forming time, and finally to predict the finished shape. Mathematical models for optimisation have been proposed.

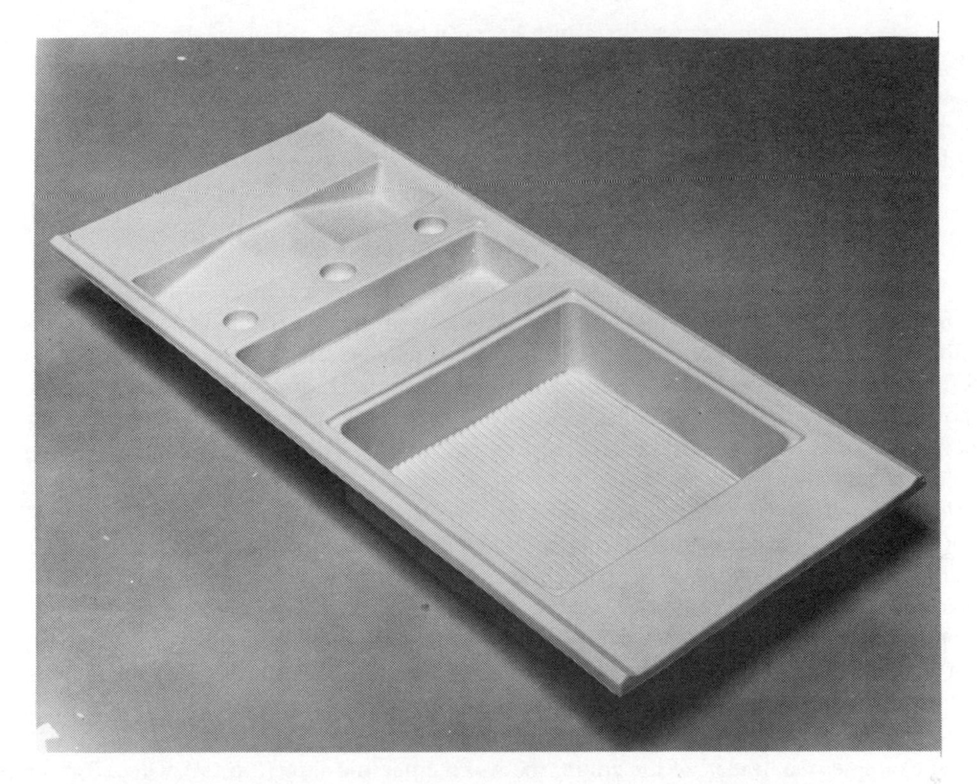

Fig. 7. Passenger amenity panel in Prestal. Pressure-formed in superplastic condition. *Courtesy of British Aircraft Corporation, Filton Division.*

Fig. 8. Stainless steel cup with corrugated base, formed in superplastic condition. Cornfield and Johnson[58].

For the sake of discussion we distinguish between high- and medium-melting-point alloys.

High-melting-point alloys. (Mainly titanium alloys.) Another advantage to mention is the possibility of obtaining a part in one step, thus eliminating the numerous expensive interstage annealings required in normal fabrication techniques. On the other side, many problems must be adequately solved, such as:

- Development of dies to stand high temperature at contained capital cost
- Contamination
- Fine temperature control and, possibly, short heating times
- Superplastic behaviour at low strain-rate range, that is long forming time

Aerospace and other advanced industries demand parts in such normally-difficult-to-form alloys as titanium. Considering the above pros and cons, it may be said that, for complex shapes with fine details very difficult to obtain otherwise, superplastic pressure-forming may play an important role. For less complex shapes only an economic assessment will show whether a superplastic or a conventional process is more convenient. This assessment should be done at the design stage.

Medium-melting-point alloys. As tooling must withstand only moderate temperature – loads are low – it can be made of such cheap materials as cast iron, aluminium and ceramics[54]. As the tooling cost is low, restyling of the part being produced may be more frequent. Electric heating in various forms may be used[61]. The above advantages should be added to those mentioned at the beginning of the section.

In the previous section the alternatives were pressure-forming versus conventional processes. In the case of medium-melting-point alloys the alternatives are there materials versus conventional materials, each with its own typical forming process.

For example, the 'leader' of the group, the Al-Zn eutectoid alloy (Prestal), whose forming and service properties are good, may be an alternative to plastics or steel in car-body construction[55]. Plastics require cheap tooling, allow rapid changes in tooling for new models, and assembly costs are low. On the other hand they are more expensive, and their shock-absorbing properties are not comparable with those of metals. This may be a good reason for limiting application in car-body construction.

Steel is a cheap material, but the cost of tooling and machinery is very high, because of the high flow stress of steel and the high number of parts required in complex assemblies. Assembly is also expensive. A comparison between steel and Prestal has been done by Hundy[54], considering material costs, production volume, labour cost etc. for construction of the body of a small sports car at 500 a week. The cost for a body in Prestal may be lower than for steel. It is also worth noticing that the cost distribution is different in the two cases. In the case of Prestal, material costs are by far the biggest element of the total cost, and labour and tool costs are reduced.

As pointed out by Hundy, the relative economies of a superplastic alloy versus other materials for particular applications vary widely and need to be studied carefully in each case. While any assessment must be done at the design stage, exploitation of superplasticity may also be completed by using design criteria similar to those for thermoplastics. The designer should realise that, in the case of superplastic alloys, he is dealing with a material exhibiting the forming properties of plastics and the service properties of metals.

8.2 *Die-less wire-drawing*

A new process, based on the use of simple tension, is die-less wire-drawing as developed by Weiss and Kot[62]. See Fig. 9. The bar to be reduced is held at one end while the other end is

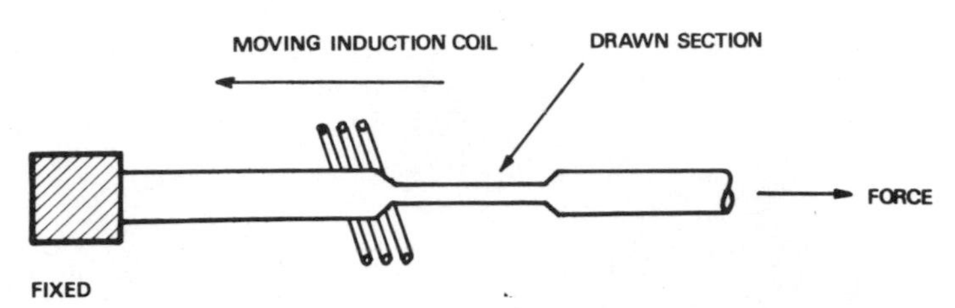

Fig. 9. Die-less drawing process in superplastic condition.

loaded in tension. An induction coil heats the bar to the desired temperature and moves along it. The method may be used to produce stepped lengths. There is no die wear and no lubrication problem. Reductions in area up to 80% have been obtained in superplastic alloys[61].

8.3 *Forging*

Very little has been published about this process. Freche[44] has forged a turbine blade from TAZ-8A, nickel-based alloy. The dies were of the same material, but not superplastic, having been heat-treated differently. See Fig. 10. Complex parts[44] that used to be cast or forged, then machined, are now being made in one superplastic forging operation. This is possible as the material flows into thin recesses to obtain details which are too fine for die casting[44].

Much work is also being done on the use of superplastic alloys for making dies for thermoplastics.

To discuss what may be the contribution of superplasticity to forging it is worth reporting, as a guide, a typical cost breakdown for a hot steel forging[46]. The various cost items are as follows: raw material 50%, overheads 21%, plant 11%, direct labour 9%, die costs 9%. Overheads, plant costs, and part of the labour costs depend on rate of production. Furthermore, at present 25% of the raw material is waste in the form of flash. A contribution to the economics of forging may come from: any savings in raw material; increase in production rate; increase in die life; reduction in the number of stages to obtain a part.

From the technical point of view high ductility and low flow strength are sought to improve die filling. Low flow strength means also low forming loads.

Superplastic characteristics give good die filling. With high ductility large amounts of deformation per operation may be obtained[46], so the number of dies is reduced. Low forming loads are required: this leads to the use of smaller, hence cheaper, equipment. But these advantages may be offset by:

- low production rate, higher cost of superplastic materials
- cost of a post-treatment
- short die life in the case of high-melting-point alloys (as the process is slow the die temperature is equal to the deformation temperature).

In the open forging of large parts, on the other hand, a reduction in load reduces the requirement for high-tonnage presses, and this may offset the low-deformation-rate problem. Keeping the workpiece for some time at an appropriate temperature may be a problem.

The future of superplasticity in forging is probably associated with difficult-to-form materials and complex parts with fine details.

8.4 *Coining*

Data on the exploitation of superplasticity in coining* are very sparse. Still, coining in superplastic condition is worth trying on difficult-to-form materials, indeed on any material when fine details are required.

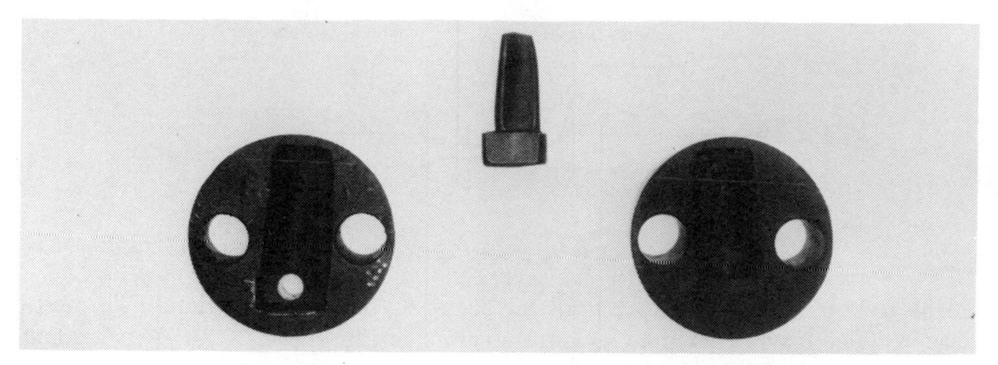

Fig. 10. Forging of a turbine blade from TAZ-8A, in superplastic condition. *Courtesy of Dr Freche, NASA.*

In ordinary metals, even if the flow stress is low, there is a threshold[64,65] below which flow does not take place. This applies also to forging. High pressures must be applied for complete filling. In superplastic materials, because of their viscous behaviour, there is no such threshold, and flow takes place under any, even very low, constant pressure[64]. Good die-filling can be obtained and the time required depends on the value of the applied pressure.

An approximate analysis to determine the pressure range for superplastic forming of a simple shape, and the time required to form under a given pressure, has been developed by Jovane and Morelli[64]. They determined a relationship linking material and geometrical parameters, time and pressure for the filling of a square groove, as shown in Fig. 11. A computer program was written to handle the relationship and forming times for different pressures, and degrees of filling were calculated. Typical curves are shown in Fig. 12, which shows that forming time is a function of the applied coining pressure. Experimental results support the analysis. Such an analysis may be of some help in assessing whether a process is economically feasible or to determine the forming pressure which minimises costs. Further research work is under way.

* Coining is the process in which 'the workpiece is completely confined in a closed die and the metal of the workpiece is made to conform to the contours of the die, without permitting any metal escape'[63].

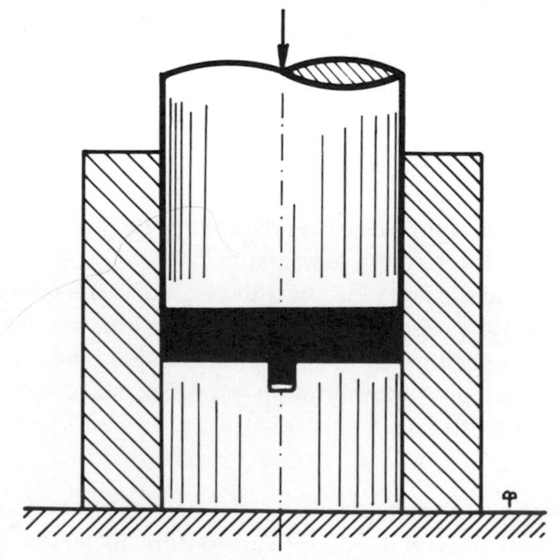

Fig. 11. Filling of a square groove.

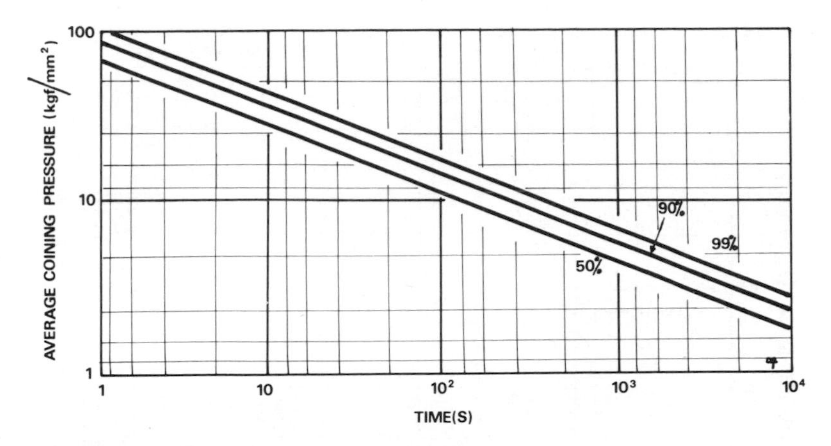

Fig. 12. Time required to obtain various degrees of filling versus coining pressure in superplastic condition. Jovane and Morelli[64].

8.5 *Extrusion*

Extrusion in superplastic condition has attracted more theoretical than development work.

Using the visioplasticity method, flow studies of the direct extrusion process in superplastic condition have been made by Jovane, Shabaik and Thomsen[66]. They showed that the velocity fields are nearly identical with those obtained using an ordinary metal. The method also allowed them to calculate extrusion loads very close to the measured values.

Mehta, Shabaik and Kobayashi[67] have presented two solutions for the detailed mechanics of the extrusion. They are supported by experimental results.

Superplastic properties may be exploited in extrusion. As reported for stainless steel[44], complex shapes may be obtained and low extrusion pressures are required when working in the superplastic range.

As the strength of superplastic alloys is sensitive to strain-rate, extrusion pressure may be lowered by reducing ram speed. This would certainly reduce the cost of tooling and machinery, but require more time for forming, that is more labour, heat cost, etc. An economic assessment of superplastic extrusion for a given alloy could be made if a relationship between extrusion pressure, material properties, geometry and ram speed were known.

An approximate analysis for the extrusion of a cylindrical bar of superplastic material through a conical die has been developed by Jovane and Della Volpe[68], who derived an equation by which to calculate the extrusion pressure for given geometry, friction and material. The equation has been used to compute the extrusion pressure versus ram speed for the Ti, 6A1,4V and A1-Zn eutectoid alloy whose $(\sigma, \dot{\epsilon})$ and $(m, \dot{\epsilon})$ curves are given in Fig. 2. They have been chosen as good example of superplasticity at low and relatively high strain-rates. The aim of this exercise was only that of showing how extrusion pressure decreases with ram speed. The curves of extrusion pressure versus ram speed for the two alloys (initial diameter 100 mm, $\alpha = 45°$, $\mu = 0{\cdot}15$, extrusion ratios 4 and 20) are given in Fig. 13. As expected, extrusion pressure is fairly dependent on ram speed. The analysis may be seen as preliminary contribution to the economic assessment of superplastic extrusion.

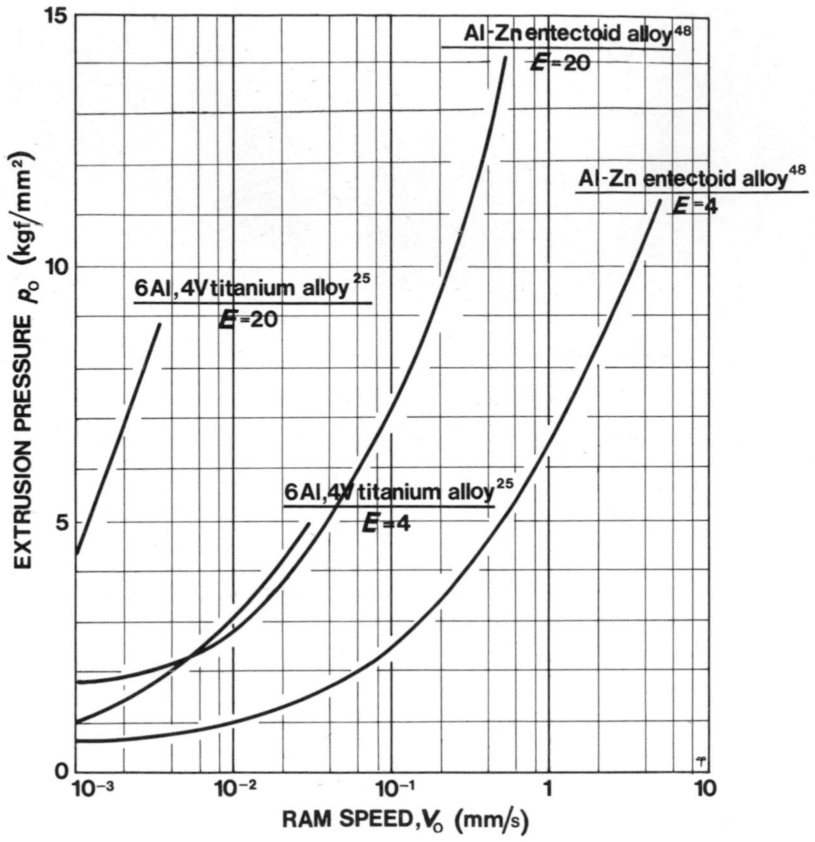

Fig. 13. Extrusion pressure versus ram speed for two alloys and extrusion ratios in superplastic condition. Jovane and Della Volpe[68].

8.6 *Miscellaneous*

Processes such as stretch-forming, creep-forming, rolling, etc. are being investigated, but very few data are available. They will not be treated here. Deep drawing has been attempted, but unsuccessfully. It was found that the cup wall elongated without drawing the flange[57].

9.0 POTENTIAL APPLICATIONS

In the foregoing sections most of the applications of superplasticity to forming have been reported and discussed. But superplasticity is not necessarily confined to them. New applications can be foreseen, particularly in such areas as the forging of large and complex ingots, the stretching of large ingots into thin sections without rolling[44], making dies for die casting and forming plastics, extruding in two stages on the same machinery (first stage: produce fine-grain structure in the alloy, second stage: make the product)[15], making prototypes (from car bodies to typewriter bodies), injection, blow-moulding and other forming processes typical of plastics[19]. These are just a few of the applications foreseeable. In most cases the greatest advantages should be related to difficult-to-form alloys.

10.0 CONCLUSION

The viscous behaviour and high ductility of superplastic alloys and the good service properties of some of them create a promising field for innovation in metalforming.

Superplasticity may be exploited in new and conventional processes. The main advantages are: reduction of number of stages to obtain a fabrication; reduction of working loads; and the possibility of obtaining complex, intricate shapes with fine details. These, in some cases, may be offset by: rate of deformation low compared with that of ordinary metalforming; high temperature, for some alloys; accurate and fine control of the process; higher cost of alloys due to thermomechanical processing needed for making them superplastic; post-treatment, necessary in some cases to improve service properties.

Superplastic forming is now moving from research to development and production, as clearly shown by the example reported (panels for a car-body or a refrigerator, turbine blades etc). The convenience of superplastic processes versus ordinary forming processes varies from case to case and must be assessed carefully in each of them. It may only be said that, at present, it looks as if superplastic forming may be competitive for short-run production, for manufacturing difficult-to-form alloys, and for producing complicated parts.

To enlarge the regions in which superplastic forming may be industrially competitive much more r. & d. work must be carried out, not only on superplastic forming but also on the production of cheaper superplastic alloys with a balanced combination of forming and service properties. Finally, full exploitation of superplasticity calls for creativity: new forming processes must be devised as well as new design approaches made. Cooperation between research and production people is a vital factor.

ACKNOWLEDGEMENT

The author would like to acknowledge experimental help from Ing. Raimondo Pasquino.

BIBLIOGRAPHY

1 ROSENHAIN, W., HAUGHTON, J. L. and BINGHAM, K. E., 'Zinc Alloys with Aluminium and Copper', *J. Inst. Metals*, 23, p. 261, 1920.

2 SAUVEUR, A., 'What is Steel? Another Answer', *Iron Age*, 113, p. 581, 1924.

3 HARGREAVES, F., 'Ball Hardness and the Cold Working of Soft Metals and Eutectics', *J. Inst. Metals*, 39, p. 301, 1928.

4 JENKINS, C. M., 'The strength of a Cd-Zn and Sn-Pb Alloy', *J. Inst. Metals*, 40, p. 21, 1928.

5 PEARSON, C. E., 'The viscous property of extruded Pb-Sn and Bi-Sn eutectic alloys', *J. Inst. Metals*, 54, p. 111, 1934.

6 UNDERWOOD, E. E., 'A review of "Superplasticity"', *J. Metals*, 14, p. 914, 1962.

7 LOZINSKY, M. G. and SIMEONOVA, I. S., '"Superhigh Plasticity" of Commercial Iron Under Cyclic Fluctuations of Temperature', *Acta Metallurgica*, 7, p. 709, 1959.

8 BACKOFEN, W. A., TURNER, I. R. and AVERY, D. H., 'Superplasticity in Al-Zn Alloy', *Trans. A.S.M.*, 57, p. 989, 1964.

9 AVERY, D. H., and BACKOFEN, W. A., 'A Structural Basis for Superplasticity'. *Trans. A.S.M.*, 58, p. 551, 1965.

10 HOLT, D. L. and BACKOFEN, W. A., 'Superplasticity in Al-Cu Eutectic Alloy', *Trans. A.S.M.*, 59, p. 755, 1966.

11 BACKOFEN, W. A., AZZARTO, F. J., MURTY, G. S. and Zehr, S. W., 'Superplasticity', *Ductility*, A.S.M., Metals Park, Ohio, p. 279 1968.

12 JOVANE, F., 'Superplasticity – Fundamental Aspects and Outlooks of Industrial Application', *Metallurgia Italiana,* 4, p. 239, 1967.

13 CHAUDHARI, P., 'Superplasticity', *Science and Technology*, p. 42 Sept. 1968.

14 SHERBY, O. D., 'Superplasticity', *Science Journal*, p. 75, June 1969.

15 WELD, H. M., 'Superplasticity', Department of Energy, Mines and Resources, Mines Branch, Ottawa; Information Circular IC 235, October 1969.

16 DAVIES, G. J., EDINGTON, J. W., CUTLER, C. P. and PADMANABHAN, K. A., 'Superplasticity: a Review', to be published.

17 JOHNSON, R. H., 'Superplasticity', *Metallurgical Reviews*, p. 115 Sept. 1970.

18 OELSCHAGEL, D. and WEISS, V., 'Superplasticity of Steels During the Ferrite Austenite Transformation', *Trans. A.S.M.*, 59(2), p. 143, 1966.

19 NICHOLSON, R. B., 'Exploitation through Metallurgical Development', *Plasticity and Superplasticity – Developments in their exploitation*, Institute of Metallurgists – Review Course Series 2-n 3, October 1969, p. 19.

20 ANON., 'New materials for car bodies', *Automobile Engineer*, p. 536, Dec. 1968.

21 JOVANE, F. and SHABAIK, A., 'On the superplastic Pb-Sn eutectic alloy, as a model material for studying the mechanics of forming processes in superplastic condition', *La Metallurgia Italiana*, (12), p. 993, 1968.

22 MORRISON, W. B., 'The elongation of superplastic alloys', *Trans. Met. Soc. AIME*, 242, p. 2221, 1968.

23 HAYDEN, H. W., GIBSON, R. C., MERRICK, H. F. and BROPHY, J. H., 'Superplasticity in the Ni–Fe+Cr System', *Trans. A.S.M.*, 60, p. 3, 1967.

24 HAYDEN, H. W. and BROPHY, J. H., 'Interrelation of Grain Size and Superplastic Doformation of Ni-Cr-Fe Alloys', *Trans. A.S.M.*, 61, p. 542, 1968.

25 LEE, D. and BACKOFEN, W. A., 'Superplasticity in some Ti and Zn Alloys', *Trans. T.M.S. A.I.M.E.*, 239, p. 1034, 1967.

26 FRECHE, J. C., WATERS, W. J. and ASHBROOK, R. L., 'Evaluation of two Nickel-base Alloys, Alloy 713C and NASA TAZ-8A, produced by extrusion of prealloyed powders', *Materials Engineering Exposition and Congress*, October 1969, Philadelphia, ASM Technical Paper P.9 – 25.2.

27 REICHMAN, S. H. and SMYTHE, J. W., 'Superplasticity in P/M In-100 Alloy', *International Journal of Powder Metallurgy*, 6, (1), p. 65, 1970.

28 REICHMAN, S. H., CASTLEDINE, B. W. and SMYTHE, J. W., 'Powder metal nickel superalloys may be formed superplastically', *SAE Journal*, 78(4), p. 59, 1970.

29 MARTIN, P. J. and BACKOFEN, W. A., 'Superplasticity in Electroplated Composites of Pb and Sn', *Trans. A.S.M.*, 60, p. 352, 1967.

30 MORRISON, W. B., 'The formability of Superplastic Alloys', *Metallurgical Journal of Strathclyde University*, 20 p. 24, 1970.

31 ZEHR, S. W. and BACKOFEN, W. A., 'Superplasticity in Pb-Sn Alloy', *Trans. A.S.M.*, 61, p. 300, 1968.

32 HOLT, D. L., 'The metallography of superplasticity in some aluminium-base alloys', *Trans. A.S.M.*, 60, p. 564, 1967.

33 GHOSH, A. and DUNCAN, J. L., 'Torsion tests on superplastic tin-lead alloy', *Int. J. Mech. Sci.,* 12, p. 499, 1970.

34 JOVANE, F., 'A new method for determining (σ, ϵ) curves of superplastic materials', *La Metallurgia Italiana*, (11), p. 923, 1967.

35 JOVANE, F. and NASO, V., 'Experimental assessment of a new method for determining (σ, ϵ) curves of superplastic materials', *La Metallurgia Italiana*, (12), p. 957, 1967.

36 JOVANE, F., 'Empirical relations to represent mechanical properties of superplastic materials', *La Metallurgia Italiana*, Atti notizie, (7), p. 191, 1968.

37 PADMANABHAN, K. A. and DAVIES, G. J., 'Numerical analysis of superplasticity data for use in metal-forming applications', *J. Mech. Phys. Solids*, 18, p. 261, 1970.

38 WOODFORD, D. A., 'Strain-Rate Sensitivity as a Measure of Ductility', *Trans. A.S.M.*, 62, p. 291, 1969.

39 JOVANE, F., 'Instability and high elongations in superplastic alloys deformed in tension', *Annali della Facolta' d'Ingegneria di Bari*, VII, 1966.

40 HART, E. W., 'Theory of the Tensile Test', *Acta Met.*, 15, p. 351, 1967.

41 AVERY, D. H. and STUART, J. M., 'The Role of Surfaces in Superplasticity', *Proc. of the 14th Sagamore Conference*, Syracuse Univ. Press, 1967.

42 HAYDEN, H. W. and FLOREEN, S., 'The Deformation and Fracture of Stainless Steels having Microduplex Structures', *Trans. A.S.M.*, 61, p. 474, 1968.

43 FLOREEN, S. and HAYDEN, H. W., 'The Influence of Austenite and Ferrite on the Mechanical Properties of Two-Phase Stainless Steels having Microduplex Structure', *Trans. A.S.M.*, 61, p. 489, 1968.

44 BACKOFEN, W. A., 'Superplasticity Enchants Metallurgy', *Steel*, p. 25, December 15 1969

45 NAZIRI, H. and PEARCE, R., 'The influence of copper adalitions on the Superplastic forming behaviour of the Zn-Al eutectoid', *Int. J. Mech. Sci.*, 12, p. 513, 1970.

46 LUNTZ, J. A., 'The exploitation of new materials and new processes in the forging industry', *Plasticity and Superplasticity-Developments in their exploitation*, Institute of Metallurgists – Review Course Series 2 – No. 3, October 1969, p. 92.

47 WEINSTEIN, D., 'The effect of Superplastic Deformation on the Ductility of a Helium-Containing Fe-Cr-Ni Alloy', *Trans. Met. Soc. A.I.M.E.* – 245, p. 2041, Sept. 1969.

48 HOLT, D. L., 'The relation between superplasticity and grain boundary shear in the aluminium-zinc eutectoid alloy', *Trans. of Met. Soc. A.I.M.E.* –242, p. 25, 1968.

49 BERNHARDT, E. C., Editor, 'Processing of Thermoplastic Materials', Reinhold Publishing Corp., N.Y. – 1959, p. 448.

50 FIELDS, D. S., 'Sheet Thermoforming of a Superplastic Alloy', *I.B.M. J. of Research and Development*, 9, (2), p. 134, 1965.

51 JOVANE, F., 'An Approximate Analysis of the Superplastic Forming of a Thin Circular Diaphragm: Theory and Experiments', *Int. J. Mech. Sci.*, 10, p. 403, 1968.

52 JOVANE, F., 'Pressure forming of Superplastic Alloys', *Fondazione Politecnica per il Mezzogiorno d'Italia*, Quaderno n. 42, 1969.

53 JOVANE, F., 'Superplasticity in Titanium', *The Science Technology and Application of Titanium*, Editors DR R. JAFFEE and N. PROMISEL, Pergamon Press – Oxford and New York, 1970. p. 615.

54 HUNDY, B. B., 'Exploitation of Plasticity and Superplasticity in the Sheet Metal Industry', *Plasticity and Superplasticity – Developments in their Exploitation*, Institution of Metallurgists, Review Course – Series 2, No. 3, p. 73, October 1969.

55 NORTH, D., 'Superplastic alloy for autobody construction', *Sheet Metal Industries*, p. 13, January 1970.

56 NEAL, M., 'New Technologies can cut aircraft production costs', *Engineering Production*, p. 921, 10 Sept. 1970.

57 AL-NAIB, T. Y. M. and DUNCAN, J. L., 'Superplastic Metal forming', *Int. J. Mech. Scie.*, 12, p. 463, 1970.

58 CORNFIELD, G. C. and JOHNSON, R. H., 'The forming of superplastic sheet metal', *Int. J. Mech. Sci.*, 2, p. 479, 1970.

59 HOLT, D. L., 'An Analysis of the bulging of a superplastic sheet by lateral pressure', *Int. J. Mech. Sci.*, 12, p. 491, 1970.

60 THOMSEN, T. H., HOLT, D. L. and BACKOFEN, W. A., 'Forming Superplastic Sheet Metal in Bulging Dies', *Metals Engineering Quarterly*, p. 1, May 1970.

61 JOHNSON, R. H., 'Superplasticity in Metals', *Sheffield University Metallurgical Society Journal*, 9, 1970.

62 WEISS, V. and KOT, R., 'Superplasticity', *Metal Deformation Processing, Vol. III,* Defence Metals Information Center, Battelle Memorial Institute, DMIC Report 243, 10 June 1967.

63 THOMSEN, E. G., YANG, C. T. and KOBAYASHI, S., 'Mechanics of Plastic Deformation in Metal Processing', MacMillan Co., N.Y. 1965, pp. 74, 230, 289, 294.

64 JOVANE, F. and MORELLI, P., 'On coining in superplastic condition', *Annali della Facolta' di Ingegneria di Bari*, 1968.

65 DUNCAN, J. L. and JOHNSON, W., 'Exploitation through engineering development', *Plasticity and Superplasticity – Development in their exploitation*, Institute of Metallurgists – Review Course Series 2, No. 3, October 1969. p. 38.

66 JOVANE, F., SHABAIK, A. and THOMSEN, E. G., 'Some extrusion studies of the eutectic alloy of Pb and Sn', *Trans. A.S.M.E. J. Eng. Ind.*, 91, p. 680, 1969.

67 MEHTA, H. S., SHABAIK, A. H. and KOBAYASHI S., 'Analysis of Tube Extrusion', *Trans. A.S.M.E. J. Eng. Ind.*, 92, p. 403, 1970.

68 JOVANE, F. and DELLA VOLPE, R., 'Extrusion in Superplastic Condition', Atti dell' Instituto di Tecnologie dell' Universita di Bari.

THE LASER CUTTING OF STEEL RULE DIES

F. W. Lunau and B. C. Doxey
BOC-Murex Research and Development Laboratories

SUMMARY

The development of gas jet laser cutting has enabled the cutting of steel rule dies to be automated. These dies, for stamping out cartons, gaskets, etc., generally consist of an 18 mm thick plywood base cut with slots in the pattern of the carton outline and desired creases, the cutting and creasing being done by sharp and rounded edge knives set in these slots. The currently accepted method of cutting these slots is by a power jig-saw manually guided. Due to the necessity constantly to interrupt the cut to leave bridge pieces to hold the die together (which means that the saw blade has to be constantly rethreaded through predrilled holes), it has been impossible to automate the process, which is desirable due to the increasing demands of modern high-speed packaging machinery. Apart from the development of the gas jet laser cutting process, a photocell line-follower had to be developed for following the complex pattern to the necessary high accuracy. Information as to the route to be followed is added to the master drawing by means of overlays. In practice, it is possible to cut the knife slots to an accuracy of ±0·003 in over an area of 65 in × 45 in The economics are discussed. These stem from the five to sixfold reduction in cutting time. Apart from this greater throughput there are the side benefits of reduced lead time and increased accuracy.

1.0 INTRODUCTION

Steel rule dies for stamping out packaging cartons generally consist of an 18 mm plywood sheet, jig-saw cut with a pattern of slots representing the outline and folding creases of the carton (Fig. 1). Steel knives, 0·7 mm thick, are set in these slots, sharpened for cutting or rounded for creasing. Alternatively the knives can be held in the correct pattern by separately cut and shaped blocks compressed together by a substantial steel outer frame. The preparation of dies by either of these methods obviously entails a considerable amount of manual work and the automation of the process using currently available cutting techniques is difficult. Such automation, however, is attractive for two reasons. Firstly, over the years a difficulty can be expected in recruiting the suitable skilled labour. Secondly, and probably more important, the steady rise in the speed of automatic packaging machines increases the need for each carton to have the same dimensions within a very small tolerance as the others of its type, which is difficult to achieve because each die can contain up to 120 carton shapes.

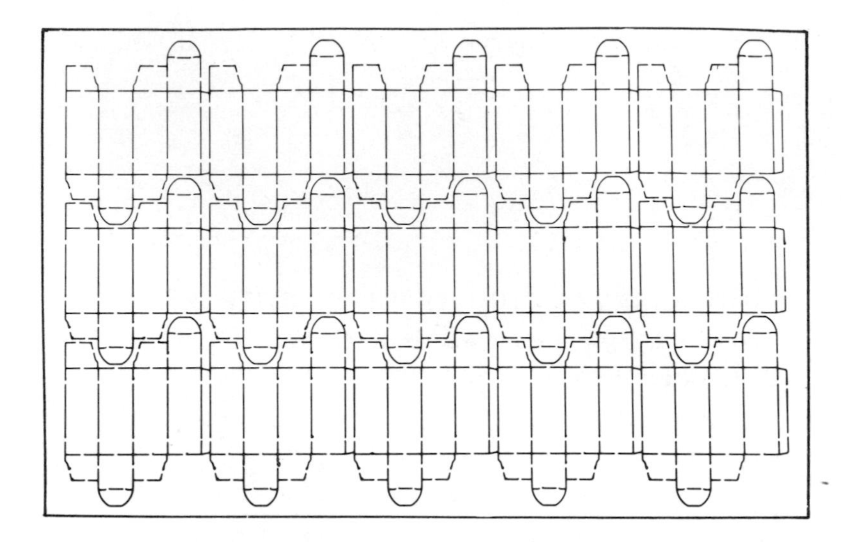

Fig. 1. Laser cut die before knifing. The die is 600 mm × 1100 mm and stamps out 15 cartons at one time.

2.0 REQUIREMENTS FOR AUTOMATION

Automation is obviously easiest with the one-piece type of die illustrated in Fig. 1, which is also a type which offers advantages in later stages of the process. The usual method of cutting by jig-saw is difficult to automate because the blade is flexible enough to wander with variations in the grain of the wood and the necessity to have a drilled hole at each stop and start position to insert or extract the blade. This occurs about every 100 mm because bridge pieces must be left to hold the die together. Thus a cutting method has to be developed which is free from these difficulties to enable the process to be automated.

Programming to obtain the desired relative movement between the cutting medium and the workpiece is another problem area. Since each die design will probably be cut only once or twice, preparation and, in particular, verification costs of the program are the most important points to be considered. An accuracy of about ± 0·01 mm is also necessary. These factors are enlarged upon in the following sections.

3.0 THE CUTTING PROCESS

It was considered desirable to retain the traditional material for the dies, which is 18 mm plywood. This material has a blend of low cost, high dimensional stability, sufficient elasticity to hold the knife, and low weight, which plastic materials cannot, at present, equal. The development of mechanical cutting methods is difficult in view of the inhomogeneities in the material and the small width of the slot, 0·7 mm, compared with its depth of 18 mm. However, at the time of this development, the authors' laboratories were engaged in assessing the potentialities of the high-power CO_2 laser and it was soon realised that this was a tool which answered the requirements for this particular application.

The CO_2 laser is a simple gas discharge tube bounded by an optical cavity. The lasing medium is CO_2, mixed with helium and nitrogen to aid lasing action. The proportions are 82% He, 12% N_2, and 6% CO_2. The pressure in the tube is about 10 torr. The radiation that is emitted is in the infrared at a wavelength of 10 600 nm (106 000 angstrom). This wavelength is invisible to the eye but behaves in most respects as visible light. The energy comes out from the laser as a parallel beam 15 mm diameter and is then focused to a spot about 0·4 mm diameter by a simple lens system. This small spot size is only possible because of the coherent and parallel nature of the laser output.

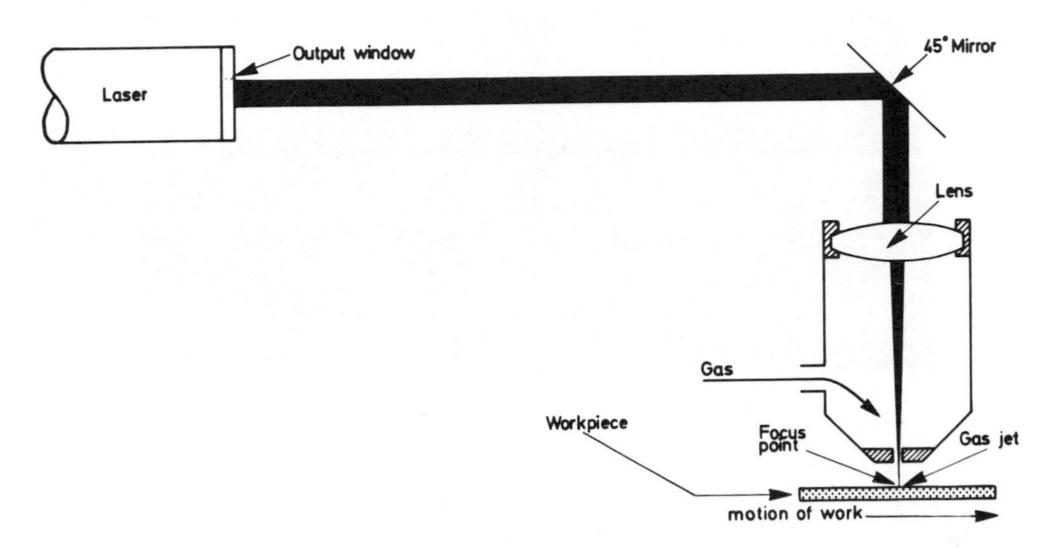

Fig. 2. Diagrammatic arrangement of gas jet laser cutting process.

Surrounding the focused laser beam is a nozzle through which is emitted a concentric jet of compressed air (Fig. 2). The effect of this gas jet is to prevent a rounding of the top surface of the cut and to increase the parallelness of it (Fig. 3).

A power output of 200 W was found to give a cutting speed of 225 mm/min, which was of the right order for a practical machine.

4.0 PROGRAMMING SYSTEM

The development of a suitable system proved one of the more difficult parts of the project. Whatever method is adopted, it consists essentially of breaking down the complex carton shape into a number of simple paths which can be traced individually.

As mentioned before, this programming, including the necessary verification, must be cheap because, unlike most other production applications, this one requires only one or two dies of the same pattern to be cut in the large majority of cases. The final choice lay between numerical control. It was finally considered that optical line-following was the better method in this case for the following reasons:

- It is considerably cheaper in capital costs than a continuous-path N.C. method, particularly if a capability for curved paths is required.
- An accurate drawing of the carton has to be prepared at some stage anyway to lay out the print and artwork.

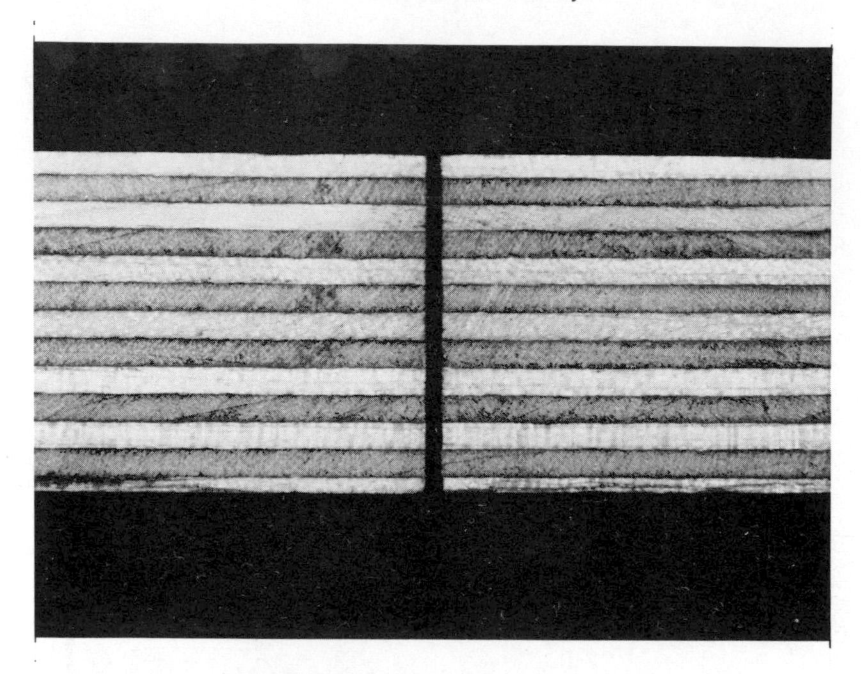

Fig. 3. Cross-section of gas jet laser cut through 18 mm thick plywood.

- The program, which is a 1:1 photographic print of the carton drawing, does not require further verification, unlike an N.C. tape which must be fully verified before being used to cut a die.
- Verification of the cut die is much more easily done by overlaying it with the transparent film master used for guiding the cutting machine, than checking against an N. C. tape.
- No additional skills or equipment are required for the optical line-following method, over and above those already possessed by a carton manufacturer.

However, line-following systems are normally used for controlling equipment such as flame profiling machines and milling machines, and the shapes which are usually cut have a closed contour without intersections. The use of line-following equipment to follow a carton blank presents a problem because of the large number of lines which may intersect any one point, thereby confusing the following system, which may track any one of the lines leading out of a junction. It was, therefore, found necessary to devise a system which would guide the follower round the carton blank in a predetermined manner.

4.1 *Masking technique*

The technique is to divide the carton into simple routes, this information being transferred on to a set of transparent masks which can be located by circular discs over the original drawing. Wherever the line follower has a choice of paths at a junction, then the line which is not to be followed is blanked out by applying white tape or paint to the mask, leaving visible the line which is to be followed. This technique is illustrated in Fig. 4, which shows a drawing of a carton blank that has been divided into three separate routes by the masking technique. It can be seen that the first mask allows the following of the horizontal lines, the second mask the vertical lines, and the third mask puts in the remaining lines. The actual number of paths or

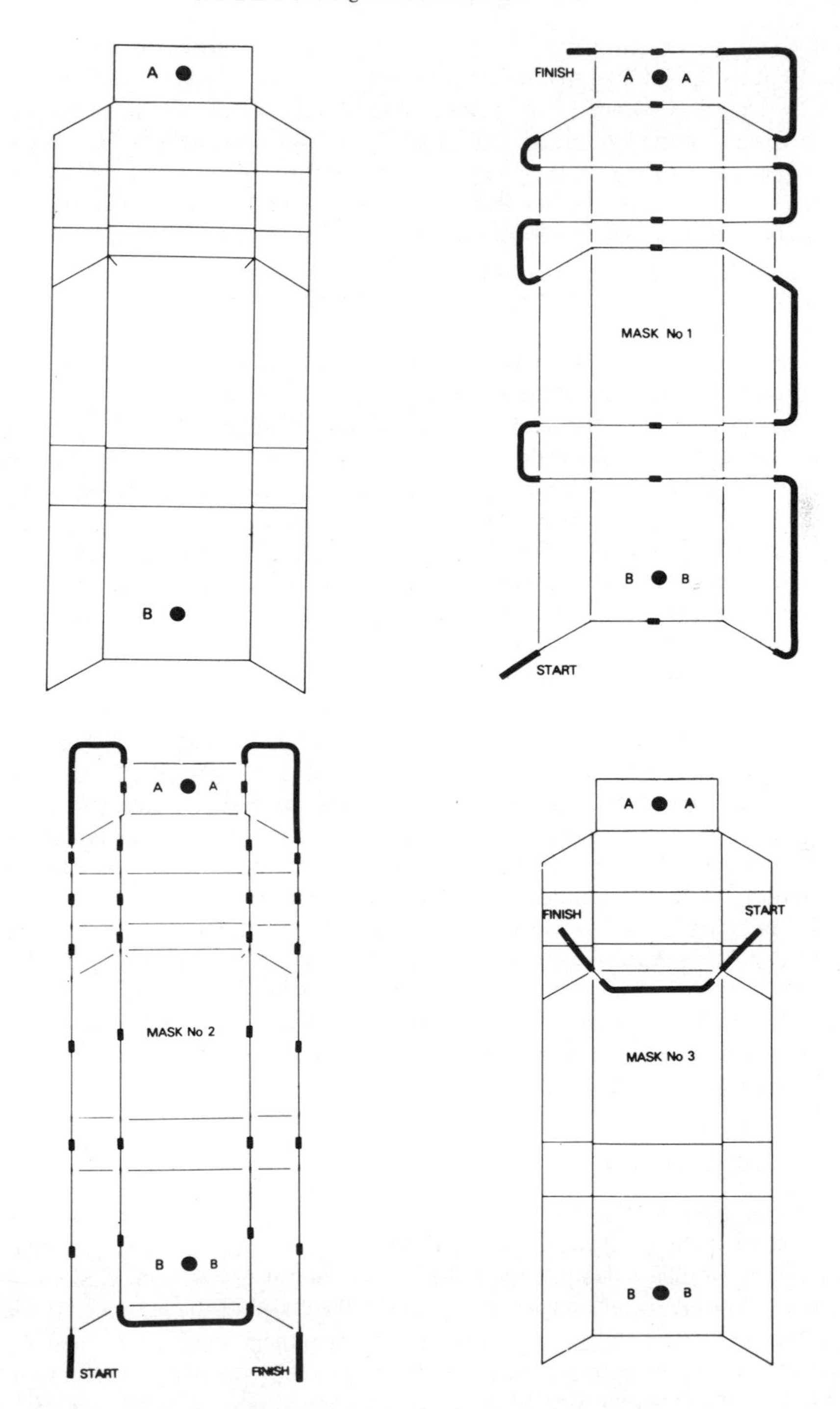

Fig. 4. Arrangement of masks splitting up a complex carton shape into three simpler shapes capable of being followed by an optical line follower.

masks required depends very much on the shape and complexity of the original carton but in practice it has been found that only two masks may be required for a simple shape whereas four may be necessary for one of the more complex shapes. The masks also carry a command signal in the form of a thick black line, which is detected by a photoelectric cell in the follower head which generates a signal causing a shutter to intercept the laser beam, thereby switching off the cutting action. This is a convenient method of leaving bridge pieces in the die; and it is also necessary where the line follower must traverse between two points on the die without any cutting action. In practice, the thick line takes the form of an adhesive backed tape which is stuck to the mask. The tape is supplied accurately cut to the required width and is easily applied or removed.

During the cutting operation, the original drawing is secured to the tracing table with adhesive tape or magnets and remains firmly in position when masks are being changed. The accuracy of location of the mask over the drawing is not of prime importance since the masks do not contain any information relating to the shape of the profile, they only inform the following head of the route to be taken and where to switch the cutting action on and off for leaving bridge pieces, etc. Where the die contains more than one carton, as is generally the case, the original master drawing is stepped and repeated by equipment already existing in carton manufacturers' plants to produce a transparent photographic positive of the multiple layout. Depending on the size of the individual cartons and the number on the multiple layout, masks can be prepared for an individual carton or for a small block or row of cartons. In general, for cartons larger than about 6 inches square, it is more economic in time to make a set of masks for a single carton and transfer the masks from one carton to another in the multiple layout. For instance, a typical carton may require a set of three masks, which means that 60 mask changes are required for a 20-up layout. This would take a total of 20 minutes during the total cutting time, experience having shown that a mask can easily be changed in about 20 seconds. The number of mask changes can be reduced by making more complex masks to cover a larger number of cartons and in some cases this may result in an overall time-saving but experience determines the masking arrangement for any particular carton and layout.

The masking technique is very convenient where common knives are used in a multiple layout as it is essential to avoid double cuts where a common knife occurs, otherwise a slightly wide cut may be produced, which results in a loose knife. This may be avoided by adding a strip of black tape, i.e. a command signal to the mask to switch off the cutting action where a cut has already been made on a previous mask. This procedure is easily carried out by the operator on the machine.

5.0 MACHINE DESIGN

The introduction of an advanced scientific piece of equipment into an industrial environment posed many problems. An initial decision was taken that the equipment should be operable by normal industrial staff with a minimum of extra training. Simplicity and lack of complication were therefore the guiding rules in the design of the machine. The laser and nozzle were kept stationary and the workpiece moved over them, since it was deemed that there were less unknowns in this approach than the alternative ones of moving the laser or its beam. Figs. 5 and 6 show the final solution. Partly due to this emphasis on simplicity, the whole development was compressed into a remarkably short time scale. The initial conception took place in the summer of 1968 and the decision to make a prototype machine was made in October 1968.

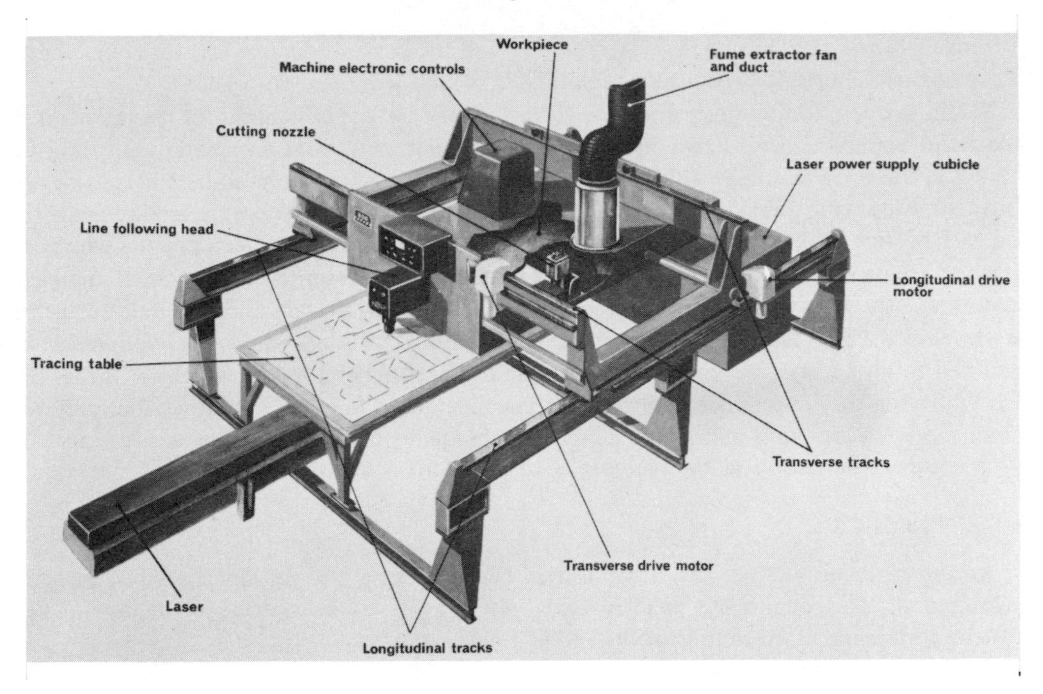

Fig. 5. Diagrammatic arrangement of Laser Falcon die-cutting machine.

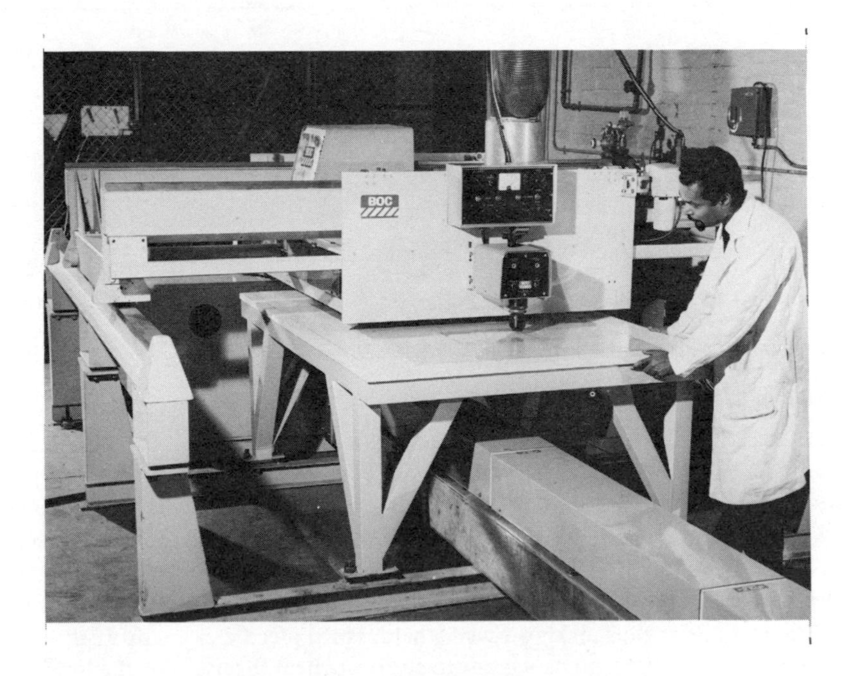

Fig. 6. General view of completed machine.

This machine was commissioned in March 1969 and the first production machine commenced its proving trials in April 1970, making the total time span less than two years.

To allow operation by normal industrial staff, a very tight specification for the laser and the associated apparatus was drawn up. The main requirements were automatic and unvarying mixing of the three laser gases (82% He, 12% N_2, and 6% CO_2), single push-button starting and, above all, long-term stability of output within ± 5% of nominal. These were requirements that had not previously been met, CO_2 lasers having been laboratory-type equipments where, for instance, frequent adjustment of the optical cavity had been regarded as all part of the game. Simultaneously with the laser development, the programming system and line follower had to be developed. The line follower itself is based upon existing commercial equipment but modified to give a high accuracy and to respond to the command signal given by a thicker line. It is gratifying that experience with the first machine out in the field has shown that only very minor modifications to it and to the design of subsequent machines have been necessary, this being another vindication of the philosophy of simplicity wherever possible.

6.0 ECONOMICS

At the time of writing, the machine has been working for six months in an industrial environment. This period has enabled some direct costs to be obtained but has not been sufficient for some of the indirect savings to be fully quantified.

The direct savings obviously come from a reduction in man-hours per die and can be evaluated from the data given in Table 1.

TABLE 1. DIRECT COST ELEMENTS IN DIE CUTTING

Direct labour cost per man hour including payroll expenses		£0·70
Direct operating cost of machine per hour		£0·50
Average programming time per die		7 hours
Average cutting time per die/laser cutting		6 hours
Average marking-out and cutting time per die (manual method)		40 hours
Direct cost of manually cut die 40 X £0·70	=	£28·00
Direct cost of laser cut die 7 X £0·70 + 6 X £0·70 + 6 X £0·50	=	£12·10
Saving in direct cost of laser cut die		£15·90

There are, however, a number of less direct savings which have not yet been fully evaluated. The higher accuracy and consistency mean the die can be put into production with no rectification. On occasion the rectification for individual cartons in a multiple die may be impossible so that these cartons have to be discarded. It is proving possible to shorten the 'make ready' time for the stamping press thus making a saving here. Multi-piece dies are difficult to store, and require between four and eight hours' labour to readjust them for use. For the long production runs sometimes required it may well be cheaper to cut a new die than to reknife an old one.

The lead time to produce a new carton is reduced considerably thus giving a selling advantage.

It is not unrealistic to assume that these incidental advantages can raise the saving per average die from £15·90 to £20·00.

The return on capital has been plotted in Fig. 7 as a function of annual throughput for the two values of saving per die. The percentage return on capital is based on a machine purchase price of £20 000 and straight-line depreciation over 10 years. It will be seen that, to achieve a reasonable return of 15%, some 250-310 dies will have to be cut per year, which is close to the single-shift capacity of the machine. If a firm is large enough to load the machine for two shifts then the return on capital, of course, rises dramatically and could be as high as 50%.

This type of calculation emphasises the sensitivity of high-capital-value machines to loading and also shows that if the cost of the machine were increased by say £8000 because it was numerically controlled, then it would be very difficult to achieve a reasonable rate of return on capital.

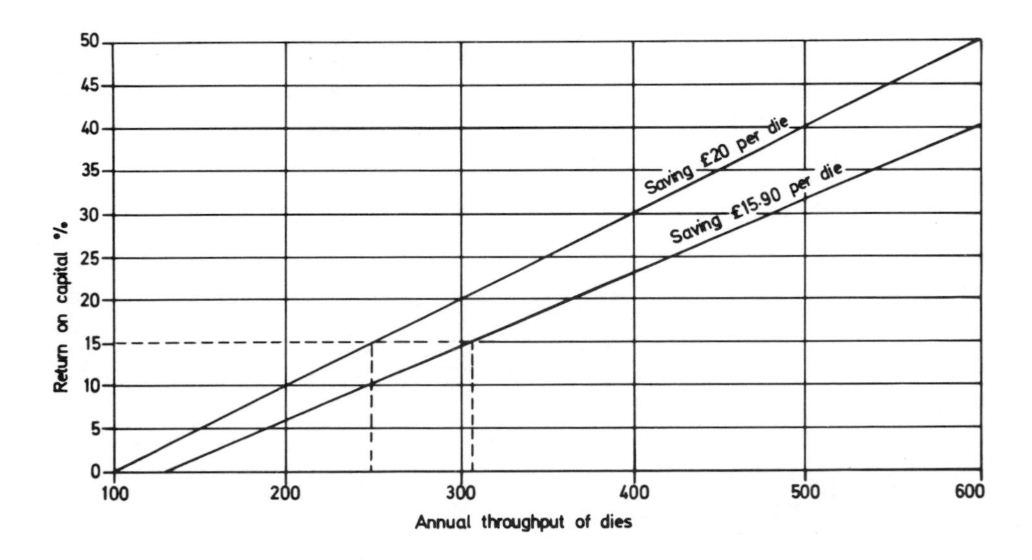

Fig. 7. Effect of annual throughput on return on capital.

7.0 CONCLUSIONS

The automating of what has essentially been a 'craft' process has been an intensely interesting undertaking. It has been necessary at all stages to consider the effects of the new process on the inter-linking processes used in carton making and to develop a system which entailed the minimum possible disturbance of the status quo. To do this, close co-operation with the user industry is required. Another very obvious difficulty in an exercise of this nature is that, unless prospective demand is fairly high, the necessary recovery of development expenses can increase the price of the machine to an uneconomic figure. It is thus necessary, in a case such as this, to aim at an international market so as to be able to spread development costs widely enough. This, in turn, raises the question of how to sell machinery into very specialised markets abroad, which may be outside the normal scope of a company's business.

Another difficulty is that of ascertaining the size of the market in the first place, for a completely revolutionary concept. A conventional type of market survey was carried out before the decision to build the prototype was made, which gave some guidance as to the possible market, but experience is proving that this survey only gave a very rough indication of what is, in actual fact, happening.

Die cutting is but one of many other likely laser cutting applications. In some of these it may be possible merely to substitute the laser beam instead of another cutting tool, but in general it appears that a total system is required of a complexity approaching this particular development.

SESSION VI

Decision Making in Design Manufacture and Marketing

Chairman: Mr J. D. Houston
Managing Director
Higher Productivity (Organisation and Bargaining) Ltd

A SYSTEMATIC PROCEDURE FOR THE GENERATION OF COST-MINIMIZED DESIGNS

P. W. Becker and B. Jarkler
Electronics Laboratory, Technical University of Denmark

SUMMARY

We present a procedure for the generation of cost-minimized designs that arise in the design of circuits and systems. Suppose a designer has decided upon the topology of his product. Also, suppose he knows the cost and quality of the different grades of the N components required to implement the product. The designer then faces the following problem: how should he proceed to find the combination of N grades which will give him the desired product reliability* at minimum product cost?

We discuss the problem and suggest a policy by which the designer, with a reasonable computational effort, can find a set of 'good' implementations.

The suggested policy is applied to the problem of implementing an electronic amplifier. The results are quite encouraging.

NOTATION

A	Amplification of transistor amplifier	M_j	Number of ratings for g_j
B	Bandwidth of transistor amplifier	N	Number of components in the product
C	Product cost	R	Product reliability
${}_iD$	Member no $(i + 1)$ of S	S	Set of sub-optimal implementations
${}_i^jF$	Fraction defined by Equations 3 or 4	S^*	Set of optimal implementations
g_i	Grade of the ith component		

1.0 INTRODUCTION

1.1 *The problem*

Suppose a designer has decided upon the topology of his circuit or system (hereafter called product) and knows the cost and quality of the possible choices for each component or subsystem (hereafter called component). Assume, for the moment, that each component is available in three grades: high, medium, and low. For example, a 100Ω resistor in the design may be realized in hardware by resistors having 1%, 5% or 10% tolerances, the lower grades presumably having lower costs than the higher grades. When the designer selects a grade for each of the N components, he can do this in 3^N different ways. With each implementation is

* By 'reliability' is meant the probability of meeting initial test specification.

associated both the cost, C, and the reliability, R, of the resulting product. Each implementation or design (both words will be used to mean a set of N component-grades) may be illustrated by a dot as shown in Fig. 1. Among the huge number of possible implementations the designer is

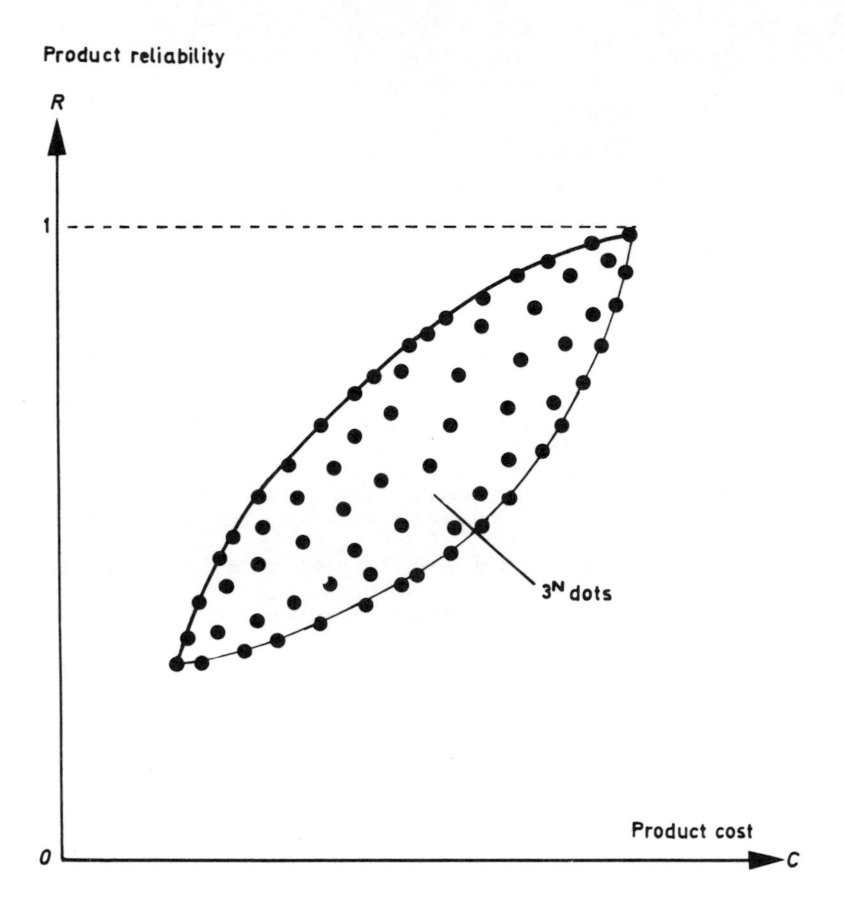

Fig. 1. The interesting implementations. If N components each are available in three different grades, the product can be realized in 3^N ways, as illustrated by the 3^N dots. The designer is only interested in the cost-minimized implementations illustrated by points on the heavy line.

only interested in those where the product reliability has been achieved at minimum cost. Such optimal implementations are illustrated by points on the heavy line in Fig. 1. It is among the members of this set of implementations that the designer will select the final design. The question to which we address ourselves in this paper may now be stated: *how can the designer find the optimal implementations without evaluating the reliability of all, or at least an exorbitant number of, implementations?*

1.2 *The organization of the paper*

The paper is organized as follows. In Section 2 we consider the case where all N grades can be changed in infinitesimal steps. It turns out that in this case (under five mild assumptions) we can specify a policy which makes it possible to compute all optimal implementations. In

Section 3 we consider the more realistic case where all N grades can be changed only in finite steps. Again we are able to specify a policy which makes it possible to compute all optimal implementations; however, the computational work becomes unmanageable. In this situation the authors suggest a sub-optimal procedure involving only reasonable computational work. This procedure, we believe, will generate a string of 'good' designs. To test the procedure, it is applied to the design of a three-transistor amplifier. We show that the sub-optimal procedure does indeed lead to a series of 'good' amplifier designs.

2.0 N GRADES, WHICH CAN BE CHANGED IN INFINITESIMAL STEPS

2.1 *Five reasonable assumptions for the case* N = 2

Let us begin by considering the case where the designer varies only the grades of two of the components. The grades are called g_1 and g_2. We now make five mild assumptions.

(1) g_1 and g_2 can both be measured in a meaningful manner by a scalar; e.g., the grade of the electrical components can be assessed by their tolerance. The scalar has the following two properties (i) the higher the g-value, the inferior the component, and (ii) all g-values must be greater than zero.

(2) g_1 and g_2 can be varied in infinitesimal steps between the highest grades, $(g_1)_{min}$ and $(g_2)_{min}$, and the lowest grades, $(g_1)_{max}$ and $(g_2)_{max}$. The case where the steps are finite will be treated later.

(3) C and R are both differentiable functions of g_1 and g_2 in the region of interest.

(4) If we determine the locus of points in the (g_1, g_2)-plane with the same C-value, we obtain a family of *convex*[1] curves as illustrated in Fig. 2. The shape of the curves is determined by the economical fact that use of the very best grades results in a dramatic rise in product cost. The radius of curvature at any curve point will, therefore, be positive. The gradient is vertical to the tangent and will be parallel to some vector from A, in Fig. 2, pointing out into the *first* quadrant.

(5) If we determine the locus of points in the (g_1, g_2)-plane with the same R-value, we obtain a family of *convex* curves as illustrated in Fig. 2. The shape of the curves is determined by the following familiar fact: the effect on product reliability of the degradation of one component's grade can only, to a small degree, be compensated for by the improvement of some other component's grade. Consequently the radius of curvature at any curve point will be positive and the gradient will be parallel to some vector from A pointing out into the *first* quadrant.

2.2 *The locus of cost-minimized implementations for* N = 2

Consider the region BFGDHK in Fig. 2. All points in the region illustrate (g_1, g_2)-combinations for which the product reliability is at least R_3 and the product cost is at most C_1. The region is convex because it is an intersection of two convex regions ANFGDHKM and EPFBKQ. If we insist that the product reliability R should be at least R_3, and we then try to minimize the product cost, C, the region will shrink first to the convex region LGDH and then degenerate to the point D. D represents the least expensive (g_1, g_2)-combination with which $R \geqslant R_3$ can be realized, i.e., $R = R_3$ at D. At D the tangents to the C_3-curve and the R_3-curve coincide due to assumption (3) meaning that the gradients will have the same direction. It should be noticed that due to the assumptions of convexity, the limiting case for the convex intersection was *one* point, point D, and not two or more separate points; two or more separate points clearly do not constitute a convex region. By repeating the procedure we can

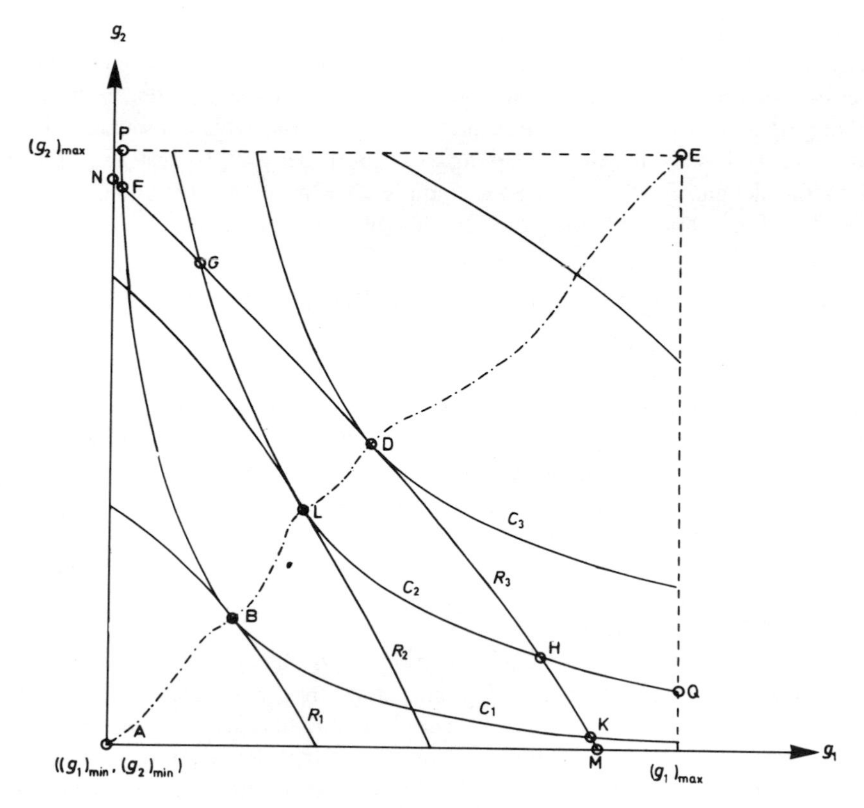

Fig. 2. The locus of cost-minimized implementations R_1, R_2, and R_3 indicate members of the family of constant-R-curves. C_1, C_2, and C_3 indicate members of the family of constant-C-curves. The line ABLDE illustrates the locus of cost-minimized implementations.

generate a string of optimal implementations A, B, L, D, E, etc. The curve through the points illustrates the locus of cost-minimized implementations which are of interest to the designer. The (C, R)-values for the points on the curve are the ones indicated by a heavy line in Fig. 1. The curve does not meander much since the tangent to the curve is always parallel to some vector from A pointing out into the first quadrant.

2.3 *Buckling and bifurcation*

In assumptions (4) and (5) we mentioned that the sets of curves should be convex. The danger of concavities is illustrated by Fig. 3 where a curve buckles and where consequently the locus of cost-minimized implementations separates into two branches. This occurrence is called bifurcation [2]. Bifurcation is highly undesirable because it necessitates a study of all alternate paths before the cost-minimizing designs can be determined. The concepts of buckling and bifurcation have lately received increasing attention [2]. Buckling, incidentally, does not necessarily lead to bifurcation. The five assumptions which are sufficient to avoid bifurcation are consequently not necessary to avoid bifurcation.

2.4 *The assumptions in the case* $N = 3$

With a few modifications the exposition from Section 2.1 applies to the case $N = 3$. The first three assumptions should now read to include g_3, the third grade to be varied. We modify

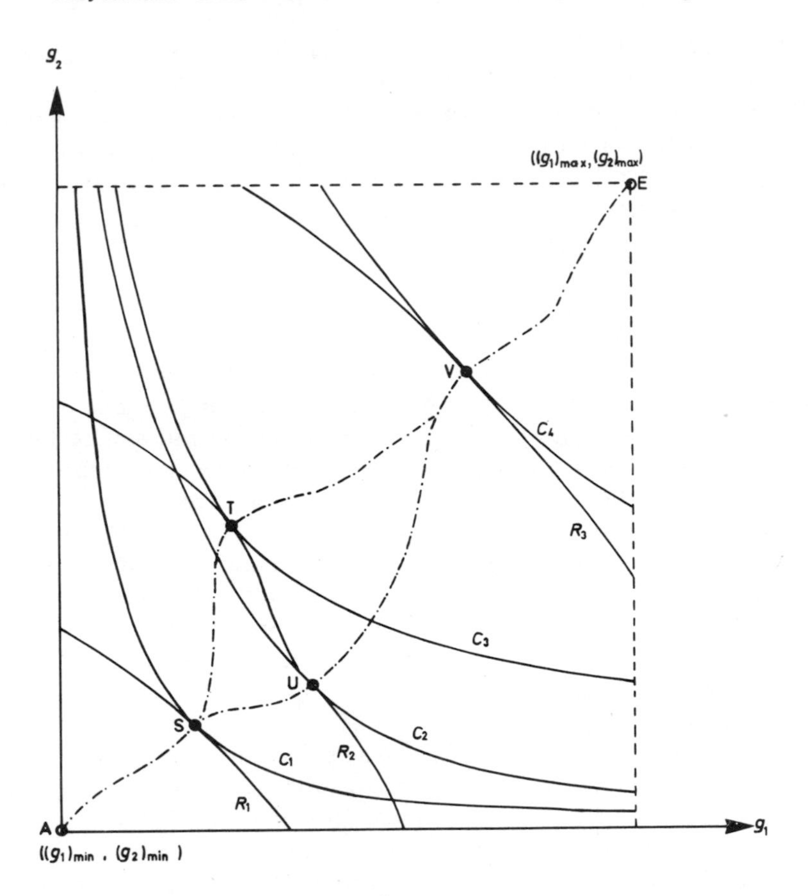

Fig. 3. Bifurcation and buckling. The constant-R-curve, R_2 buckles. This gives rise to bifurcation of the locus of cost-minimized implementations. The implementations SUV are inferior to the implementations STV.

assumption (3) to state that points with the same C-values or R-values are located on families of differentiable surfaces in the first octant. It should be noticed that the curvature at a point on a surface now is two-dimensional; the curvature may be described by the two main radii of curvature.

Assumption (4) now reads as follows. The locus of points (g_1, g_2, g_3) with the same C-value constitutes a convex surface. The shape of the surface is determined by the economical fact that use of the very best grades results in a dramatic rise in product cost. As before, we assume the constant-C-surfaces at all points to have (a) main radii of curvature which are more than zero, and (b) a gradient which is parallel to some vector pointing from (0, 0, 0) out into the first octant. For example, a possible family of C-surfaces may be obtained by using a line through (0, 0, 0) and (1, 1, 1) as axis for rotation of hyperbola branches. Equation (1) illustrates such as possible family of C-surfaces.

$$g_1^2 + g_2^2 + g_3^2 = (g_1 + g_2 + g_3)^2/2 - 1/(\text{Product cost})^2 \tag{1}$$

Assumption (5) now reads as follows. The locus of points (g_1, g_2, g_3) with the same R-value constitutes a convex surface. The shape of the surface is determined by the following familiar fact: the effect on product reliability of the degradation of one component's grade can only, to a small degree, be compensated for by the improvement of some other component's grade. As before, we assume the constant-R-surfaces at all points to have (a) main radii of curvature which are more than zero, and (b) a gradient which is parallel to some vector pointing from (0, 0, 0) out into the first octant. For example, a possible family of R-surfaces is given by Equation (2) where a_1, a_2 and a_3 are constants.

$$a_1^2 g_1^2 + a_2^2 g_2^2 + a_3^2 g_3^2 = 1/(\text{Product reliability})^2 \tag{2}$$

2.5 *The locus of cost-minimized implementations for* N = *3*

Under the modified assumptions it is seen that the region where the R-values of the points all exceed or equal a certain value R_0 is a convex region. It is also seen that the region where the C-values of the points all are less than some value C_0 is a convex region. If the intersection of the two regions exists, it will also be a convex region. We can now rêpeat the limiting argument from Section 2.2 and make C_0 smaller and smaller while maintaining the value of R_0 In the limit the convex intersection is reduced to *one point*; the assumption of all radii of curvature being more than zero insures us against the possibility that the intersection in the limit becomes a straight line rather than a point. At the point where the convex intersection reaches its limit, the tangential planes to the R_0-surface and the C-surface coincide and the two gradients have the same direction. By repeating the procedure we can find other cost-minimized implementations. Due to the assumption of differentiability, the locus of cost-minimized implementations will, as before, form a curve connecting the two points $[(g_1)_{min}, (g_2)_{min}, (g_3)_{min}]$ and $[(g_1)_{max}, (g_2)_{max}, (g_3)_{max}]$. The (C, R)-values along the curve constitute the locus of cost-minimized implementations illustrated by the heavy line in Fig. 1.

If a constant-R-surface for, say $R = R_a$, intersects with a constant-C-surface for, say $C = C_b$, the intersection becomes a closed curve in space. All points on the curve correspond to only one point in Fig. 1, the point where the lines $R = R_a$ and $C = C_b$ intersect.

Buckling and bifurcation obviously are possible also in three dimensions; our assumptions about convexity, however, insure us against such phenomena.

2.6 *The extension to the case* N > *3.*

In Section 2.2 and 2.5 it was demonstrated for $N = 2$ and $N = 3$, an even and an odd number of dimensions, that the locus of cost-minimized implementations constitutes a path from the point illustrating the implementation combining maximum reliability and maximum cost to the point illustrating the implementation combining minimum reliability and minimum cost. The proof was based on five mild assumptions.

When the dimensionality exceeds three, $N > 3$, essentially the same thing holds true. The extension of the five mild assumptions to N grades, $g_1, \ldots, g_N$, is trivial. For reasons stated earlier, the implementations with reliability $R > R_0$, or with cost $C < C_0$, are illustrated by points located in two convex N-dimensional volumes. If any implementations at the same time have $R > R_0$ and $C < C_0$, they will be illustrated by points located in a convex N-dimensional volume (this is readily seen by approximating both the R_0- and the C_0-volume by a convex polytope [3] of sufficient complexity and then showing that the intersection becomes a new and smaller convex polytope). Due to the assumption of all curvatures being positive (which insures us against saddle-points) the convex region degenerates to *one* point in the limit when C is being steadily reduced. Due to the assumption of differentiability, the locus of cost-minimized designs will be a one-dimensional curve from $[(g_1)_{min}, \ldots, (g_N)_{min}]$ to $[(g_1)_{max}, \ldots, (g_N)_{max}]$. For example, the equation for the locus could, in principle, be parametric and consist of N equations of the form $g_j = h_j(t)$, where $j = 1, ..., N$, and t is the parameter. When t changes monotonically from t_{min} to t_{max}, a running point on the curve will follow the curve from one end of the locus to the other.

2.7 *The optimal policy*

We can now, under the stated assumptions, formulate an answer to the question which was propounded in Section 1.1. The locus of cost-minimized implementations is obtained by (a) starting with the implementation which combines maximum reliability and maximum cost, and (b) changing the N grades in such a manner that the cost-gradient and the reliability-gradient always are proportional; when the gradient vectors have the same direction we obtain maximum cost savings as we gradually reduce the reliability. The locus will end with the point illustrating the implementation which combines minimum reliability and minimum cost.

3.0 N GRADES, WHICH CAN BE CHANGED ONLY IN FINITE STEPS

3.1 *Sub-optimal designs*

When the N grades can be changed in finite steps only, the possible combinations of grades may be illustrated by lattice points in N-dimensional space. Apart from the end points, $[(g_1)_{min}, ..., (g_N)_{min}]$ and $[(g_1)_{max}, ..., (g_N)_{max}]$, no lattice point will (generally speaking) fall on the locus of cost-minimized implementations. This means that we must, in practice, select our implementations among sub-optimal implementations. The degree of sub-optimality of an implementation with (Cost, Reliability) = (C_0, R_0) may be described in one of two ways. One can evaluate the excess cost $(C_0 - C^*)$ or the reliability deficiency $(R^* - R_0)$ of the implementation where (C_0, R^*) and (C^*, R_0) are the two corresponding cost-minimized implementations. When selecting a sub-optimal implementation, we are faced with the problem: which of two sub-optimal implementations (C_1, R_1) and (C_2, R_2) is the better? If, at the same time $C_1 < C_2$ and $R_1 > R_2$, clearly (C_1, R_1) is superior to (C_2, R_2); but if $C_1 < C_2$ and $R_1 < R_2$ it is an open question whether or not the decrease in reliability, $(R_2 - R_1)$, was worth the cost-saving, $(C_2 - C_1)$. This question can only be answered after a study of the particular case.

3.2 *The suggested policy*

When the N grades can be changed only in finite steps, the optimal policy consists of the following two steps. (i) Discard all implementations which are inferior to some other implementation. This leaves us with a set of implementations, S^*, with the following property. If two implementations (C_a, R_a) and (C_b, R_b) both belong to S^*, and $C_a > C_b$, then it is also true that $R_a > R_b$. (ii) Among the admissible implementations S^*, select the one which seems best for other reasons (e.g. marketing reasons).

Clearly the above procedure is unrealistic because it involves the computation and comparison of a horrendous number of implementations since we cannot obtain S^* in practical cases. We can, however, with a reasonable amount of work, obtain another and smaller set of implementations, S, which the authors believe constitutes a 'good' set of implementations. The procedure is based on the policy described in Section 2.5 where we maximize cost-saving while reducing the product reliability. This time, however, we must take steps of finite length in N-dimensional space. The suggested policy is a procedure consisting of the following steps.

(1) The first member of S is $[(g_1)_{min}, ..., (g_N)_{min}]$ for which C and R both are maximized, $(C, R) = (C_{max}, R_{max})$.

(2) Evaluate (C, R) for the N implementations obtained by reducing the grade g_j, $j = 1, ..., N$, by one rating while maintaining the $(N - 1)$ other grades at maximum level. The result is called $({}^j_1C, {}^j_1R)$.
Compute the N values of the fraction ${}^j_1F$ defined by Equation (3).

$$ {}^j_1F = (C_{max} - {}^j_1C)/(R_{max} - {}^j_1R) \quad (3) $$

The value of j which maximizes ${}^j_1F$ is determined and the corresponding reduction of g_j by one rating is made permanent; the implementation is called ${}_1D$. ${}_1D$ is included in S which by now consists of two implementations. The corresponding value of $({}^j_1C, {}^j_1R)$ is renamed $({}_1C, {}_1R)$.

(3) Starting with ${}_1D$, evaluate (C, R) for the N implementations obtained by reducing the N grades, one at a time, by one rating; the result is called $({}^j_2C, {}^j_2R)$, $j = 1, ..., N$. Compute the N values of ${}^j_2F$ defined by Equation (4).

$$ {}^j_2F = ({}_1C - {}^j_2C)/({}_1R - {}^j_2R) \quad (4) $$

Determine the value of j which maximizes ${}^j_2F$; the corresponding reduction of g_j by one rating is made permanent. The resulting implementation called ${}_2D$ is included in S. (C, R) for ${}_2D$ is renamed $({}_2C, {}_2R)$

(4) ${}_2D$ is used as starting point for the next set of N implementations, etc. By repeating the procedure sufficiently many times, we will in the end arrive at the design $[(g_1)_{max}, ..., (g_N)_{max}]$ for which $(C,R) = (C_{min}, R_{min})$. This design is also included in S.
If the number of ratings for g_j is called M_j, it is seen that the number of implementations in S equals M.

$$ M = 1 + (M_1 - 1) + (M_2 - 1) + ... + (M_N - 1) \quad (5) $$

Clearly M is much less than the total number of possible implementations which is $M_1 \cdot M_2 \cdot ... \cdot M_N$.

4.0 AN ILLUSTRATIVE EXAMPLE

4.1 *The three-transistor amplifier and its components*

To test the suggested policy, we decided to try it out on a product of reasonable complexity. We decided to use an electronic amplifier, a device with which we had some familiarity. The amplifier configuration which was selected for the study is shown in Fig. 4, and it has been described in the literature [4]. The component values listed in Tables 1 and 2 were selected to achieve at the same time (i) a gain, A, with value between 19·5 and 22·5, and (ii) a 3 dB bandwidth, B, exceeding 84·5 MHz.

A preliminary study of the effect of component grade changes on the output specifications showed that some of the components obviously should have a particular grade to obtain a cost-minimized product design; eight such components are listed in Table 1. For some of the other components several grades seemed to be reasonable choices; five such components are listed in Table 2. The resistors listed in Table 2 are assumed to have constant probability densities; e.g., 10% resistors with nominal value 470Ω are assumed to have true values evenly distributed between 423Ω and 517Ω. With each transistor are associated values of r_{bb}, $r_{b'e}$, r_{ce}, g_m, $C_{b'e}$ and $C_{b'c}$ which are statistically dependent. Fortunately, corresponding transistor

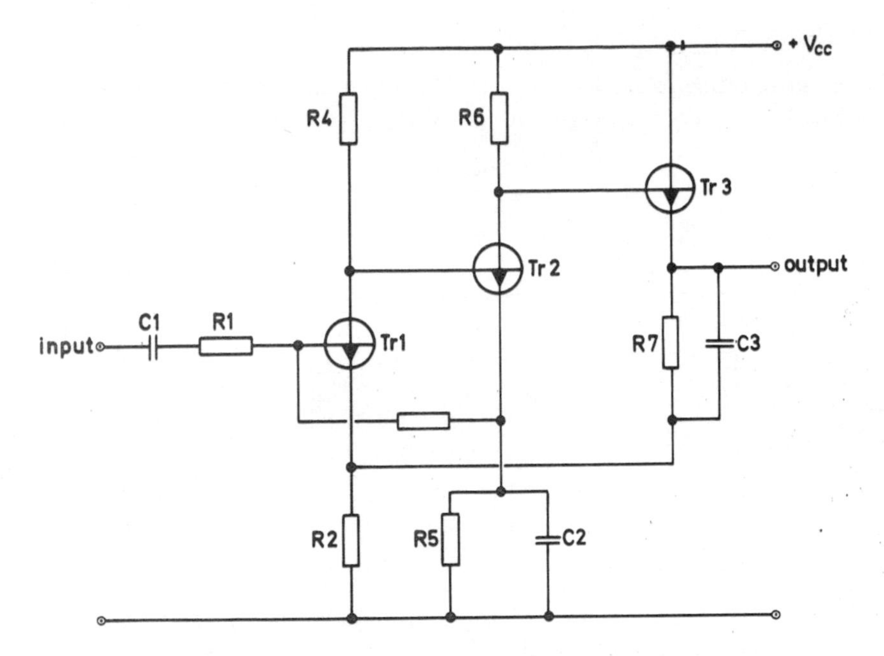

Fig. 4. The product. The three-transistor-amplifier was used to test the suggested policy.

parameter values for 100 2N918 transistors as well as for 100 2N2369 transistors have been tabulated in the literature[5]. The tabulations cover nine operating points per transistor. By '2N918 selected' is meant that 2N918 transistors have had the $C_{b'c}$ value measured and only the best half is used. The selected transistors had $C_{b'c}$ less than 615 nF.

4.2 *Exhaustive search* v. *the suggested policy*

An inspection of Table 2 shows that the total number of ways of implementing the product is $4{\cdot}4{\cdot}2{\cdot}2{\cdot}2 = 128$. The number is moderate and we decided to perform an exhaustive search for good implementations. For each possible implementation we determined two quantities, C and R. The first quantity, C, is the total cost for the components from Table 2. Due to grade variations, C is the variable part of the product cost so we are only concerned with changes in C-values. The second quantity, R, is the joint probability of the product simultaneously meeting the two output specifications, $19{\cdot}5 < A < 22{\cdot}5$ and $B > 84{\cdot}5$ MHz. The (C, R)-values for the 128 implementation methods are illustrated as points in Fig. 5. Twenty-four of the points are not shown as they coincide with some of the remaining 104 points. Fig. 5 clearly shows the danger of getting an inferior implementation. The excess cost, $(C - C^*)$, can easily be 5p or more for an implementation which should have a specified reliability and which was selected carelessly.

It is evident from Fig. 5 that we have obtained a 'good' set of implementations as anticipated. From Equation (5), we know that there are only ten members in the set, S. Consequently, we miss out on several good implementations, e.g., the one with $(C, R) = (20{\cdot}05, 92{\cdot}0)$. In only one case does the method really let us down: we miss out on the implementation $(18{\cdot}36, 86{\cdot}0)$. The implementation has the following grades:

$$(R_2, R_7, C_3, \text{Transistors no 2, Transistor no 3}) = (\tfrac{1}{2}\%, 5\%, 1\tfrac{1}{2}\%, \text{2N918, 2N2369})$$

5.0 CONCLUSIONS

The problem of selecting an implementation, which minimizes cost for a desired reliability, has been studied. A policy has been suggested for finding a 'good' set of implementations with a

TABLE 1

Component	Grade	Price (Danish kroner)
R_1 = 470 Ω	5%	0·29
R_3 = 10 kΩ	5%	0·29
R_4 = 22 kΩ	10%	0·07
R_5 = 1 kΩ	5%	0·29
R_6 = 4·7 kΩ	5%	0·29
C_1 = 10 μF	−10%, +50%	2·00
C_2 = 50 μF	−10%, +50%	2·00
Transistor no 1	2N918	6·30

A preliminary study shows that the eight components listed above should have a particular grade. [One Danish krone is approximately equal to 5p].

TABLE 2

Component	Possible grades	Price (Danish kroner)
R_2 = 22 Ω	½% tolerance	1·07
	1% ,,	0·85
	5% ,,	0·29
	10% ,,	0·07
R_7 = 470 Ω	½% tolerance	1·07
	1% ,,	0·85
	5% ,,	0·29
	10% ,,	0·07
C_3 = 15 pF	1½% tolerance	0·50
	5% ,,	0·40
Transistor no 2	2N918	6·30
	2N918 selected	8·30
Transistor no 3	2N918	6·30
	2N2369	3·90

A preliminary study shows that the five components listed above could reasonably be used in several grades. [One Danish krone is approximately equal to 5p].

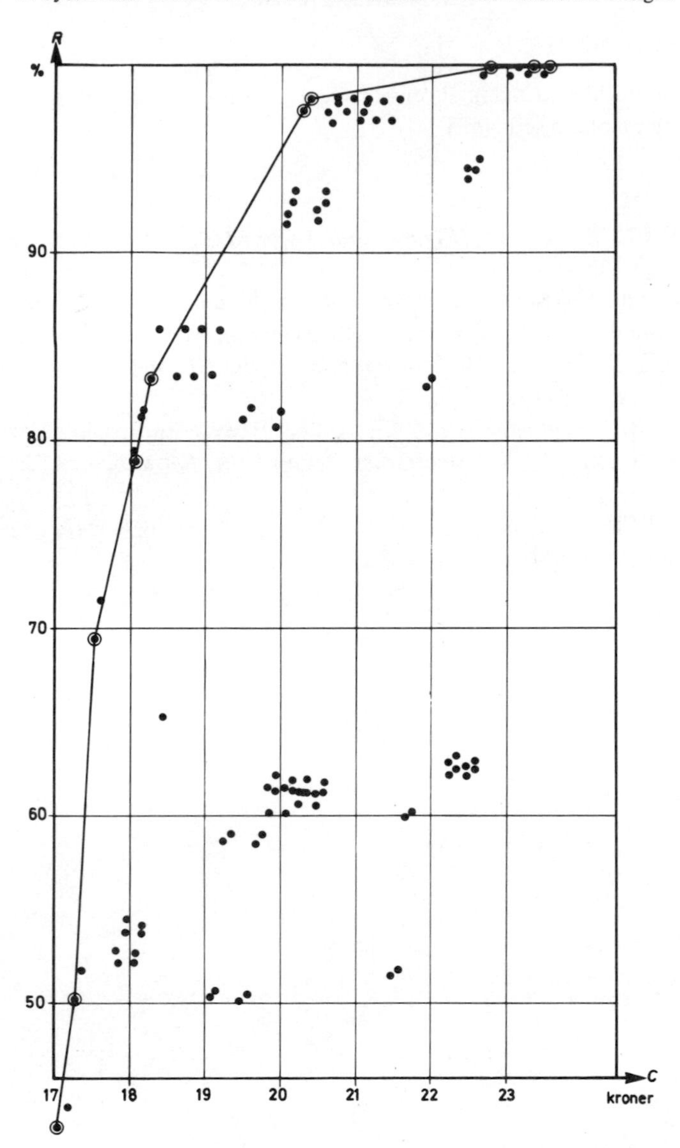

Fig. 5. A 'good' set of implementations. The points illustrate 128 possible implementations. By following the suggested policy, we find the ten implementations illustrated by encircled points. The ten implementations obviously constitute a good set of implementations. (Danish kroner equal approximately 5p).

reasonable amount of computational work. The designer can either settle for one of the 'good' implementations or use the 'good' set as a starting point in a search for a more suitable implementation.

6.0 ACKNOWLEDGEMENTS

The authors would like to thank Professor Georg Bruun, Mr. Finn Jensen and Dr. Frank C. Karal, Jr. for many helpful discussions.

REFERENCES

1 LYUSTERNIK, L 'A., *'Convex Figures and Polyhedra'*, Article 7.7. New York: Dover Publications, Inc.; 1963.
2 KELLER, J. B., and ANTMAN, S., *'Bifurcation Theory and Nonlinear Eigenvalue Problems'*; New York: Courant Institute of Mathematical Sciences, New York University, N.Y.; 1968.
3 SOMMERVILLE, D. M. Y., *'An Introduction to the Geometry of N Dimensions'*; New York: Dover Publications, Inc.; 1958.
4 FERRANTI, Ltd., 'Applications of the E-line Epoxy Encapsulated Transistor; U.K.: *Application notes issued by the Electronics Department, Ferranti, Ltd.*; 25–27, November 1967.
5 RACA RESEARCH Ltd., 'REDAC'; U.K.: 'Racal Electronics Design and Analysis by Computer; Measurements Manual, 1, 1968.

THE DESIGN OF A PRODUCTION CONTROL SYSTEM: IN PARTICULAR THE REQUIREMENTS PLANNING SEGMENT

A. W. Buesst
National Cash Register Co. Ltd

SUMMARY

The paper discusses the design of the central element of a production control system suitable for an average-sized, mixed-production (usually metalworking) manufacturer. The facilities of a straightforward disc storage business computer were used in designing a system to carry out the vital *requirements planning* function, that is to determine the need for, quantity of and due date of a continual stream of replenishment orders to meet the user's production plan. Having mentioned the main objectives and constraints, the paper continues with the matters about which the design revolved: the file of outstanding demands and replenishment orders, its processing during a level-by-level explosion, the necessary input and the principal resulting output. The design decisions involved are discussed with the object of showing how a system was derived to perform a practical and much-needed task.

1.0 INTRODUCTION

Our overall brief was to design a production control system for the batch production of engineering goods in a conventional factory. It was assumed that these goods would be assemblies of some sort and that many of the parts would be cut from metal, the latter being more a description of common practice than a point of real significance.

The system was to be suitable for operating on the NCR Century computer, which is a fairly straightforward commercial machine with disc storage, allowing access to files at random. This computer happens to be quite a good one and relatively inexpensive, but, more important, it is an adequate and suitable machine for this task. Our problems were, in general, not due to the computer as such.

These terms of reference covered a lot of ground and would, if successfully met, provide a system of value to a very large number of manufacturing companies.

2.0 DESIGN PHILOSOPHY

Naturally, like most designers, we had the benefit of earlier work by NCR and other companies available to us. Almost instinctively we decided to take a straightforward approach – to do precisely what we understood had to be done. We would not copy existing manual systems for their own sake; nor, if possible, would we resort to tricks to produce an 'improved manual system'. This major decision may sound trite, but so many earlier computer systems we studied seemed to skirt around the real issue. Perhaps this was because of the restriction of punched-card equipment or the limitations of magnetic tape computers.

The central element of all the computer-assisted systems which we knew of was either:

- the calculation of what was wanted; or
- the control of work in progress.

The second major decision was to tackle the calculation of what parts were wanted and when, since we felt that it would, in due course, be far easier to schedule a properly initiated work-in-progress situation. Further, the control of the stock position which would result should give a substantial financial reward fairly quickly, possible paying for the new system.

3.0 THE INDUSTRIAL SYSTEM

The overall system or environment in which requirements planning may operate is shown in the diagram which we call the industrial system, Fig. 1. The central element is the box marked IMMAC, standing for Inventory Management and Material Control.

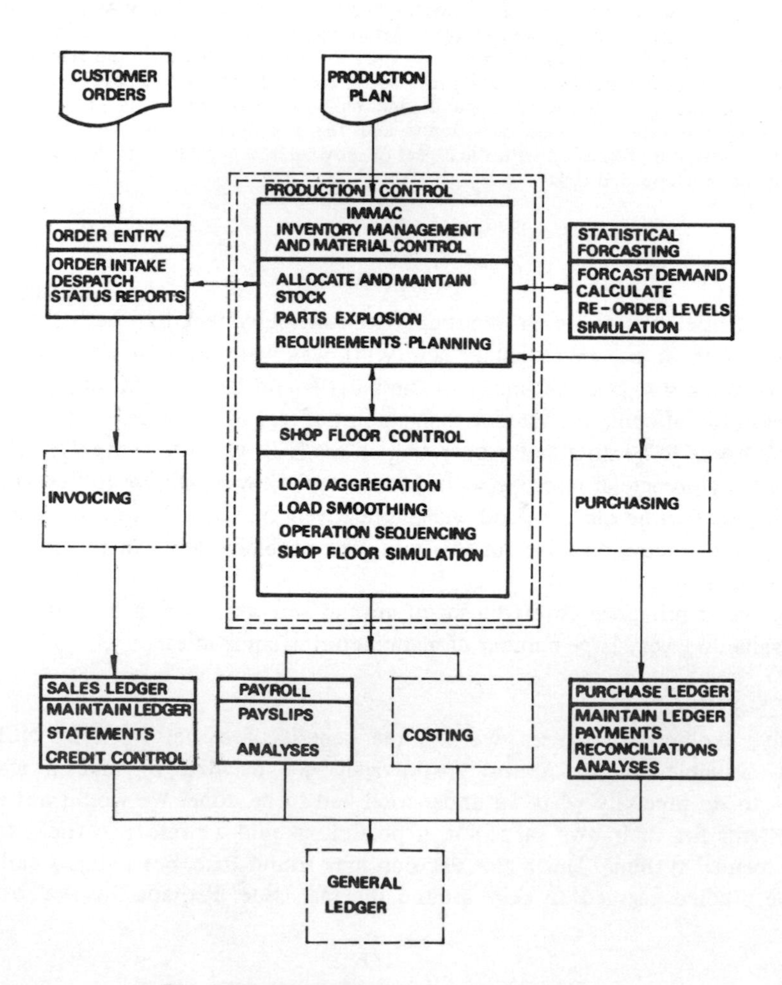

Fig. 1. The N.C.R. Century industrial system.

The figure shows that it is possible to grow out in various directions from this system, according to particular needs. For example, an order entry system may handle the acceptance of customer orders in relation to planned product stocks and initiate their despatch at the proper time. Statistical forecasting may be applied to certain items. Shop floor control may be developed steadily, first aggregating the load of individual batches, then smoothing this load, then resequencing operations for more efficiency, and finally conducting simulation exercises to assist management decision-making. Purchasing and subcontract control could be added.

4.0 OBJECTIVES OF THE REQUIREMENTS PLANNING SYSTEM

To be judged satisfactory, the material planning phase must:

- ensure that materials are available when they are required
- allow the production facility to function efficiently
- keep stocks to the minimum that is consistent with ordering strategies.

The prime objective is to set out all those items which must be made or bought, showing the quantity and due date. This can be done in list form, arranged by item, by start date, by supplier or otherwise. Alternatively it can be set out in the form of a schedule by some arbitrary time periods. Obviously before this can be done, we must know, for any item:

- how many there are now
- how many more are already on order and when they are due
- how many are required and when

In a manual system it is usually too difficult to keep track of all the individual orders and requirements, so they are lumped together into three totals: *on hand, on order* and *allocated.* These are often further reduced to one figure, the *free balance.* To do otherwise within a reasonable time and with a realistic number of people would be impossible.

It seemed, however, that a sound computer-assisted system should be able to cope with this situation, and following this line of thought we considered that it should:

- maintain an inventory control system
- maintain a file of outstanding orders and requirements

Thus the necessary analysis of the outstanding order situation, to provide for order expediting and so on, should logically be provided as well.

Note that we defined for convenience:

an *order* is placed for replenishing stock; these goods are needed BY Production Control

a *requirement* is for goods which Production Control must supply TO others; parts to make higher-level orders, products for customers.

It follows from these definitions that a 'customer order' is a 'requirement'.

To work out what is required, and when, the make-up of every part must be known. So the system must also:

- maintain a parts list or bill of material file

This, besides providing for 'breakdown' or 'explosion' of products, also allows for printing assembly and material requisitions if desired.

4.1 CONSTRAINTS

There are many constraints on a system, such as this, which has considerable interaction with its environment, but the significant restrictions are (in no particular order):

- The system must be stable

- The system must be flexible
- The user must be able to control or even override the system
- The system must cater for both *planned* and *re-order level* stock control
- The system must be able to raise orders covering single requirements as well as to groups requirements so as to raise 'economic' orders
- The computer files should be as large as possible
- Data processing time should be kept to a minimum
- Should the volume of data being processed rise the system must not seize up or 'stall'.

4.2 THE REQUIREMENTS FILE

Having decided to maintain details of all outstanding orders and requirements, we had to make a decision on the form in which this should be done. As already mentioned, there could merely be summary totals stored with each item's description and on-hand balance. These totals could be *on order* and *allocated,* or just *free stock* or *free balance,* or perhaps these could be broken down into arbitrary time periods. Our experience suggested that such blanket figures would quickly become incorrect if an allocation were cancelled but not advised, or if an issue went unrecorded, and, since they are not self-checking, we decided to avoid them and to keep details of each order and requirement separately, with its due date expressed in working days.

Another major decision was to make or buy any item at the last possible moment, which in practice means that the start date of an order is its due date less that item's lead time. By our definition the *lead time* of an item is the number of working days which usually elapse between deciding an order is needed and the goods being available in stores. Obviously things go wrong and orders run late; equally, it is frequently possible to get an order completed in half the average time. We assume also that the lead time is unaltered by the quantity ordered.

It should of course be allowable to make or buy items in larger 'economic' quantities in time to meet the first of a series of requirements. The remainder would then be stocked until needed.

This file of orders and requirements was called the *requirements* file and arranged in part-number sequence. There are arguments for using other sequences but the obvious one prevailed after due consideration. From this file we could now extract the current situation of any part. For example, if X were physically in stock, Y more due next week and Z due out in two weeks, then the position in 1 week could be $X + Y$ and in 2 weeks $X + Y - Z$. Note that this is the position as known at the present instant, relating to a moment one or two weeks hence.

It might be thought that the use of individual-working-day dating could lead to myriads of records on the file. This is not so. A record must be kept of every order placed whatever the system used. Similarly, an allocation is usually made for parts or materials to make any order, and no extra information is therefore demanded. If lead times are given in round weeks (5 working days) events will occur in an orderly manner. Also, if grouping together a number of requirements and supplying them with one replenishment order is allowable, the file remains quite tidy. The use of working days for timing also permits individual users to set up their own time-periods and drop the appropriate figures into them without any difficulty.

4.3 MASTER FILES

The most dynamic file of data in the system having been settled the other files were fairly self-evident. These master files were named:

- The *requirements* file, just described
- The *inventory* file, containing details of each item, its control parameters and the quantity on hand
- The *parts list* file, in which the make-up of any item made from other items is recorded
- The *where-used* file, which is the *parts list* reversed
- The *calendar of working day numbers*

A major decision, of little importance to the user, was to have a perfectly straightforward parts list file, showing the parts needed to make any next-higher-level part. From such lists, in which any part appears only once as an 'assembly' but as often as necessary as a constituent of assemblies, we may construct bills of material. These, by arbitrary definition, are the entire contents of any assembly, right down to raw materials and bought-out parts. An alternative method, using chaining techniques, seemed after careful consideration and calculation to offer no particular advantages, nor did it contribute anything extra to the primary purpose of the system.

The where-used file is a bonus generated automatically and will not be discussed further in this paper; nor need the facility to handle 'versions' of parts lists, alternatives and multiple changes concern us particularly. Items in the inventory file, the parts list and the where-used file may be obtained at random or in part-number sequences as desired. We also decided to allow for a user's existing part numbers, up to as many as eighteen characters long.

4.4 ITEM CLASSIFICATION

The various ways that an item could be classified seemed to be unmanageable. Our first simplification covered:

1. ABC
2. Type of item: forging, bar, assembly
3. Made, bought, other
4. How held: no stocks, re-order level, two bin, min-max etc.
5. How ordered
6. How issued

It was then realised that 4 and 5 were really the same, and, from the point of view of the requirements planning system, the most important. We therefore devised a control code to specify how an item would be ordered and how it would be issued. The other classifications would be the user's responsibility. There would be three main types of ordering:

- Stock control i.e. that is on re-order level
- Explosion item with an E.O.Q.
- Explosion item, but each requirement provided by a separate order.

4.5 THE EXPLOSION CYCLE

Having settled the files of data and basic concept, it was necessary to decide upon the method of breakdown or 'explosion' to be used, and to provide for the proper generation of replenishment orders – the heart of the material planning function.

There are several ways of doing this explosion of products down to all their constituent parts. Having done it, and collected together all known demands and replenishments outstanding, one can calculate the need for, quantity of, and target date of fresh orders in many ways again. We chose to use a level-by-level explosion, netting off at each level. This works in the following manner.

Products have level zero, and all demands (whether actual customer orders or forecasts submitted by the sales department) are considered in turn. If stock exists to satisfy them nothing further need be done. If not, the requirements may be grouped, if desired, and a potential order placed to meet them.

This potential order is 'exploded' by reading its constituent assemblies, sub-assemblies and parts off its parts list. Each of these potential requirements is dated to arrive in time for the start of the potential product order, which is determined by its lead time. Some parts may 'enter' their parent during its build and so are dated accordingly. All these 'potential lower-level requirements' are then sorted to part number sequence and merged into the requirements file.

The cycle then starts again, this time considering level-one items (mainly major assemblies). As the cycle continues down through the levels, all demands for items are added to the file and, once any particular item's lowest level has been reached, they are all considered together to see what orders might be needed to satisfy them.

The lowest level of any item is determined automatically, when the parts list file is set up, by finding the maximum number of assembly stages between the item and any level-zero finished product. Raw materials and bought-out parts may be sent to an arbitrary lowest level of thirty to save computing time.

At the end of the explosion the requirements file contains a mixture of firm requirements and orders currently in progress, and many potential ones stretching out into the future.

Some of these potential orders should be put in hand before the next review and these are listed in the *net requirements* report. Both the firm and projected situation for any part may be displayed in the *projected requirements report.* These are both described in section 4.7.

During the explosion a routine must be followed to determine the need for, quantity and due date of replenishment orders. A lot of time was spent on this aspect, but the overall design of the system is such that a user could alter many details of these calculations to suit himself. For example, a dynamic order quantity calculation might be incorporated. Stock control items are excluded from these calculations.

After completing the explosion and printing the two reports the file is edited to remove future potential requirements and orders. The new review will recreate these from the exact product demand pattern then existing, since each review is a complete one, not just the new items. This gives the user the opportunity to change his production plan completely between reviews, although it would clearly be foolish to do this if the requirements of various long-lead-time raw materials were also radically altered.

4.6 INPUT

However the explosion had been designed, the prime input would still be customers' orders. The nature of these will, of course, depend on the business. In some cases specific customer orders are entered directly as requirements. In others the only 'customer' of this system is the sales department, which places 'stock orders' on the factory. In most companies a mixture of these will exist, with some spare part demands as well.

It must be made clear that there have to be demands of one sort or another, dated from the present out into the future, or the system cannot operate properly. The most forward demands should be dated at least the overall lead time of the finished product ahead. In this case the overall lead time is measured from the stage at which stocks are maintained to meet demand. This is frequently the raw material stage.

Where customers do not place orders for delivery up to and beyond this overall lead time, the sales department must do so instead. This is done by stock orders or sales forecasts, or, more generally, by a formally agreed production plan.

The other input of the system would be mainly stock movements and requests entered on punched cards or equivalent media.

4.7 OUTPUT

The output of the system consists of printed lists, analysis, instruction and reminders, collectively known as *reports*. Two basic principles were decided upon: first that the computer's instructions were not mandatory but were suggestions which could be altered by the user; second that as far as possible only exceptions would be shown.

In due course, as further systems are added, there will be direct links, with data output by one being worked on by the next. Thus the content of particular reports may differ and they may well be produced at different points, but they remain the user's essential link with the system. Every detail on every report was therefore carefully examined.

The two fundamental outputs of the system have already been mentioned. The projected requirements report contains all known future demands and both firm and proposed replenishment orders. This could be arranged across the page by arbitrary periods or down the page for as far as necessary. The latter course was chosen, as can be seen in Fig. 2. The due dates are in sequence downwards, and the cumulative totals of demand and supply may be compared to determine the situation at any date. Individual users could relatively easily alter the print program to give a demand schedule in their traditional format.

DATE 18/ 4/69 DAY 104   P R O J E C T E D / C H A N G E D R E Q U I R E M E N T S   PAGE 7

					REQUIREMENTS			ORDERS OUTSTANDING			
				DATE	RFCE	BAL REQD	CUM REQS	O/NO	BAL DUE	O/H BAL + CUM ORDS	ORDER ACTN
PART 014	5A12456	STRS LOCN EAST	4							37.000	
		LEAD TIME	44	175/69				A001249	150.000	187.000	
HUB ASSY		% SCRAP	7.5	180/69	1246	200.000	200.000				
		ORDER QTY	150.000	190/69				A001302	150.000	337.000	B/F
		RE ORD LVL	12.000	192/69	SPARES	25.000	225.000				
		UNIT		204/69	A001253	110.000	335.000				
				227/69		285.000	620.000		300.000	637.000	

Fig. 2. This is an optional report showing the pattern of demands and replenishments, as now known, projected into the future. There are three blocks of information: (i) facts about the item on the left, (ii) dated allocations in the middle, and (iii) stock plus orders on the right. This includes existing orders, any to be started before the next review and possible future orders. Suggestions to bring orders forward (B/F) or to cancel them may appear. The *current stock position* report is similar.

This report could be very lengthy, so it is given for specific part numbers on request only, unless the computer finds that an item now appears to be under- or over-supplied. This may occur if an existing order is closed short, extra requirements are suddenly introduced, or cancellations occur. In this case any outstanding orders may be signalled for bringing forward or cancellation.

It might be said that the projected requirements report will be useless because the workshops or suppliers will always let you down. Does this mean that, in turn, you always let your customers down? While the answer may be to increase lead times to their true length and to expedite effectively, the report remains valid in that it shows the effect of the promises made to customers.

The net requirement (orders to be placed) report is a list of suggested manufacturing and purchase orders whose start date occurs before the next explosion (Fig. 3). Any items which should have started already are highlighted for special attention. The user is expected to vet this list and initiate the necessary paperwork in the usual way. Perhaps one or two of the orders need to be altered for one reason or another; the others are confirmed and the computer may then print assembly or material requisitions if desired. This emphasises the philosophy of suggesting action on an exception basis. The user then has the final say, after appraising both the facts marshalled by the computer and any external influences known to himself.

DATE 18/ 4/69 DAY 104 NET REQUIREMENTS (ORDERS TO BE PLACED) PAGE 23

GROUP	PART NO	DESCRIPTION	QUANTITY TO ORDER	UNIT/MEASURE ORDER USE	START DATE	DATE DUE	LEAD TIME	USUAL SUPPLIER	CTL COD	POTNL SHORT	ACTION	DATE	SIG
120	1004159	.092 CRS S.SEL	32.000	SQFT (SQIN)	108/69	138/69	30	46126	10				
120	1006517	.092 CRS COIL	50.000	FOOT (INCH)	97/69	117/69	20	46126	10	***			
120	1021414	19/64 CDS NTRGN	10.000	FOOT (INCH)	113/69	153/69	40	20012	10				

Fig. 3. This report is printed after a review or 'explosion' and lists all new orders which should commence before the next review. Items asterisked will have to be made or bought in less than the usual lead time to avert shortages. This is the main 'action' list of the system.

The system could be considered as having four functions:

- Planning
- Monitoring performance
- Information
- File maintenance

Under *Planning* come the two major reports just discussed. Also, to cater for what we called stock control items, controlled by re-order level, there is a stock action report which requests fresh supplies of these items and also contains reminders, warnings and discrepancy messages.

To exercise control, the production controller must be able to monitor the performance of the system and thus of his personnel. We therefore added a daily list of reported shortages, and at the end of the week a shortage summary, listing those items which have not yet been cleared.

From the detailed requirements file the computer can check each issue and receipt and report any deviations in quantity and time which are outside individually stated tolerances. Another useful report is a list of outstanding orders, arranged by due date.

To answer the traditional question, 'how many of such and such are there?', we decided to provide on request a report identical with the projected requirements report. This would show supply and demand details during the lead time only, and was called the *current stock position.*

All these reports show, where appropriate, all other relevant facts available in the computer. The recipient is thus better able to take suitable action.

Within the system's files is stored a large amount of valuable information and, mainly on a request basis, we provided for reports such as:

- Stock valuation

- Purchase commitment
- Usage listing
- Inventory analysis

The file maintenance section provides:

- Parts lists
- Where-used lists
- Bills of material

5.0 PRACTICAL DIFFICULTIES

During the explosion cycle the requirements file, the inventory file and the parts list are all needed in succession once for every level. There seemed to be little point in squeezing these files into restricted space; they simply had to be large enough to cater for a growing concern. There was no way of estimating with certainty the length of the requirements file by the end of an explosion, since this would involve too many factors besides the average parts-list length.

These considerations had been apparent throughout the design and had affected our approach to many points of detail which need not be recorded here. We believe, however, that they are considerations of vital importance to the user, as are such mundane matters as file security, ease of operation and recovery from mishaps.

6.0 IMPLEMENTATION

The implementation of a computer-assisted production control system is undoubtedly a big task, so any assistance the system designer can provide will be valuable. We tried to make the various sections discrete so that they could be implemented separately and in differing sequence.

The first, and in many ways biggest, hurdle is to create the files of data, but this is a study in itself. Having set this in train the user may then commence by one of the following methods:

- *Stock recording*. All movements are recorded but only re-order-level control items are so controlled by the computer.
- *Stock control*. As above, but all items are treated as re-order-level control until explosion can be effected.
- *Explosion* is effected, but either for a small range of products or for only one or two levels to start with.
- *Gross explosion* of the production plan but no netting off against stock.

Each of these starting methods has certain advantages but, whatever the method, four basic factors are vital:

- Top management must be behind the scheme
- Disciplined communication of data must be established
- The master file data must be made accurate
- Recognition that this sort of system will not run itself

7.0 BATCH PROCESSING

The whole system relies on data being processed in batches. One might expect the stock and shortage subsystems to be run at least once a day so that, allowing for punching, the computer would be 24 or more hours behind the event. This sounds slow, but is it likely that all movements get posted any faster in a manual system? No other portion of the system is required as often so batch processing is quite suitable.

Looking to the future, the possibility exists of reporting stock movements to the computer from on-line terminals and receiving immediate displays on video units for action. This can in fact be done today, but at a cost which most users would not consider worthwhile. We are quite certain that these improvements will come in time, but equally sure that they will be built onto the present design, not make it obsolete. By the time this occurs the typical user will probably have developed, through shop-floor control, to a fairly well established total company control system, and will have the experience and skill to make the best of the equipment and methods then available.

8.0 CONCLUSION

This paper is deliberately more a description of the design process than of the resulting system. Underlying this process were a number of decisions in which every effort was made to choose the method allowing the most flexibility, since every user has a slightly different situation.

The system was designed to undertake the requirements planning function of a batch manufacturing company and to provide the basis of a complete, computer-dependent, production control system in due course. This function, at the heart of production control, itself requires a number of detail decisions to be taken regularly about every item. In addition, the user is required continuously to make decisions for controlling the system. Many of these decisions will have far-reaching financial effects. It may have been apparent that the computer itself has hardly been mentioned. We set out to do what we, as production people, knew was needed, not to play with machinery. We set out to bridge the gap between the computer and one of its potentially most important users, the production controller. If this paper encourages previously sceptical production executives to accept that such a system might indeed be practical, even useful, we will be well satisfied.

ACKNOWLEDGEMENTS

The writer wishes to thank The National Cash Register Co. Ltd for permission to publish information about the IMMAC system. He also places on record thanks to his colleagues on this project, and in particular Mr P. R. Thomas, the project leader, and Mr P. R. Hazeldene.

DECISIONS IN TECHNICAL SELLING

D. J. Leech and D. L. Earthrowl
Division of Industrial Systems Engineering, University College of Swansea

SUMMARY

Aircraft accessory manufacturers spend considerable sums of money in preparing tenders for the supply of new equipment to aircraft and engine manufacturers. Tenders are for components which have not been designed in detail and the money is at risk in that (a) tenders may not be accepted and (b) when a tender has been accepted, subsequent design and manufacturing costs exceed the agreed selling price. This paper describes an investigation of the documentation, information flow and decision making in the system by which a number of selected companies submit tenders. A system common in principle to all the companies is described. Investigation of the functioning of this system indicates that it is not operating usefully in controlling expenditure. Decisions to be made are, typically, whether to enter a new market, whether to tender when invited, how much to spend on any tender and whether to modify decisions in the light of overspend. Generally, these decisions are either made or are not made in spite of the control system.

1.0 INTRODUCTION

This paper is concerned with the decisions made in technical selling and has been prepared from information collected during a larger study of the design procedures of high-technological-content organizations. To ensure comparable levels of engineering sophistication the study has been confined to the aircraft component industry. In this field, work is initiated by a customer's invitation to tender, followed by a company proposal which may or may not be successful. The vendor is committed by his proposal to meeting very stringent specification requirements for an agreed sum of money and in an agreed time, despite not having solved all the attendant problems. Because of this commitment, companies have evolved decision-making processes to be applied during selling.

Two companies have been studied in depth and a basically common procedure was found. More superficial information was obtained from a number of other companies to check variations from this common procedure. All companies were generous in giving access to documents. The documents reproduced here have been obtained from three companies but have been modified to some extent in order to preserve anonymity.

Company budgeting, based on the Sales Director's estimates of income, is the initial step in the selling procedure. The estimate of income derives from sales already agreed and the assessment of probable income from current proposals. In the companies studied, annual turnover varied from about £2 000 000 to about £5 000 000 and the amounts allowed to be written off against proposal spend ranged from about £20 000 to about £100 000. In every company, it was estimated that about 10% of all proposals are successful but this figure is probably low.

When the amount to be spent on proposals is decided, it is the Sales Director's responsibility to see that it generates maximum turnover.

In every company, the procedure ensures that an enquiry is examined and the decision to propose is made after consideration of the financial, technical and production feasibility of the problem. An amount of money is allocated to each proposal, governed by the value of the prospective order, the probability of success, the work necessary to complete the proposal, the submission date and the budgeted total proposal spend.

If work is authorised then it is general for a control to be applied to keep the spend within the authorised amount. Information on spend/achievement is supplied weekly and decisions to continue or not are made. Although there appears to be agreement that a control system for spend/achievement should be used, the degree of sophistication of the system in operation varies quite widely. The purpose of this paper is to highlight some of the decisions necessary in a proposal system.

2.0 THE PROPOSAL

2.1 *Proposal requirements and constraints*

A number of customer requirements and corresponding proposals have been examined, but again, for reasons of commercial confidence, we cannot generally quote directly either from requirements or proposals. The significant points are that the customer is asking the vendor to commit himself to:

- A performance.
- A first cost.
- The running costs.
- A delivery programme.

By his proposal the vendor is accepting those commitments. Only a small fraction of proposals result in orders, so money spent in making a proposal is spent in uncertainty or, on occasion, risk. Further, this money must be spent, not only in the hopes of getting an order but also to ensure that predicted performance, costs, and delivery dates are acceptably accurate. Clearly, inaccurate predictions will lead to loss of money by the vendor – the loss resulting from redesign to meet performance, penalties for late delivery, the cost of rectification of faults in service etc. Increased expenditure on making a proposal will result in increased technical accuracy but this must be weighed against the probability of failure to get the order, in which case all expenditure is wasted.

The constraints within which performance must be achieved may be seen in such a general specification as BS.2G100, which commits the vendor to demonstrating, by expensive tests, that the required performance will be met in specific environmental conditions (such as temperature, contaminants, vibration, humidity). Prices quoted must allow for the difficulty of this design problem, for the cost of the testing which demonstrates that the problem has been been solved, and also for the probability that there may be several attempts – and hence several test sequences – to solve and demonstrate the solution of the problem.

2.2 *Preparation of a tender*

The need to establish a profitable selling price is obvious. The ability to do so requires that the Sales Director knows the market and the competition, and also that sufficient design be done before tendering for accurate estimation of manufacturing and development costs. More difficult to predict are the customer's running costs but a typical specification says

The standard of reliability required is that the equipment failure rate shall not exceed 0·1 per 1000 unit hours. The vendor must guarantee that the failure rate shall not be more than 0·4 per 1000 unit hours.

Or

The item shall achieve a mean time between failures (MTBF) of not less than 60 000 hours and the estimated troubles rate for the item shall be less than 10 per 10^5 hours.

Even if costs can be estimated with accuracy, the Sales Director has to decide, where the customer permits, whether design and development costs will be amortized over some probably hypothetical production order or whether they will be claimed separately. Direct payment for

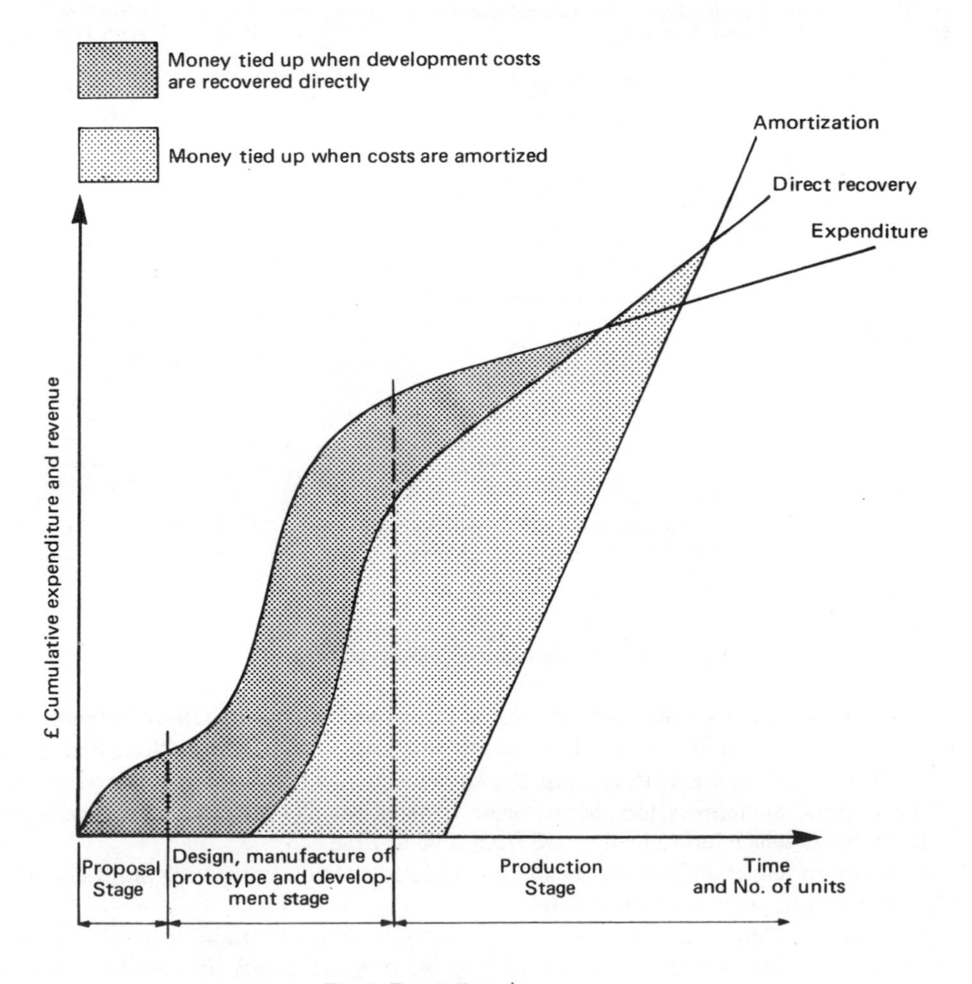

Fig. 1. Expenditure/revenue curves.

non-recurring costs is obviously better for the vendor (Fig. 1) but nowadays the customer frequently expects the vendor to share his risk.

Item	Quantity	Description	Delivery required by
			month year
1.1	1 set	Wooden mock-up fully representative of proposed flight model.	6 – 68
1.2	1 set	Metal installation model with all external features fully representative of flight cleared units for prototype aircraft.	12 – 68
1.3	1 set	As item 1.2 above for production aircraft.	9 – 70
2	1 set	Test rig units.	12 – 68
3.1	6 sets	Units to Flight Cleared Standard for installation in Development Phase Aircraft (Nos. 1–6).	2 sets by 9–69 2 sets by 1–70
3.2	6 sets	As item 3.1 for aircraft (1–12).	2 sets by 1–71 2 sets by 4–71 2 sets by 7–71
4.1	4 sets	Units to a Flight Cleared Standard for use as Development Phase spares (Aircraft 1–7).	2 sets by 6–69 2 sets by 11–70
4.2	2 sets	As items 4.1 (for Aircraft 8–12).	2 sets by 4–71
5.1	20 sets 20–50 sets 50–100 sets 100–200 sets	Units to a Production Flight Cleared Standard. Note: Prices quoted for production units should indicate the minimum economical batch ordering size against each quantity.	Delivery will commence on 1–72 building up to 4 sets/month within 6 months, 8 sets/month within 12 months. Max. rate there-after 10 sets/month.

Fig. 2. Extract from required delivery schedule.

Delivery programmes are important and this is recognized in any invitation to tender. In a proposal, the vendor will be expected to supply and justify such information. Fig. 2 is an abstract from a typical requirement, and Fig. 3 is an example of a delivery programme submitted as part of a proposal. Sometimes, too, the customer will insist that the delivery programme is given as a PERT network which has to be updated from time to time.

An invitation to tender will sometimes give an indication of how competing proposals will be judged, as the following quotation illustrates.

> *If the technical proposal falls significantly short of the specification, it is unlikely that it will be given further serious consideration. If the proposal meets the specification in the opinion of our technical authority, the factors which are then most likely to affect its*

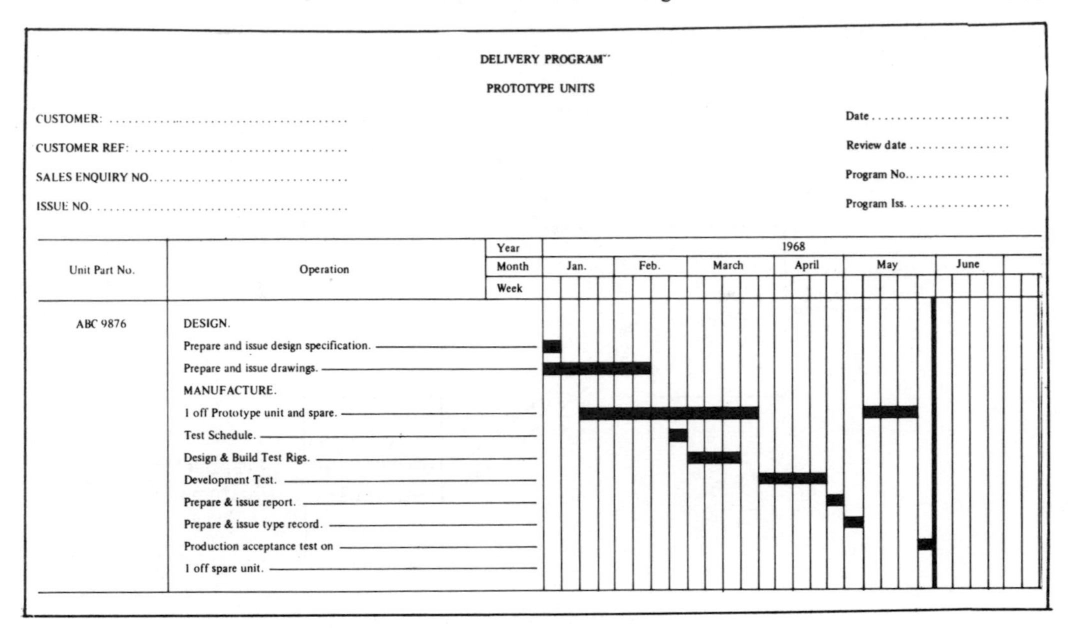

Fig. 3. Typical delivery programme for prototype units.

success in competition with other proposals are price, and the adjudged ability to complete the task within your acknowledged capabilities and on time.

A typical proposal will contain the folllowing information:

- Design scheme.
- Installation details.
- Description of the proposed unit, its construction and operation.
- Performance, stress, weight and life analysis to demonstrate compliance with every clause of the specification.
- Failure analysis.
- Reliability analysis.
- Servicing description.
- Design, development and manufacturing program (PERT or GANTT).
- Cost details – development costs, model and prototype costs, production costs, repair costs.
- Capability of vendor, i.e. demonstration that vendor's resources will permit programme to be met.

3.0 NETWORK OF ACTIVITIES TO PRODUCE A PROPOSAL.

3.1 *Preparation of proposals*

The general system for the preparation of proposals is shown in Fig. 4. Technical Sales obtains details of a customer requirement, the Technical Department produce a design scheme and the Commercial or Contracts Department complete the proposal by adding the selling price. The flow of information between departments is illustrated in Fig. 5, and Fig. 6 shows the procedure chart for proposal preparation. Table 1 lists types of documents and should be used with both figures.

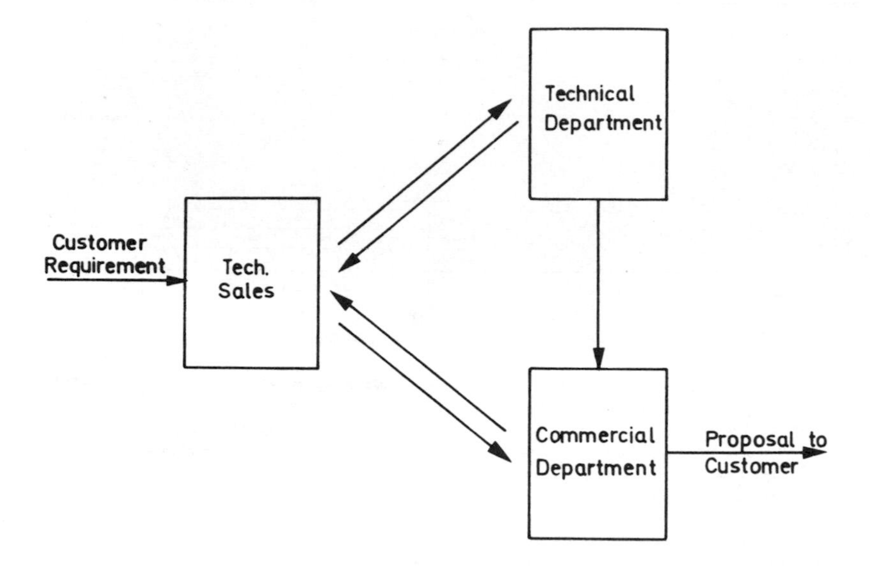

Fig. 4. Basic system for the preparation of a proposal.

Information about the requirement is first given to the Sales Director, who decides whether the requirement is worth investigation. If, on examination of the Initial Sales Information (1)*, he decides that he wants to tender for the job, he will:

(a) raise a sales enquiry form (2) against which time may be booked;

(b) circulate to the appropriate departments the sales enquiry, issue 2, which carries a request for an estimate of the time needed to complete the proposal.

Clearly, some work is required, even to make this estimate, since a possible method of meeting the customer's requirement must first be postulated. Typically, it is permissible to spend up to 100 hours against the enquiry number while preparing the estimate.

The sales enquiry form and specification (3), which may be supplied either by the customer or compiled by the Technical Salesman, are passed to a group of engineers concerned with the formulation of an answer to the problem.

3.2 *Design structures*

Generally, there seem to be two distinct design structures in operation. The first uses a group of designers, development engineers and stressmen under a product or engineering manager, who produce a proposal and continue to be responsible for the detail design if and when an order is received.

The other structure has a project engineer responsible for the design scheme stage only. He directs a team of designers and co-ordinates the effort of this team in the production of a proposal. On receipt of an order, a designer is given the responsibility of the detail design and may work with the same personnel that produced the proposal or with a completely new team. The 'product group' in both types of organization make considerable use of many specialist departments, as illustrated in Fig. 7. Some of the department's work can be directly costed to the proposal while others are costed within a general overhead.

The 'product group' discusses, and formulates a solution to the problem. The estimate of

* Numbers in parentheses refer to the documents listed in Table 1. A selection of the more important documents is given in the Appendix.

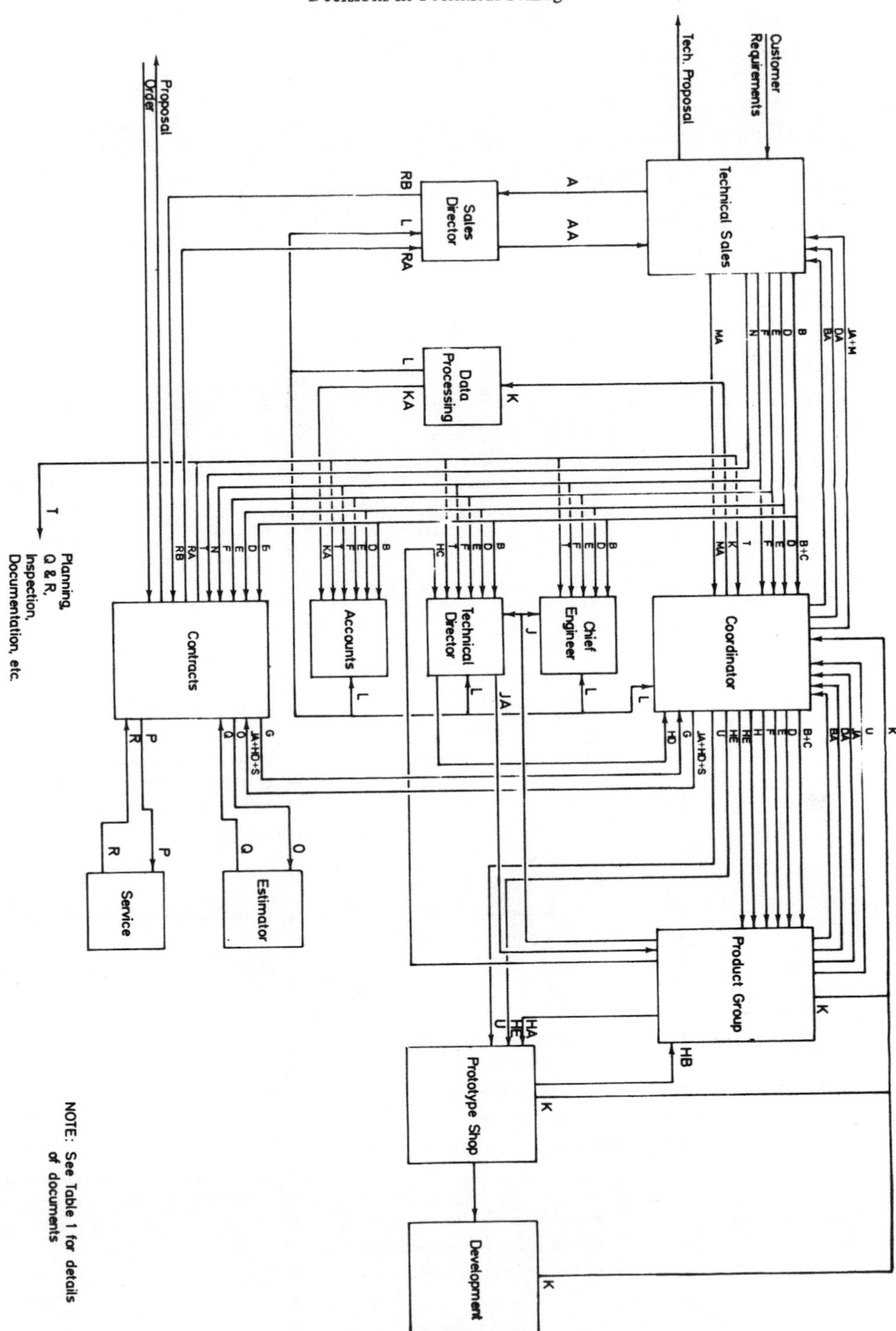

Fig. 5. Information flow diagram.

TABLE 1.

Document No.	Code	Document	Remarks
1	A	Initial Sales Information.	The basis on which decisions to propose are first made.
	AA	Approval for consideration.	Authorisation for the use of a specified time to only consider the requirement.
2	B	Sales enquiry form. Iss. 1.	Details of authorised time for consideration and estimate form for time to complete proposal.
	BA	Sales enquiry form. Iss. 1.	Estimate for time to complete the proposal.
3	C	Initial Specification	Supplied by the customer or compiled by T.S. T.S. responsibility until order received.
4	D	Sales enquiry form. Iss. 2.	Authorisation for work to commence on proposal to estimated time on BA.
	DA	Sales enquiry form. Iss. 2.	Request for extended time to complete scheme.
5	E	Sales enquiry form. Iss. 3.	Authorizing extended time.
6	F	Closure of sales enquiry.	
7	G	Request for technical estimate.	
8	H	Technical Estimate.	Estimate for design, manufacture, development and type approval testing materials and rigs.
	HA	Technical estimate completed by Engineering Manager of Product Group.	
	HB	Technical estimate completed by Prototype Shop Manager.	
	HC	Completed technical estimate.	
	HD	Approved technical estimate.	Contingency may be added by TD.
	HE	Copies of HC.	
9	J	Design scheme for approval.	Part sectioned GA with notes on construction and manufacture Installation drawings.
	JA	Approved design scheme.	Signed by C.E. and/or T.D.
10	K	Time sheets and work completed details.	
	KA	Cost details.	
11	L	Weekly print out cost/achievement.	
12	M	Technical clearance form.	Details drawing No. and issue.
	MA	Technical clearance form complete.	
13	N	Brochure.	May be technical content only sent to technical contact.
14	O	Request for Production Estimate.	
15	P	Request for Publication and Service estimate.	
16	Q	Production Estimate.	
17	R	Publication & Service Estimate.	
	RA	All estimates.	
	RB	Price details.	
18	S	Program	Generally a GANTT chart of design manufacture development of prototype.
19		Proposal (commercial & technical).	
	T	Notification of order and works number.	
	U	Details drawings, schedules of parts, test schedules, etc.	

*** The Appendix contains several typical documents.**

time needed to complete the design scheme is completed and returned to the Technical Salesman. If the time to complete the proposal is such that the submission date is hard to meet, a decision is made whether to proceed with or reject the invitation to tender. The sales enquiry is closed (6) if it is decided not to proceed and the cost to that time is written off. If the estimate allows the submission date to be met, authorisation for work to begin to estimate is given by raising the issue of the sales enquiry (4). Very often the estimate is too low and an extension of time is sought. Again, a decision is taken and the sales enquiry issue raised (5).

3.3 *Commencement of work on the scheme*

The authorisation for work to begin on the design scheme triggers the Contracts Department who send a Request for a Technical Estimate (7) to the 'product group'. This estimate is the basic document for apportioning money to the Technical Department when and if an order is received.

In conjunction with the 'product group', and while they are producing a design scheme, a brochure (13) is being prepared by the Technical Salesman or Project Engineer. The Technical Estimate (8) is completed by the Product or Engineering Manager who then passes it on to the Prototype Shop Manager so that the prototype manufacture estimate can be completed.

The Technical Director approves the completed estimate and may add a contingency allowance, which is really a measure of confidence in the Product Manager.

Design Scheme drawings (9) are supplied to the Coordinator, who sends one copy to Technical Sales. Sometimes a Technical Brochure, together with a Technical Clearance Certificate (12), is supplied to the customer's engineering division (with no price or delivery information) for technical acceptance. If this is given, then the Technical Clearance Certificate (12) is completed with the drawing number and issue. This gives an agreed technical basis on which subsequent commercial decisions may be based. The modern trend, particularly where inter-

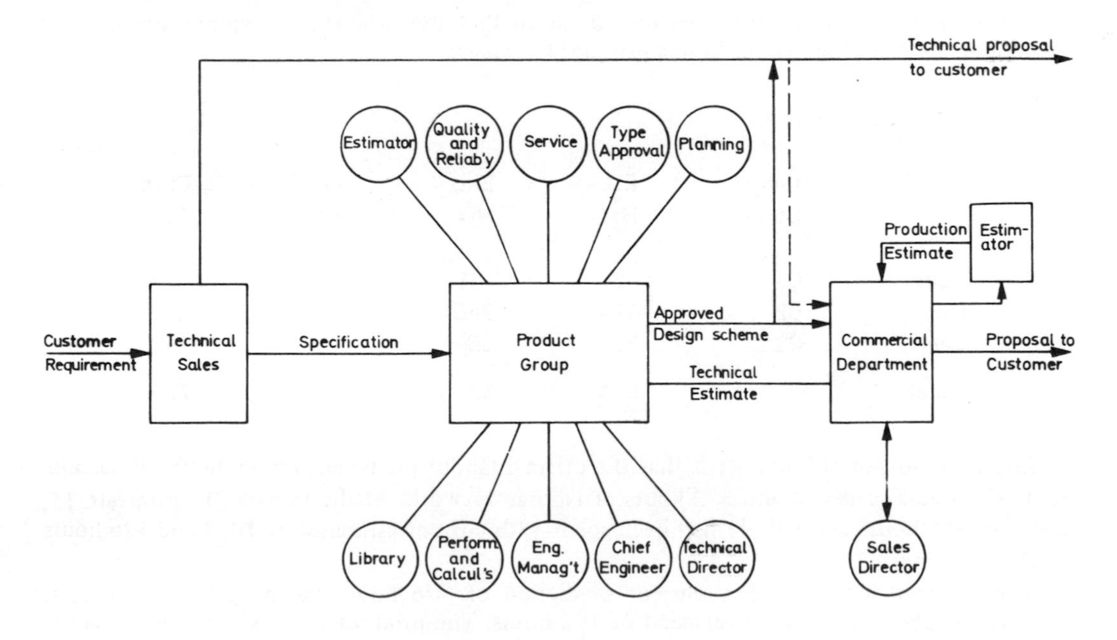

Fig. 7. Expanded design system.

national liaison is involved, is for the whole proposal, technical and commercial, to be submitted at one time.

Meanwhile, the Contracts Department is supplied with the approved Technical Estimate, Design Scheme and a Delivery Programme (18), produced by the Coordinator, so that a Production Cost Estimate (16) can be made. Publication and Servicing Cost Estimates (17) are also supplied to Contracts.

All estimates are put together and then shown to the Sales Director, who decides what price is to be quoted, and a Complete Proposal (19) is despatched to the customer. The time spent on proposal preparation is recorded weekly on Time Sheets (10) so that a cost/achievement record can be computed by Data Processing and printed on a weekly printout (11).

4.0 MONITORING EXPENDITURE

The monitoring of expenditure is based on information obtained from Time Sheets (10). Each week, each directly accountable engineer uses a Time Sheet to say what job he has been working on, how many hours he worked on each job during the past week, and *how many more hours* he expects to work before his part of the job is finished. It is necessary for the forecasting aspect of this task to be vetted by a section leader to ensure coordination of the forecasts of all members of the same labour code.

In theory, it is simple to sum hours spent by any labour group against any job, to sum the hours yet to be spent, and to compare the current prediction of total labour spend (hours already spent plus hours yet to be spent) with the originally predicted spend. Normally this is done by computer, and while only simple arithmetic and logic is involved, the number of jobs current (usually several hundred), the number of sub-divisions of expense (different labour codes, different types of monetary expenditure etc.), the number of men involved and the need for maintaining and up-dating from records mean that the programme requires considerable development. A typical extract from a printout (11) reads:

JOB NO. 4321.

	Lab Code	Est Hrs	Bkd Hrs	Plus Hrs	Curr Total
Calcs	01	38	15	10	25
Des	02	174	266	3	269
Lab	03	372	356	126	482
Total		584	637	139	776

Interpretation of this extract is that the original labour predicted was 38 hours of calculation, 174 hours of design and 372 hours of laboratory work. At the time of the printout, 15, 266 and 356 hours respectively had been spent with current estimates of 10, 3 and 126 hours work remaining.

The printout thus shows a current prediction of 776 hours spend against an original estimate of 584, a predicted overspend of 192 hours. The printout from which this extract is made also converts hours into cost, shows expenditure on materials and shows overspend or underspend as a variation from the previous accounting period. When a programme has been developed, time-sheet information given on a Friday can be analysed and be available to the managers on the Monday.

All firms admit to having either developed or at least considered a cost control system similar to the above. One company uses it fully but others believe that it is not justified at the proposal stage since many apparently required decisions are not made and work proceeds in spite of any attempt to control it. Further, the latest date for proposing is itself a control.

5.0 DISCUSSION

5.1 *The Sales Director's first decision*

The first decision that the Sales Director thinks he makes is how much to write off against the Technical Department's efforts to make sales. In fact, there is very little evidence that this decision is taken seriously by any company. The basis on which this amount is decided was given variously as 'What we can afford' and 'The average proposal spend in the last year multiplied by expected number of proposals'. The usefulness of budgeted write-off is very dubious. In a declining market, a company will permit itself to spend more than it intended, to ensure that it gets an adequate share of the market and if any spend results in future sales, it can be afforded. There is conflict between long- and short-term plans since the return for this year's spend may be many years off and overspend may be unjustifiably regarded as a loss. Whatever the reasoning behind budgeting sales write-off, where records were available they showed considerable overspend, in one case as much as 100%.

5.2 *The Sales Director's second decision*

The second decision that the Sales Director makes is whether, when informed of a possible sales opportunity, he will make a proposal to the potential customer. Occasionally, the Technical Salesman acts on his own initiative and does not pass on an enquiry, but generally there is no question of not proposing in a field in which the vendor is interested.

The areas of interest to the companies investigated were fairly clear and for all but a few jobs there was no question but that a company would compete for an order. Only in those few cases which might be considered to take the firm into a new line of business would the Sales Director have seriously to weigh up whether to propose or not.

Most companies have attempted to diversify their product range, but there has not been much success in diversifying outside the aircraft field and this has led to some timidity. It is often thought that 'over design' is the main stumbling block in the general engineering field but there is no clear justification for this opinion. An equally plausible view is that the customer does not appreciate his own problems.

It was generally thought that about a tenth of all proposals resulted in orders but no-one could back this guess with figures from records. It seems that in some fields, where there is little competition, a company will judge its chance of success as being of the order of one in three, but where the company is trying to enter a new market it accepts a one in a hundred chance of success. In fact, for an aircraft accessory company to enter a new field usually involves a policy decision rather than an ad hoc decision by the Sales Director.

The nature of customer contact varied considerably from one company to another. At one extreme, technical staff were screened from all contact with the customer by a sales staff who regarded themselves as technically competent, while at the other extreme the salesman served only to act as the initial contact and to introduce the technical staff to the customer. Whatever

Fig. 6. Procedure chart for proposal preparation.

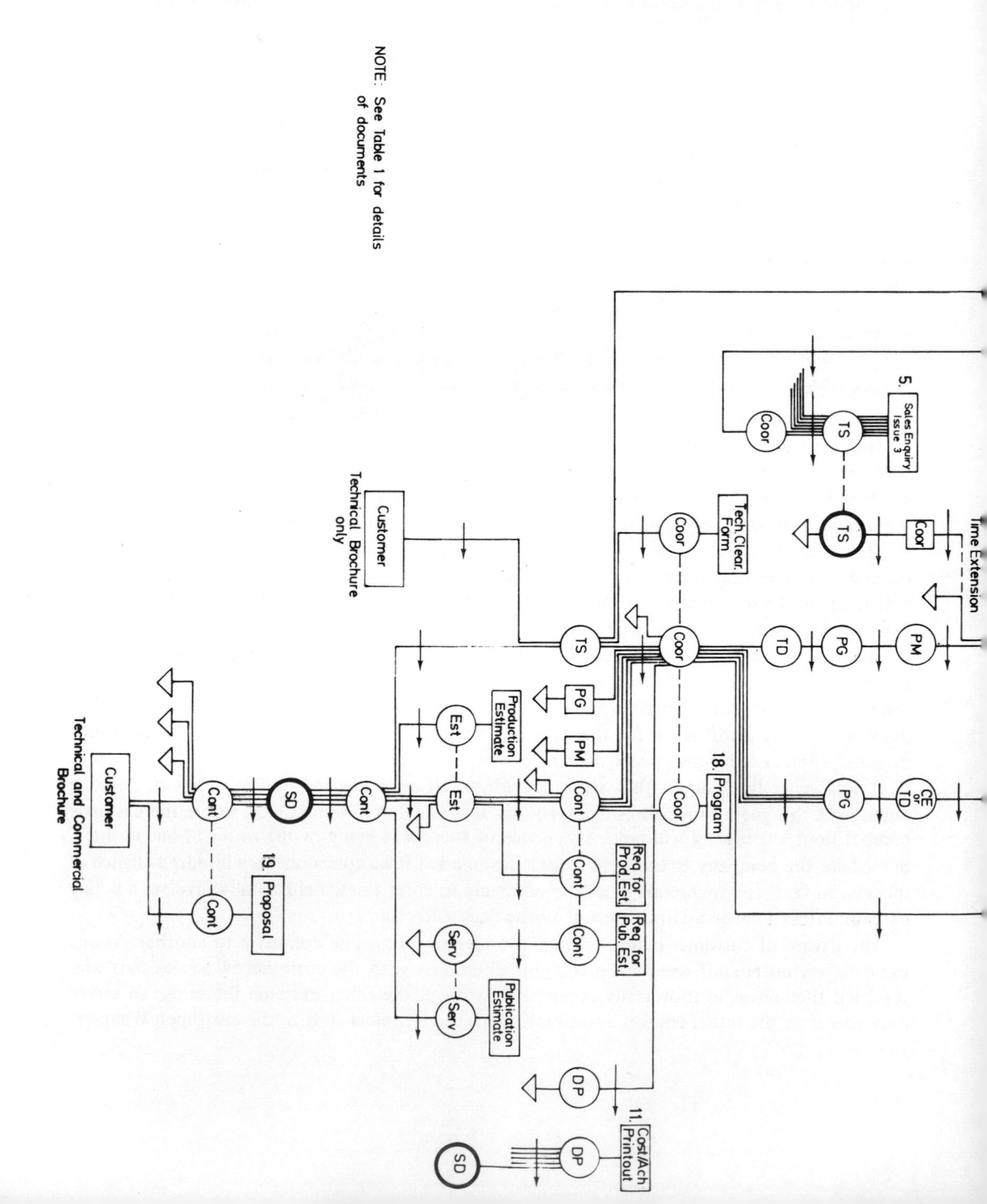

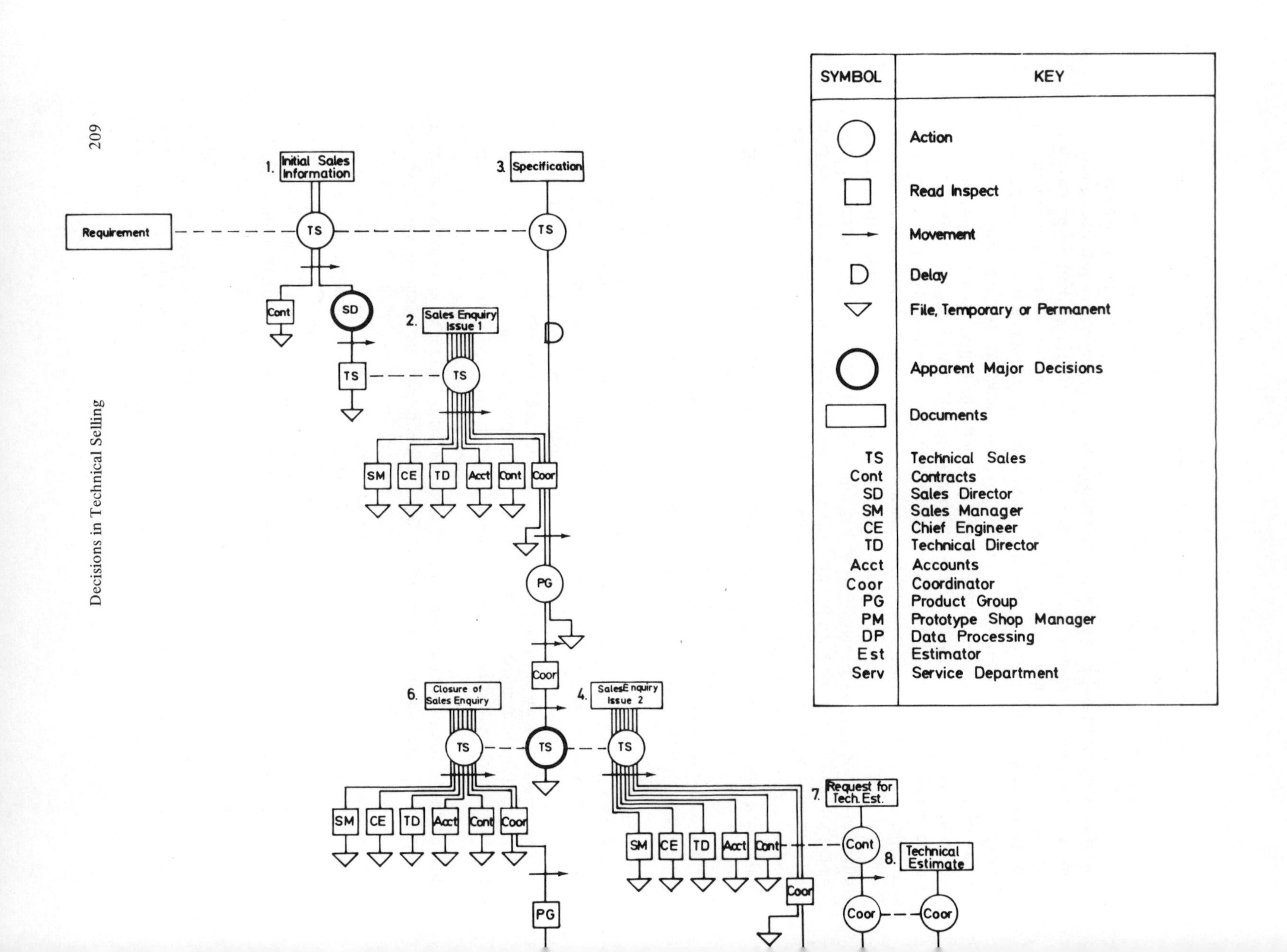
Requirement
1. Initial Sales Information
TS
Cont
SD
TS
2. Sales Enquiry Issue 1
TS
SM
CE
TD
Acct
Cont
Coor
3. Specification
TS
PG
Coor
6. Closure of Sales Enquiry
TS
SM
CE
TD
Acct
Cont
Coor
PG
TS
4. SalesEnquiry Issue 2
TS
SM
CE
TD
Acct
Cont
Coor
7. Request for Tech. Est.
Cont
Coor
8. Technical Estimate
Coor
SYMBOL
KEY
Action
Read Inspect
Movement
Delay
File, Temporary or Permanent
Apparent Major Decisions
Documents
TS Technical Sales
Cont Contracts
SD Sales Director
SM Sales Manager
CE Chief Engineer
TD Technical Director
Acct Accounts
Coor Coordinator
PG Product Group
PM Prototype Shop Manager
DP Data Processing
Est Estimator
Serv Service Department

view the company took of representation however, there was universal reluctance to allow engineers to be involved in questions of cost. This separation at high level of technical and commercial problems may contribute towards the apparent lack of cost conciousness of some designers.

5.3 *The Sales Director's third decision*

The third apparent decision for the Sales Director is whether to make a proposal when he knows how much it will cost to do so. The control system was designed to allow this decision to be taken and it is worth asking what the object is. Presumably, the Sales Director wishes to make the best allocation possible of the money written off but there is no direct measure of goodness. In its own line of business the company is almost certain to make a proposal and, naturally, if resources are limited, priority will be given to those proposals likely to result in the greatest profit. In almost all cases, however, once jobs are ranked, the company will spend as much as available labour permits to get the most lucrative jobs and spend will be limited only by the staff available and the date of submission.

If the Sales Director is asked whether the work is to continue when the money authorised has been spent without yet completing the proposal, the original estimate of the proposal time needed is incorrect. The cause of overspend should be discovered and information fed back to the estimator so that his accuracy of prediction can improve.

One reason for overspend on proposal work is the incorrect booking of the time spent on various jobs. Considerable memory, skill and manipulation are used to account for exactly 37½ hours, especially when the time sheet is completed at the end of a week. Often, the slow closure of sales enquiries enables lost time to be accredited to other jobs.

Money already spent cannot be returned and unless the proposed overspend is much greater than the originally authorised spend, there can be no decision but to persist with the proposal.

5.4 *Deciding on the quoted price*

Determining the price to be quoted in the proposal is clearly a serious decision. Part of the work on the proposal is the estimation of design, development and manufacturing costs and, knowing these, the competition and what the customer is prepared to pay, the Sales Director might be expected to determine the most profitable selling price (and also the most profitable price breakdown). The most serious difficulty is the estimation of design, development and manufacturing costs of an, as yet, unspecified component.

It is common knowledge that new aeroplanes frequently cost many times what was predicted but all the companies investigated show pride in their accuracy of estimating the design, development and prototype costs. Universally it was thought that an accuracy to within 10% was being achieved. A limited number of records examined showed that, although actual cost varied from half that predicted to a 50% overspend, there were quite a number of cases where actual cost was within 10% of prediction. The sample of records was only representative to a limited extent, however, since we are aware that in every company a few expensive rogue jobs were many times overspent. It appears, then, that a price based on an estimate of non-recurring costs will be not far out on most occasions but will be very expensively wrong on a few occasions (see Fig. 8).

Even where records showed the estimate to be approximately correct, it was noticeable that the breakdown of any job showed great inaccuracies which, often by chance, tended to cancel out. The possibility of failure is, of course, much enhanced when the production order may be too small to cushion losses at the development stage.

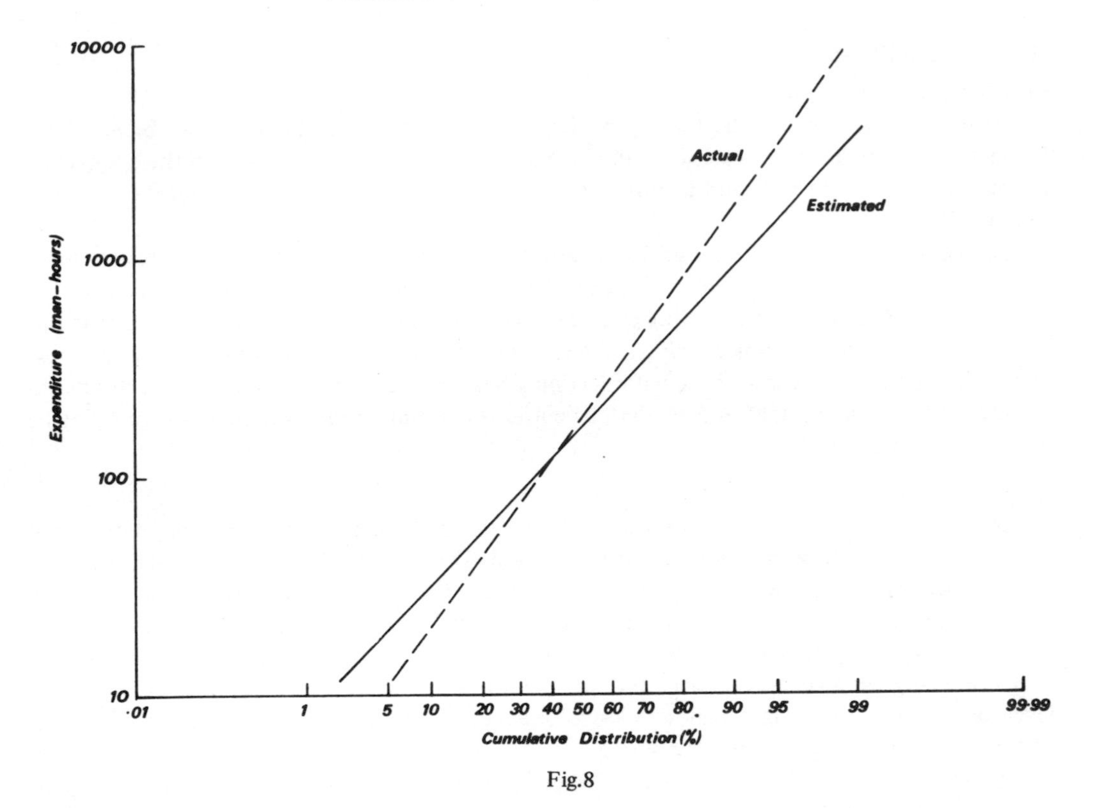

Fig.8

5.5 *The computer–aided control system*

The computer-aided control system is of questionable value in controlling expenditure on proposals. Urgency is such that work will go ahead in advance of approval or estimated proposal cost so that, week after week, the computer prints out only total spend and we see zero entries in predicted cost columns. Even if the control system worked efficiently, however, we still have the problem that, once work is initiated, it is hardly possible for the Sales Director to call a halt.

The control system is so complex that its usefulness has been questioned by more than one company. At least one company uses the whole system while another has rejected any attempt to estimate the work remaining on any job. Other companies have special staff to work on proposals so that spend is limited by the number of man-hours available.

One aspect which seemed related to the size of turnover was the manner in which the salesman saw his work. In the firms with smaller turnovers, there seemed to be a fatalistic acceptance of the small market, whereas in the other companies there was conscious effort to push up turnover.

6.0 CONCLUSIONS

6.1 *Necessity for decisions*

Decisions clearly required by the companies' procedures and documents are not being taken. In some cases, no decision is really required (as in whether to tender for work in the company's accepted range of activities) and in others no real decision is possible (as in whether to permit overspend).

The Sales Director might reasonably be expected to decide whether or not to tender for a job in a new field but, in fact, such a decision should not be made on a job-by-job basis.

The permissible write-off in selling is always exceeded and there is no real evidence that this write-off has been derived objectively. There is some evidence that those companies which most restrict this write-off are also those with declining business. All the companies were subsidiaries of much larger organizations and so the difficulties of raising money for selling would not be great if it could be justified.

6.2 *Assessment of computer control systems*

Computerized control systems produce information with considerable efficiency but do not work. Input is late or wrong and output is not used. This is partly because the control systems were devised when there was a larger market and when the luxury of refraining from tendering was possible. Also, however, the declining market has led to cuts in technical staff and in some cases the effort of operating controls is a large part of the technical effort available.

Records suggest that a very large rate of spend occurs during design. Faults shown during development may necessitate major redesign and hence a large increase in cost, so that if an estimate is wrong it is probably very wrong.

One has to ask whether, in a small market, any control system can be useful in view of the probability of underestimating the costs of design and development. Records do not confirm the belief that estimating is reasonably accurate.

APPENDIX 1

TYPICAL DOCUMENTS – Document numbers refer to Table 1.

INITIAL SALES INFORMATION

CIRCULATION: NO.

ORIGIN: ISS.

CUSTOMER: DATE
Contact and position.

APPLICATION:
System
Type of unit
Qty per system
Related documents

PROPOSAL CLOSING DATE:

PROPOSAL TO INCLUDE:

WORK REQUIRED:
Technical proposal by:–
Technical estimate by:–
Production estimate by:–
Program.................. by:
Quotation.............. by:

TARGET COSTS:

SALES POTENTIAL:
Estimated sales
Probability factor
Known competition

DESIGN & DEVELOPMENT COST RECOVERY:
PRODUCTION TOOLING COST RECOVERY:
ADDITIONAL FACILITIES REQUIRED:
ADDITIONAL COMMENTS:

Document 1. Initial sales information

FROM: Technical Sales

ENQUIRY NO.

ISSUE NO.

TO:

DATE:

SALES ENQUIRY

Please carry out a preliminary examination of the specification attached and assess the number of hours necessary to put forward a scheme to the customer by:

CUSTOMER:

APPLICATION:

SYSTEM:

TYPE OF UNIT:

QUANTITY PER APPLICATION:

PRELIMINARY SPECIFICATION:

HOURS AUTHORISED TO TECHNICAL DEPARTMENT	01 02	Calcs. Design.	
HOURS BOOKED TO ISSUE			
FURTHER HOURS REQUIRED BY TECHNICAL DEPT.	01	Calcs.	
(State what is required, e.g. proposal, scheme etc.)	02	Design	
	03	Laboratory	
		Total	

Document 2. Sales enquiry issue 1.

FROM: Technical Sales

TO:

ENQUIRY NO.

ISSUE NO.

DATE

SALES ENQUIRY

This enquiry covers the necessary technical work required to provide the Sales Department with:–

(State what is required, e.g. proposal scheme, technical estimate, etc.)

01	CALCULATIONS	
02	DESIGN	
03	LABORATORY	
HOURS AUTHORISED : Total		

CUSTOMER:

APPLICATION:

SYSTEM:

TYPE OF UNIT:

QUANTITY PER APPLICATION:

PROVISIONAL SPECIFICATION NO.

PROPOSAL DATE TO CUSTOMER:

EXTRA TIME REQUIRED:

01	CALCULATIONS	
02	DESIGN	
03	LABORATORY	

REASON FOR EXTRA TIME:

NOTE: If it becomes apparent that allowed time is insufficient, this form must be returned with extra time columns completed. No hours beyond those authorized may be booked until a further issue of this form has been raised.

Document 4. Sales enquiry issue 2.

JOB NO. .... DATE. ....

TECHNICAL ESTIMATE

ENQUIRY NO. .... ISSUE ....

COMPONENT/SYSTEM ....

CUSTOMER ....

AIRCRAFT/ENGINE ....

DRG. NO. .... ISSUE ....

SPEC. NO. .... ISSUE ....

Labour Code	Work	Man Hours	Remarks
01 (Calculations)	Performance Calcs. Liaison Stress		
	Sub Total		
02 (Drawing Office)	Scheme Details & Assemblies Schedules Liaison Rectification		
	Sub Total		
03 (Laboratory)	Development Environ. Test (Summary) Rig Design Rig Build Reports		
	Sub Total		
04	Records D.D.P. Type rec.		
	Labour Total		

Cost Code	Material	Cost (£)	Remarks
13	Rig Materials Laboratory Consumables		
	Sub Total	£	
15	Units for Development Units for Environ. Test		No. off No. off
	Sub Total		
	Total Cost		

Document 8.1. Technical estimate Page 1.

ENVIRONMENTAL TEST ESTIMATE

JOB NO. Date

ENQUIRY NO. Issue No.

COMPONENT/SYSTEM

AIRCRAFT/ENGINE

DRAWING NO. Issue No.

SPEC. NO. Issue No.

Work	Man Hours	Restriction on Clearance		Remarks
Preliminary Testing				
Type Testing		Spec. e.g. BS. 2G100	Para.	
1. Temperature 1.1 Performance 1.2 Endurance 1.3 Derangement 2. Pressure 3. Climatic 3.1 Combined pressure-temp-humidity 3.2 Tropical exposure and storage 4. Icing 4.1 Formation 4.2 Resistance 5. Contamination 5.1 Sand & Dust 5.2 Salt spray 6. Waterproofness 7. Vibration 7.1 Resonance search 7.2 Endurance 8. Acceleration 8.1 Normal 8.2 Crash 9. Explosion proofness 10. Fire resistance 11. Flame-proofness 12. Radio interference 13. Magnetic Interference 14. Transient volts				

Document 8.2. Technical estimate: environmental test estimate.

JOB NO. .... DATE ....

PROTOTYPE MANUFACTURE ESTIMATE

ENQUIRY NO. .... ISSUE NO. ....

COMPONENT/SYSTEM ....

CUSTOMER ....

AIRCRAFT/ENGINE ....

DRAWING NO .... ISSUE NO. ....

SPEC. NO. .... ISSUE NO. ....

NO. OFF ....

LABOUR CODE	WORK	MAN HOURS		REMARKS
		PER UNIT	TOTAL	
105	Machining			
205	Assembly			
305	Tooling			
405	Modifications			
05	Total Labour			
COST CODE	Material	COST		
		PER UNIT	TOTAL	
115	Raw Materials			
215	Bought out parts			
15	Total Cost			

Document 8.3. Technical estimate: prototype manufacture estimate.

NAME:	CLOCK NO.	LABOUR CODE.	WEEK ENDING:

	DIRECT				INDIRECT		DAILY TOTAL
JOB NO.							
MONDAY							
TUESDAY							
WEDNESDAY							
THURSDAY							
FRIDAY							
SATURDAY							
SUNDAY							
JOB TOTAL							
HOURS TO COMPLETE							

Document 10. Time sheet.

SESSION VII

Manufacturing Systems and Automation

Chairman: Mr W. Gregson, C.B.E.
Assistant General Manager,
Ferranti Ltd,
Edinburgh

MANUFACTURING INFORMATION SYSTEMS AND THE EXCEPTION PRINCIPLE

N. J. Radell
Cresap, McCormick and Paget Inc.,
Chicago

SUMMARY

This paper discusses the process and approach to instituting an effective management information system as it pertains to establishing meaningful management control over manufacturing operations. Special emphasis is given to the approach needed to implement the exception principle of management and how it can be woven into schemes of management reporting. A case study, complete with a sample management report, illustrates the techniques and concepts described in the paper.

1.0 INTRODUCTION

Since the end of World War II, there have been significant changes in the management processes of United States companies. These changes have particularly affected manufacturing management and the information systems used to make management decisions. Generally, it appears that the changes have favourably influenced the efficiency and productivity of the manufacturing functions. The basic purpose of this paper is to describe several of the changes having to do with management reporting and decision-making, and to illustrate the key elements involved.

Before discussing the subject matter directly, it would be useful to understand some of the management pressures that have affected United States industries during the past 25 years. First, there has been unparalleled growth, diversification and competition among United States companies. New products have been introduced at an increasingly rapid rate. As a result, many new industries have been developed.

Second, as you might imagine, the demands upon managers have grown at an exponential rate. The situation has been aggravated by a shortage of experienced, capable managers because of the effects of World War II, and by the availability of more management positions due to the substantial growth rate of industry. This shortage still exists in the age bracket of 45 to 55 year-old managers.

To some extent, the management crisis has been eased by the introduction of high-speed computers, management science and operations research techniques. In recent years, management information technology, which combines elements of these techniques, has contributed to lessening the burden on management.

One of the key benefits derived from management science techniques has been the gradual development of a 'systems approach' to understanding the management processes needed to manage a business effectively. More and more careful studies of input, output, feedback and control information requirements have been undertaken in United States companies. These studies have resulted in increasingly logical and objective descriptions of business operations in terms of engineering principles. The studies have been particularly helpful in the manufacturing area of business which typically presents the most complicated and complex management problems.

Although significant advances have been made in applying the concepts of scientific management, the fact remains that most United States managers are still overburdened. This situation is even more critical because it results in younger men being forced into receiving significant management responsibility before they are fully ready to understand and cope with the associated problems of management.

To help combat the problems of inexperience, emphasis has been given to developing management information technology and corresponding management reports that stress the use of the exception principle in the management process. Emphasis has also been given to identifying the key factors and indicators that signal the direction in which a business is progressing. The end result has been an increasing effort to help the manager direct his attention to those subjects needing his attention, and, also to help him amplify himself through his subordinates.

During the last five years, there has been a gradual emergence of more sophisticated management reporting techniques combined with the exception principle to provide the information needed to implement the systems approach to management. Management by exception and the corresponding management reporting will be the key subject of this paper.

2.0 THE EXCEPTION PRINCIPLE

The current notion of 'management by exception' originated in the early 1900s when the proponents of scientific management (Taylor, Gantt, Gilbreth) were formulating, discussing and implementing its concepts. At that time the 'exception principle' was embodied in the scientific management approach. The primary concern of the scientific management advocates was to make managers more effective and prevent them from overmanaging. There is a possibility that ways were also being sought to develop the management team approach in order to effectively handle the anticipated expansion of industry.

Statements attributed to two of the early proponents of scientific management and the exception principle are paraphrased as follows:

Frederick W. Taylor

Under the 'exception principle,' a manager should receive only condensed, summarized, and invariably comparative reports which leave him free to consider the broader lines of policy and to study the character and fitness of the important men under him.

Frank B. Gilbreth

The personal work of the executive should consist as much as possible of making decisions and as little as possible of making motions. This can be done in varying degrees according to the kind of work done by the executive and how well he realizes the possibilities of eliminating waste through the use of the 'exception principle' in management.

Each advocate contributed to the development of the exception principle in his own way and for his own reasons. Taylor continually asserted that the extension of the principle to every function of a company was vital if scientific management was to become a reality. Gantt laboured continually to make the exception principle effective through developing techniques of visualizing results and conditions. Gilbreth envisioned its extension to minor executives, declaring that it made good executives of mediocre ones, and gave good ones more time for constructive planning.

3.0 MANAGEMENT BY EXCEPTION

The early exception principle has steadily evolved until it now carries the title of 'management by exception' and can be defined in its simplest form as a system of identifying problem areas and signalling a manager when his attention is needed and, conversely, remaining silent when his attention is not required.

The primary purpose of such a system is to simplify the management process itself by permitting a manager to identify the problems needing his attention and those that are better handled by subordinates.

Developing and using a system of management by exception implies several assumptions that are important to understand.

- A management process requiring a high degree of delegation of responsibility and authority is an acceptable form of management for a company.
- Seventy per cent or more of the key indicators that reflect the condition of the company follow a predictable path.

Furthermore, a successful use of the management-by-exception principle requires that three things be accomplished from a systems standpoint.

- First, a basis for planning and controlling the company's activities can be determined over a one- to five-year period.
- Second, a set of performance indicators can be developed which reflects the conditions of the company.
- Third, a method of visualizing the status of the performance indicators can be developed and periodically presented to management.

4.0 MANAGEMENT INFORMATION SYSTEM DEVELOPMENT

Most of the early management information systems consisted of an amalgamation of data collection and data processing procedures already in use and emanating directly from basic operational systems. These basic systems included cost accounting, budgetary control, sales order entry, purchasing, production control, inventory control, and the like.

Unfortunately, in many cases these basic information systems produced data of minimal value to top and middle management. Overall perspective and management direction were difficult to achieve from the many detailed reports provided by these various specialized systems. They could even be detrimental to sound operations, since they encouraged top management's 'second guessing' of operating-level decisions. Whatever the case, the result was an increased burden on key managers because of the proliferation of detailed and unnecessary information.

A successful management information system must meet the communication requirements of top and middle management both as individuals and as members of a general management team. To do this, the system must possess these essential characteristics, which were often missing in early attempts to develop a system:

- It must emphasize visibility over events that may significantly affect future financial and operating results.
- It must employ reporting techniques that enable key management to review the basic company strategies and plans underlying anticipated financial and operating results.
- It must provide a means for formal communication among members of management regarding the way in which each member interprets variances from plans.
- It must provide a status report on key indicators affecting the business.

Recent experience has shown that the development and implementation of a complete management information system takes several years. However, to ensure that the most important needs are met in a minimum amount of time, the initial focus should be on top management reporting. Development of the system for middle and other levels of management should follow. In essence, top management requirements form the guidelines for lower-level management information requirements.

The following case study shows the results of a management reporting concept that combines the exception principle and the sound principles of management information technology. To the extent possible, the impact upon the top manufacturing executive will be used to illustrate the points.

5.0 CASE STUDY

The subject organization of this case study is a division of a large multi-product company. The division has an annual sales volume of approximately $100M. Its primary product is a sophisticated hydromechanical transmission for vehicles of various types for commercial and government markets. The division has a unique position of being the world's sole supplier of this type of equipment.

The development of the management information system for the division started with the top management group, i.e. Vice President and General Manager, Vice President – Manufacturing, Vice President – Engineering, and Vice President – Finance. A brief description follows of each of the six phases of the development activity performed by the project team.

5.1 *Fact-finding and documentation*

By fact-finding and documentation the project team gained a thorough understanding of the existing management information and reporting system. Management reports currently used by top and middle management were collected, reviewed and analysed. These reports were catalogued showing their title, frequency of issue, contents and recipients. A general flow diagram was developed to depict the sequence of reporting events through the organization. Through interviews with management personnel, the key factors that had to be measured and reported upon were identified.

5.2 *Articulation of division objectives*

A statement of division objectives allowed division management to re-examine the principal objectives of the organization to establish a proper base for determining the key management indicators and the design of the related management reports. The overall division objectives were identified and formally stated as shown in Table 1.

TABLE 1. DIVISION OBJECTIVES

Nature of the business	To be in a business which capitalizes on existing strengths in marketing, engineering and manufacturing. This implies that the products must be: • Automotive- and vehicle-oriented • High technology • Hydromechanical or turbomachinery
Relationship within the corporation	To be responsible for the research and development of new products related to present product lines as well as unique and unrelated products.
Market position	To maintain pre-eminence in the transmission market. To penetrate other automotive and vehicle markets.
Customer satisfaction	To optimize customer satisfaction with the division's products with respect to: • Quality • Reliability • Delivery
Earnings	Within the above, to optimize earnings on a year-to-year basis. To achieve stated dollar earnings.
Sales	To develop and prepare a sales forecast that is accurate. To meet or exceed the sales forecast.
Return on investment	Not an ongoing measure; used to evaluate capital expenditures only.

Several unusual, if not unique, objectives resulted from this effort. The division had responsibility for research and development work for the entire corporation. In addition, annual earnings needed to be optimized from a long-term viewpoint because of the sole source supplier position the division enjoyed. Finally, high return on investment was not a key objective.

5.3 *Analysis of objectives and decision-making responsibilities*

This analysis specified the particular division objectives and decision-making responsibilities related to each top management position. Where possible standards of management performance were identified. A responsibility checklist was developed to ensure complete understanding of top management responsibility assignments. Special care was given to identifying those situations in which each executive had sole responsibility or shared responsibility for a certain area of the business. Table 2 lists the sole and shared responsibility of the Vice President – Manufacturing.

TABLE 2. MANAGEMENT RESPONSIBILITIES OF THE VICE PRESIDENT – MANUFACTURING

A – Shared responsibilities	B – Sole responsibilities
Company image	Manufacturing and assembly
Sales plans and performance	Manufacturing engineering
Inventory control	Operations control
Expense control	Manufacturing cost performance
Earnings	Plant engineering
Management development	Product delivery
Promotion and transfer	Production planning and control
Staffing	Productivity
	Purchasing
	Quality assurance
	Systems and data processing

5.4 *Definition of management information and reporting needs*

These definitions determined the information needed by top management personnel to reach their objectives and discharge their decision-making responsibilities. Responsibilities were translated into specific information needs. Information needs were translated into reporting requirements. These responsibilities, performance standards, information needs and reporting requirements were documented and reviewed by each member of top management. Table 3 illustrates this information for the Vice President – Manufacturing in some of his areas of responsibility.

5.5 *Determination of the adequacy of present information and reporting systems*

This defined the information needs not currently satisfied, such as information needed but not available, information needed but not provided, and information provided but not needed. Further opportunities for improving present management information and reporting procedures were also identified.

5.6 *Selection of indicators and design of management reports*

Included here were the complete design of a top management reporting system and the selection of indicators which are important to follow. The key indicators selected for continuous 'tracking' are shown in Table 4. From these a family of integrated management reports was designed for members of top management.

Special emphasis was placed on condition reports which illustrated the condition of key indicators as compared to plan. The condition report received by the Vice President and

TABLE 3. VICE PRESIDENT – MANUFACTURING

Functional responsibilities Performance requirements	Information needs	Reporting requirements
Total manufacturing cost (TMC) control		
Forecast accurately the cost of goods sold, based on: Historical costs Sales forecast Delivery schedule	Historical TMC by program Order backlog Sales forecast	Actual TMC compared to forecast TMC Recurring sales Sales of new products
Measure the actual cost of goods sold, controlling deviations from the forecast	Inventory investment Raw materials Work in process Finished goods	TMC forecast for 1 and 3 year periods Cost breakdown of program overruns
Update forecasting basis	Cost accounting data by program for: Direct labour Indirect labour Material Overhead	
Inventory control		
Maintain optimum inventory	Inventory investment Raw materials Work in process Finished goods	Inventory coverage of shipments for 6 and 24 months Actual and anticipated shipment delinquencies due to stock shortages

Table 3 continued

Functional responsibilities Performance requirements	Information needs	Reporting requirements
Maintain adequate information on inventory composition and value Maintain inventory turnover at the desired level	Inventory input Inventory depletions Inventory stockouts Shipment schedule	Inventory turnover ratios, by category Trends and implications Recommended actions
Production		
Provide optimum capacity required to meet sales commitments Operate production facilities in accordance with production plan in terms of: Production schedule Manufacturing cost	Production schedule Production capacity Program production schedule Program completion Labour utilization Manufacturing costs	Production schedule compared to capacity Labour efficiency Forecast of potential program problems Delay Cost overrun
Expense budgeting		
Translate TMC forecast into operating budget for year ahead including: Direct labour Material Factory overhead	Operating budget Actual costs Overtime Factory overhead	Actual expenses compared to budget Budget forecast to year-end Burden variances

Quality control		
Translate customer quality requirements into test procedures	Product acceptance rates Spoilage rates	Quality levels by program, area, etc. Spoilage applied against orders to be shipped
Implement test procedures to measure product acceptance		
Reduce spoilage		
Product delivery		
Deliver product within 30 days of scheduled delivery date specified in contract	Delivery schedule Delinquencies	Value of delinquencies (actual and forecast) by program
Exceptions allowed for purposes of capacity balancing		Trends and implications
Reduce delinquent deliveries		

TABLE 4. KEY DIVISION INDICATORS

Business strategy	Operating and sales		Financial
General economic conditions	Sales programs	Sales volume	Sales
Market demand	Process improvement	Sales mix	Cost of sales
Competitive position	Administrative systems improvement	Price structure	Expenses
Company resources	Supervisory training	Labour efficiency	Profit
Product development	Facilities	Material usage	Return on assets
		Labour and material prices	
		Inventories	
		Product quality	

General Manager is shown in Table 5. Favourable and unfavourable variances from plan were further explained by narrative discussions, as shown in Table 6, which covers the unfavourable year-to-date variance for cost of sales. Each narrative discussion was written in a standard format and consistently described the situation, cause, action and impact for each variance. In addition to the narrative discussion, supporting graphs, charts and tabulations could also be used to amplify the condition reports.

6.0 CONCLUSIONS

The approach to developing a set of top management reports described in the case study has consistently proved to be quick and effective. In addition, it provides the emphasis on obtaining the information needed to effectively run a business, instead of accepting the information already available to lower-level management and attempting to operate the business on that basis. For the manufacturing executive, this approach has particular value. Because of the emphasis on tracking key indicators against the plan and gaining visibility on the corresponding trends, the manufacturing executive has the kind of data he needs to react to adverse conditions

TABLE 5. SAMPLE CONDITION REPORT

Key indicators	Actual year-to-date	Projected year-end	Paragraph reference number
Business strategy			
General economic conditions	–	–	–
Market demand	–	–	–
Competitive position			
Technology	–	–	–
Price	●	●	1
Delivery and service	–	–	–
Company resources			
Assets	–	–	–
Personnel	–	–	–
Product development			
New product program	○	○	2
Product improvement	●	–	3
Operating and sales			
Sales programs	–	–	–
Sales volume	–	–	–
Sales mix	–	–	–
Price structure	●	●	4
Process improvement	–	–	–
Product quality	–	–	–
Labour efficiency	–	–	–
Material usage	–	–	–
Labour and material prices	●	–	5
Inventories	–	○	6
Administrative systems improvements	–	–	–
Supervisory training	–	–	–
Facilities	–	–	–
Financial			
Profit	●	●	7
Sales	–	–	–
Cost of sales	●	–	8
Expenses	–	–	–
Return on assets	●	–	9

Legend: ○ Favourable variance; – On plan; ● Unfavourable variance

TABLE 6. COST OF SALES – NARRATIVE EXPLANATION

Situation	Year-to-date cost of sales is six per cent over plan caused by higher than planned manufacturing costs. These costs are expected to be on plan for the balance of the year, therefore, the year-end position will be only slightly over plan.
Cause	During the last month, unusually high startup costs were experienced on the new product line in Denver. These high startup costs resulted from overtime premiums and temporary addition of second shift operating personnel. These moves were required in order to make up for time lost due to the late delivery of lathes from our supplier.
Action	The loss of time in our Denver startup has been compensated for by the overtime and second shift additions. These will not continue after this month. Little opportunity is seen for recovering the additional costs incurred before year-end.
Impact	Manufacturing cost of sales will not increase further during the balance of this year. The over-plan costs at year-end are estimated to be only two per cent.

before they occur. Without a doubt, this approach gives the manufacturing manager a chance to compensate for errors caused by sales or engineering managers before the full impact is felt, rather than on a 'crash basis' after a problem has surfaced.

In general, the management reporting concepts presented have many advantages. They save time and reduce the trivial distractions of managers reviewing the reports. Managerial coverage is increased, and it is possible to identify both crises and opportunity situations readily. Furthermore, providing yardsticks of performance and understanding their relationships to the various functional areas of the business stimulates communication among the various executives in a company. This last advantage is of great value in itself.

Of course, there are limitations involved in pursuing the foregoing approach to management reporting. It does require a comprehensive knowledge and understanding of the interactions of a business. It does assume a certain degree of stability in the operations of the business, which may or may not exist. Executives can also be given a sense of false security by overlooking critical indicators that have been prejudged to be noncritical. Finally, it is often hard to obtain reliable initial data to measure the indicators that are considered critical. In many respects, this last limitation is the most difficult to overcome.

In summary, the exception principle combined with new approaches to management information reporting offer an important way to relieve the already overburdened manager. Although the ideas have been available and to some extent ignored for many years, they have great relevance to business situations today. If a company will take the time to carefully develop and implement this approach, without a doubt the benefits derived will be well worth the efforts and time expended.

REFERENCES

BITTEL, L. R., *Management by Exception*, McGraw-Hill, New York, 1964, 320 pp.

HEYEL, C., *Management for Modern Supervisors*, American Management Association, New York, 1962, 255 pp.

MILLER, E. C., *Objectives and Standards of Performance in Production Management*, American Management Association, New York, 1967, 112 pp.

PAYNE, B., *Planning for Company Growth,* McGraw-Hill, New York, 1963, 316 pp.

ROSE, T. G., and FARR, D. E., *Higher Management Control*, McGraw-Hill, New York, 1957, 290 pp.

SCHLEH, E. C., *Management by Results*, McGraw-Hill, New York, 1961, 266 pp.

MITURN, A COMPUTER-AIDED PRODUCTION PLANNING SYSTEM FOR NUMERICALLY CONTROLLED LATHES

J. Koloc
Metaalinstituut TNO, Netherlands

SUMMARY

A technology-orientated programming system for the planning of production of turned parts on numerically controlled lathes is described in detail, the approach being based on 'group technology', principles. It is claimed that the system has potential for future implementation in manufacturing systems using programmable direct numerical control. The programme is fully conversational and may be used directly by production planning personnel.

1.0 INTRODUCTION

Investigations from the point of view of group technology have shown regular incidence pattern of basic machining methods and the same regularity has been found in the incidence pattern of certain workpiece characteristics in the seemingly endless variety of machinery components.

These facts are now generally accepted, which, after all, is not surprising after decades upon decades of producing this endless variety of workpieces on a large but certainly not unlimited number of basic machine tool types and with a large percentage of standard tools.

It has been shown repeatedly that technologically significant workpiece characteristics do conform to certain regular incidence behaviour, which can definitely be termed deterministic but certainly not stochastic. Examples for this can easily be quoted from technical literature. Centrally organised reviews of whole industrial branches in Eastern Europe have yielded convincingly uniform information on this (Fig. 1). Comparisons with similar reviews in he West have shown very good agreement.

Many conclusions may be drawn from this, but one relatively meagre one can be drawn with great safety: this emerging incidence pattern of workpiece characteristics in the metalworking industry cannot be and is not accidental. It must be the result of universally valid factors which exercise objective constraints on the creative freedom of any designer and of any production planner. The influence of these factors must be very stable, shaping the workpiece assortment into what it is.

The actual laws of these influences are not yet known. Anyway, configuration, dimensions, accuracy and surface finish of any component of a machine is not the result of free creative contemplation by its designer, and these characteristics are not dictated by the ultimate function in the assembled machine alone.

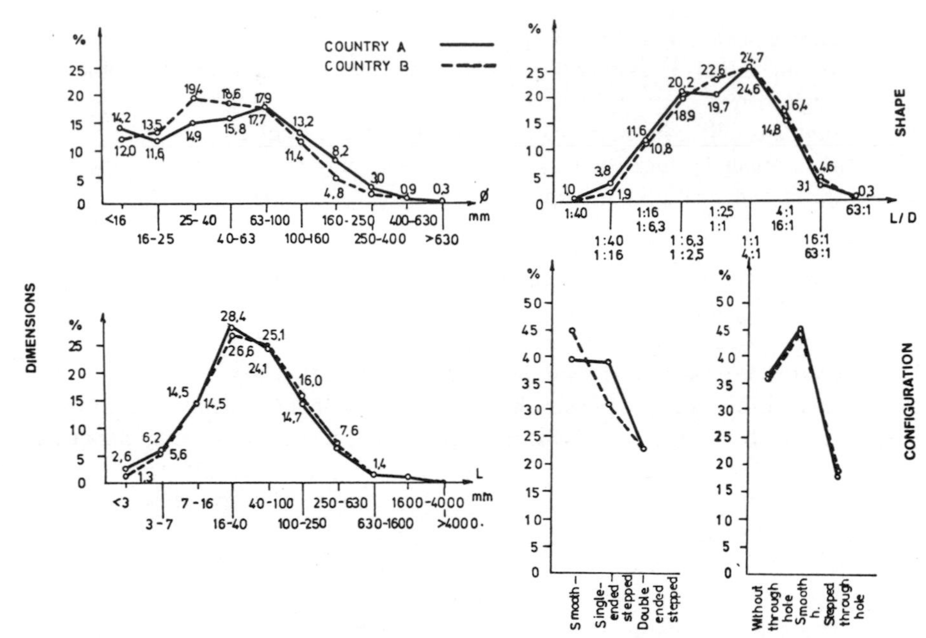

Fig. 1(a). Comparison of typical dimensions. Country A, full line. Country B, broken line.
Fig. 1(b). Comparison of shapes in countries A and B.
Fig. 1(c). Comparison of configurations in countries A and B.

Generally, the active limitations belong to one or more of the following groups:

- Component's ultimate function
- Materials from which component is manufactured
- Technological methods used for making the part
- Level of expertise and skill of the personnel

What has been said about workpiece characteristics is applicable also for most stages of machining, where similar constraints are at work. This results in a situation where again a clear incidence pattern of machining sequences, tooling sets, typical cycles etc. emerges.

This has been recapitulated only to set a foundation for a discussion of a technology-oriented programming system.

2.0 STRUCTURE OF THE MITURN-PROGRAMMING SYSTEM

The MITURN-programming system provides for a high degree of automation of production planning work. It:

- Provides the cutting sequence
- Computes and implements the feeds, speeds and depths of cut
- Implements the programmed tolerances and surface finishes
- Selects the appropriate tool for each cut
- Prepares printed instructions for machine set-up
- Computes the expected machining time
- Produces the n.c. punched tape in the appropriate code
- Prepares the printed text of the program with important information in open (not coded) form, if requested.

The system is implemented on a time-sharing basis so that:

- There is no initial investment and there are no communication problems with a computing centre
- Programming is fully conversational, with some measure of diagnostics
- The programming system is easily accessible and may be used directly by production planning personnel without any 'middle-men'.

2.1 *Input-format*

The input-format is treated in some detail to show the part-programming grammar and the workshop-orientation of the terms used.

The part programme consists of the following:

- Heading with the programmer's and program's identifications
- Set-up specification with important basic parameters of the machine used, overall workpiece size and configuration, clamping method, major positioning parameters and machinability
- Workpiece specification
- Comment for production

Both rough (blank) and finished workpiece contours are separately divided into consecutive longitudinal segments, or elements. First the rough, then the finished contour is described.

Elements available at present are (Figs 2, 3): cylinder, taper, circular arc, and threaded cylinder. Radial grooves are also available, but are not considered longitudinal elements. They may be superimposed on any cylinder or threaded cylinder.

Part programming consists merely of listing the elements consecutively from the tail-end of the workpiece to the headstock end. That is, in usual cases, from right to left. Consecutive diameters from right to left must be non-decreasing for the external contour and non-increasing for the internal contour, with the exception of grooves.

In programming the part, one first lists the number of the external elements, then the string of elements codes of the external contour. After this one lists the parameter string of the first element, then of the second, and so on. After all external elements are listed, one supplies the number of the external radial grooves followed by parameter strings defining these grooves.

After the external contour is complete the above procedure is repeated for the internal contour. To document the procedure a very simple example is shown in Fig. 4, while input for this workpiece is given in Fig. 5. A real part drawing is given in Fig. 6, and its three operation drawings are in Figs 7, 8, 9. The pertinent part programs are in Figs 10, 11, 12.

There are a few intentionally redundant input items, which are used by the system diagnostical routines, which detect serious programming errors.

The combination of conversational input with efficient information coding makes it possible to limit the input size to the essential one. Only meaningful information has to be programmed as the addresses are implicitly sequential-coded. The readability of the program is safeguarded by the computer requests for input, which are printed out with appropriate explanations of requested items.

In Fig. 13 is a comparison of input for the same simple part in two different programming systems: an APT-like system offering also automatic technological planning and MITURN. The volume of input measured by number of characters is approximately 1:5, and the difference in the number of defined words makes the difference in grammar also pronounced.

The required programming effort in MITURN may thus be termed low.

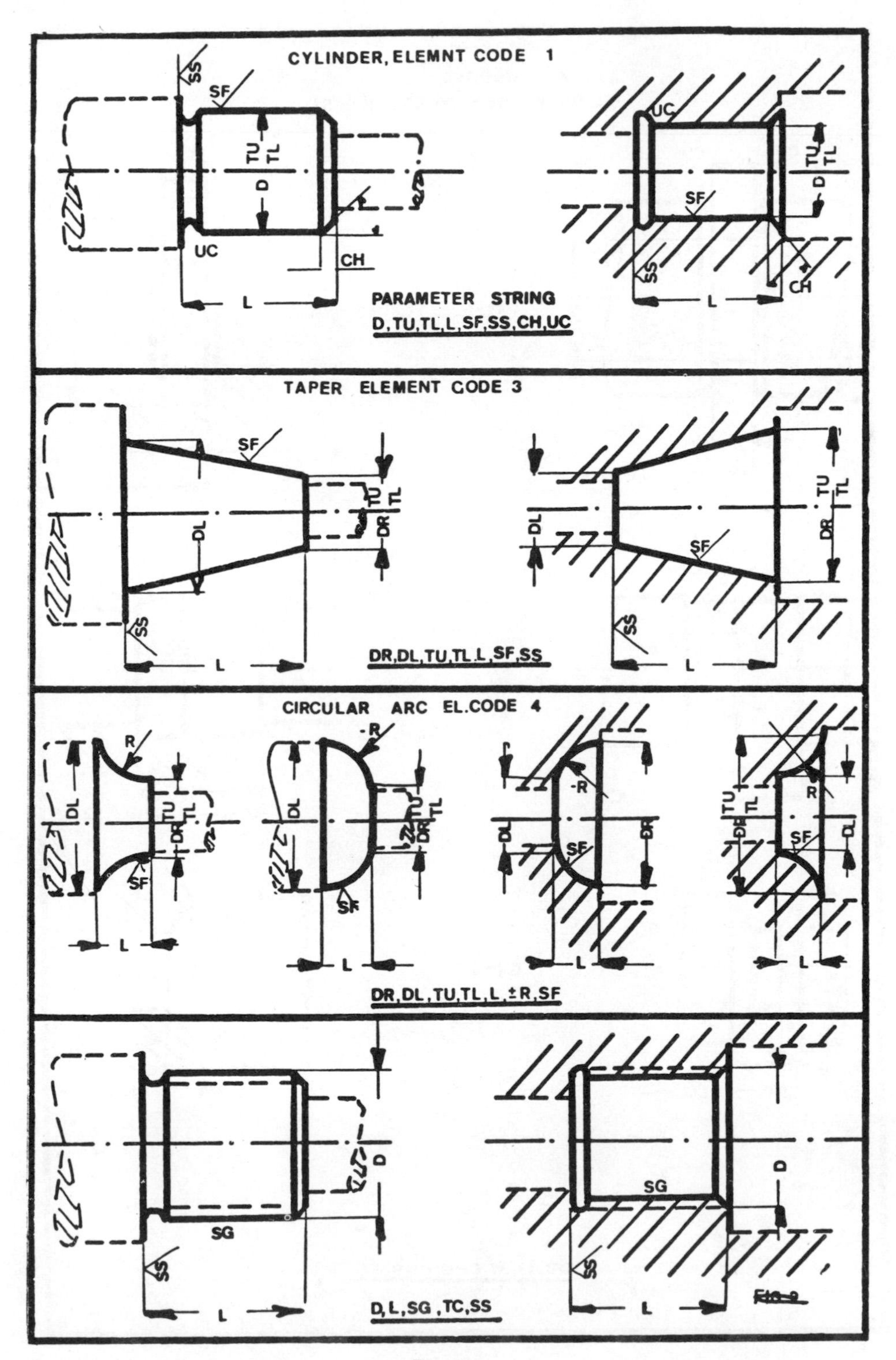
CYLINDER, ELEMNT CODE 1
SS
SF
UC
TU
TL
D
CH
L
PARAMETER STRING
D, TU, TL, L, SF, SS, CH, UC
TAPER ELEMENT CODE 3
SF
DL
DR
TU
TL
SS
L
DR, DL, TU, TL, L, SF, SS
CIRCULAR ARC EL.CODE 4
R
-R
DL
DR
TU
TL
SF
L
DR, DL, TU, TL, L, ±R, SF
SG
D
SS
L
D, L, SG, TC, SS

Fig. 2.

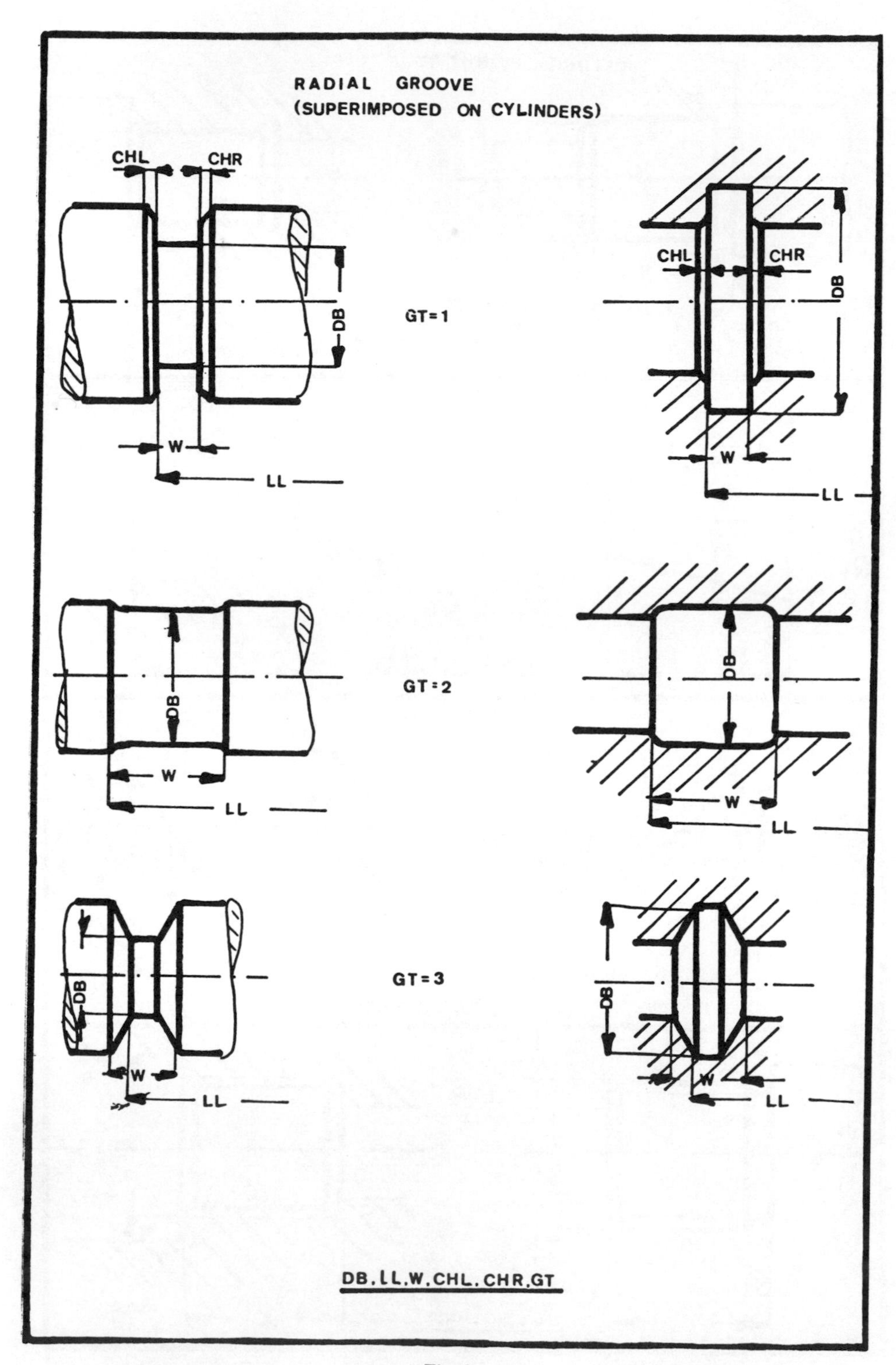
RADIAL GROOVE
(SUPERIMPOSED ON CYLINDERS)
CHL
CHR
DB
W
LL
GT=1
GT=2
GT=3
DB, LL, W, CHL, CHR, GT

Fig. 3.

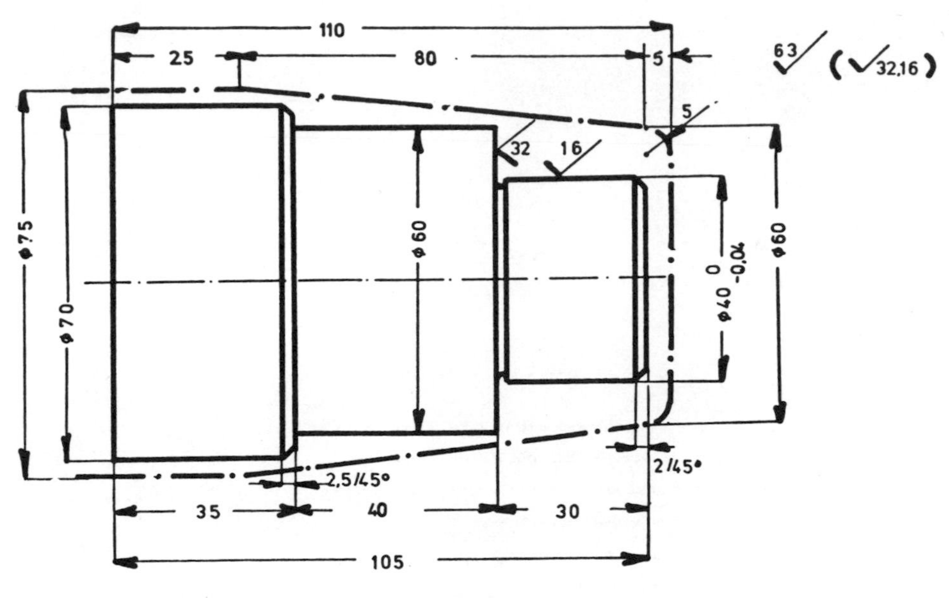

Fig. 4.

2.2 *Technology*

In the programming system presented there is an integral logical part of the compiler which produces the necessary decisions during the automatic run.

Generally speaking, the description of the basic set-up, of the rough and finished contour of the workpiece, is sufficient to initialise a computer run resulting in a control punched tape for the n.c. lathe.

This, however, could not be applicable in many practical cases without serious limitations. There must be provision for various specific aspects of machining technology which may differ from case to case, and user to user.

The activities in planning a turning operation may be classified as routines:

- which are mathematically formulated and which may be applied exactly, because they are based on known physical laws
- which may be formalised into mathematical relationships but which may be applied only as empirical values based on experience
- for which there is no possibility, now or in the forseeable future, of mathematical formulation.

Because of this, the MITURN system defines input information in three levels (Fig. 14):

A Workpiece-dependent information, which constitutes the part program. This information has to be programmed separately for each new workpiece.

B Information, relating to tool-dependent, machine dependent and control-system-dependent parameters, relating to the machining hardware environment. However, there is also the know-how, objective and/or subjective preferences specific for every user, so to say machining 'software'. So that the system does not become unacceptably rigid, there must be a channel for conveying varying amounts of information to the system, thus influencing the automatic decision-making. However, data relating to such items do not

```
RUN

MITURN      0154 EDT    09/10/70

   MITURN PRØGRAMMA VØØR NC DRAAIMACHINES

********** HØØFDGEGEVENS *****

PRØGRAMMEUR EN DE DATUM?J.ANØN 10/9/1970

TEKENINGNUMMER?EXAMPLE 1

TEKENINGSTANDAARD?MM

WERKSTUKMATERIAAL?ST 60

********** SETUP GEGEVENS *****

MACH,TYPE,KLEN,SAFD,BLEN,PLEN,HLEN,VHA
?1 10 207 5 115 105 0 3

MAT,MAC,MAS,RIC?1 .75 10 1

********** RUWE VØRM *********

AANTAL ELEMENTEN VAN DE RUWE BUITENVØRM?3

 3 EL.CØDES?4 3 1

EL 1
RAD?50 60 5 -5

EL 2
TAP?60 75 80

EL 3
CIL?75 25

AANTAL ELEMENTEN VAN DE RUWE BINNENVØRM?0

********** WERKSTUK **********

AANTAL ELEMENTEN VAN DE BUITENCØNTØUR?3

 3 EL.CØDES?1 1 1

EL 1
CIL?40 0 -0.04 30 16 32 2 1

EL 2
CIL?60 0 0 40 63 63 0 0

EL 3
CIL?70 0 0 35 63 0 2.5 0

AANTAL GRØEVEN ØP DE BUITENCØNTØUR?0
```

Fig. 5.

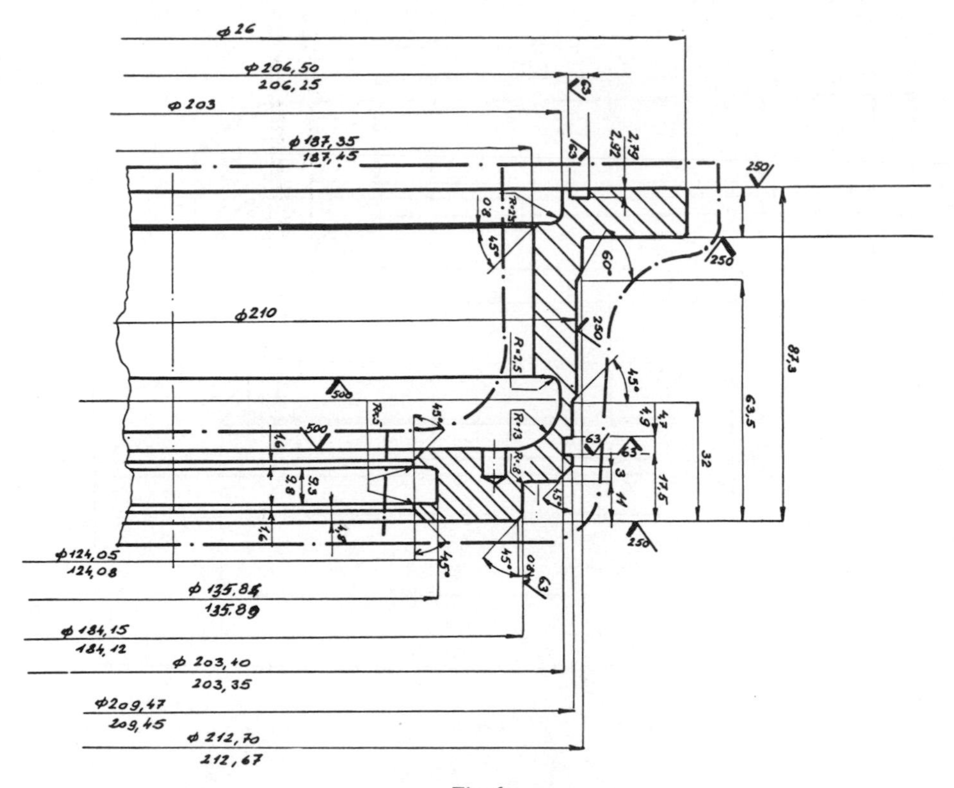
φ26
φ206,50
206,25
φ203
φ187,35
187,45
2,79
2,92
250
φ210
87,3
63,5
32
500
4,9
4,7
17,5
φ124,05
124,08
φ135,86
135,89
φ184,15
184,12
φ203,40
203,35
φ209,47
209,45
φ212,70
212,67

Fig. 6.

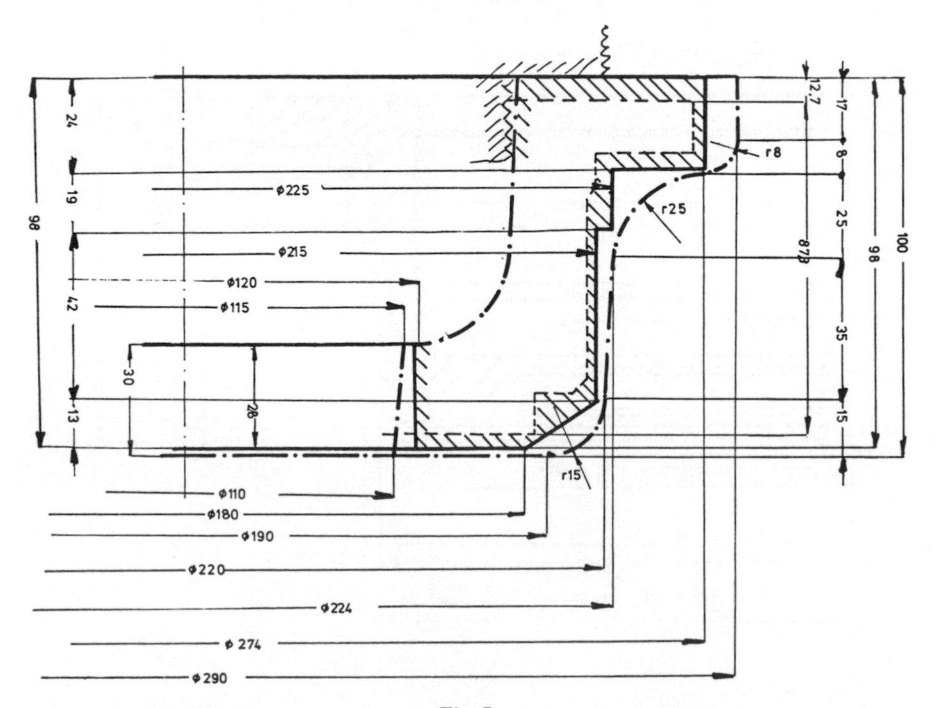
24
19
98
42
13
30
28
φ225
φ215
φ120
φ115
φ110
φ180
φ190
φ220
φ224
φ274
φ290
12,7
17
8
25
35
15
87,3
98
100
r8
r25
r15

Fig. 7.

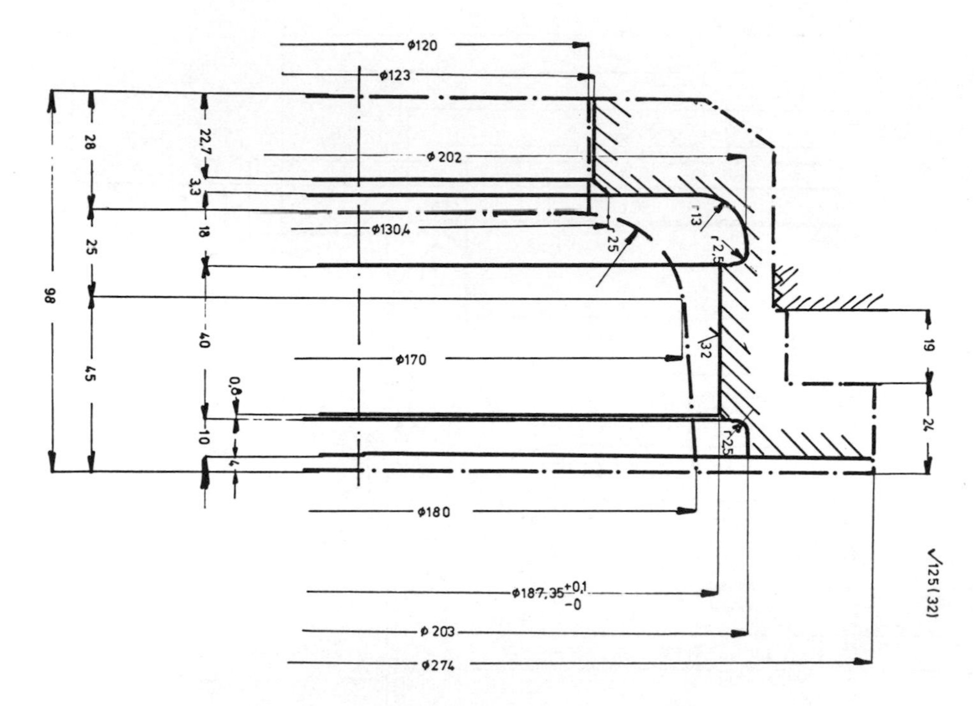
φ120
φ123
φ202
φ130,4
φ170
φ180
φ187,35 +0,1 -0
φ203
φ274
28
22,7
3,3
18
25
98
40
45
0,8
10
4
r13
r2,5
r25
32
19
24
√125 (32)

Fig. 8.

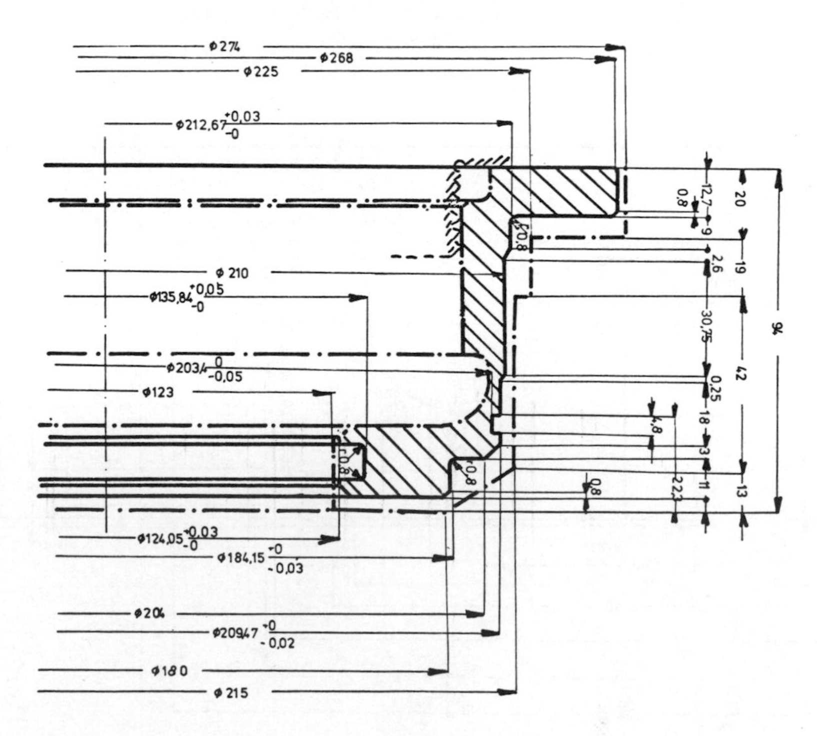
φ274
φ268
φ225
φ212,67 +0,03 -0
φ210
φ135,84 +0,05 -0
φ203,4 0 -0,05
φ123
φ124,05 +0,03 -0
φ184,15 +0 -0,03
φ204
φ209,47 +0 -0,02
φ180
φ215
12,7
0,8
20
9
19
2,6
30,75
94
42
0,25
18
4,8
3
22,3
11
13
0,8

Fig. 9.

```
RUN

MITURN      0559 EDT     09/10/70

   MITURN PRØGRAMMA VØØR NC DRAAIMACHINES

********** HØØFDGEGEVENS *****

PRØGRAMMEUR EN DE DATUM?J.ANØN 10/9/1070

TEKENINGNUMMER?THØM-1

TEKENINGSTANDAARD?MM

WERKSTUKMATERIAAL?RØESTVRIJ STAAL

********** SETUP GEGEVENS *****

MACH,TYPE,KLEN,SAFD,BLEN,PLEN,HLEN,VHA
?1 10 190 3 103 98 28 3

MAT,MAC,MAS,RIC?1 .6 10 1

********** RUWE VØRM *********

AANTAL ELEMENTEN VAN DE RUWE BUITENVØRM?5

 5 EL.CØDES?4 3 4 4 1

EL 1
RAD?190 220 15 -15

EL 2
TAP?220 224 35

EL 3
RAD?224 274 25 25

EL 4
RAD?274 290 8 -8

EL 5
CIL?290 17

AANTAL ELEMENTEN VAN DE RUWE BINNENVØRM?1

 1 EL.CØDES?1

EL 1
CIL?110 30

********** WERKSTUK **********

AANTAL ELEMENTEN VAN DE BUITENCØNTØUR?4

 4 EL.CØDES?3 1 1 1

EL 1
TAP?180 215 0 0 13 63 0

EL 2
CIL?215 0 0 42 125 125 0 0

EL 3
CIL?225 0 0 19 125 125 0 0

EL 4
CIL?274 0 0 24 125 0 0 0

AANTAL GRØEVEN ØP DE BUITENCØNTØUR?0

AANTAL ELEMENTEN VAN DE BINNENCØNTØUR?1

 1 EL.CØDES?1

EL 1
CIL?120 0 0 2 -8 125 0 0 0

AANTAL GRØEVEN ØP DE BINNENCØNTØUR?0

PRØGRAM STØP AT 1750

USED      .89 UNITS
```

Fig. 10.

```
RUN

MITURN      0606 EDT     09/10/70

   MITURN PRØGRAMMA VØØR NC DRAAIMACHINES

********** HØØFDGEGEVENS *****

PRØGRAMMEUR EN DE DATUM?J.ANØN 10/9/1970

TEKENINGNUMMER?THØM-2

TEKENINGSTANDAARD?MM

WERKSTUKMATERIAAL?RØESTVRIJ STAAL

********** SETUP GEGEVENS *****

MACH,TYPE,KLEN,SAFD,BLEN,PLEN,HLEN,VHA
?1 10 207 3 43 36 94 3

MAT,MAC,MAS,RIC?1 .7 10 1

********** RUWE VØRM *********

AANTAL ELEMENTEN VAN DE RUWE BUITENVØRM?1

 1 EL.CØDES?1

EL 1
CIL?274 40

AANTAL ELEMENTEN VAN DE RUWE BINNENVØRM?3

 3 EL.CØDES?3 4 1

EL 1
TAP?180 170 45

EL 2
RAD?170 120 25 -25

EL 3
CIL?120 28

********** WERKSTUK **********

AANTAL ELEMENTEN VAN DE BUITENCØNTØUR?0

AANTAL ELEMENTEN VAN DE BINNENCØNTØUR?6

 6 EL.CØDES?1 4 1 4 3 1

EL 1
CIL?203 0 0 7.5 63 0 .2 0

EL 2
RAD?203 198 0 0 2.5 -2.5 63

EL 3
CIL?187.35 .1 0 56 32 0 .8 0

EL 4
RAD?187.35 168 .1 0 2 -13 32

EL 5
TAP?130.4 123 .1 0 3.3 63 0

EL 6
CIL?123 .1 0 22.7 63 0 0 0

AANTAL GRØEVEN ØP DE BINNENCØNTØUR?2

GR 1?210 55 5 10 0 3

GR 2?210 61 11 12 0 3

PRØGRAM STØP AT 1750

USED      .90 UNITS
```

Fig. 11.

```
RUN

MITURN      0819 EDT    09/10/70

   MITURN PRØGRAMMA VØØR NC DRAAIMACHINES

********** HØØFDGEGEVENS *****

PRØGRAMMEUR EN DE DATUM?J.ANØN 10/9/1970

TEKENINGNUMMER?THØM-3

TEKENINGSTANDAARD?MM

WERKSTUKMATERIAAL?RØESTVRIJ STAAL

********** SETUP GEGEVENS *****

MACH,TYPE,KLEN,SAFD,BLEN,PLEN,HLEN,VHA
?1 10 192 3 97 87.3 19 3

MAT,MAC,MAS,RIC?1 .7 10 1

********** RUWE VØRM *********

AANTAL ELEMENTEN VAN DE RUWE BUITENVØRM?4

 4 EL.CØDES?3 1 1 1

EL 1
TAP?180 215 13

EL 2
CIL?215 42

EL 3
CIL?225 19

EL 4
CIL?274 20

AANTAL ELEMENTEN VAN DE RUWE BINNENVØRM?1

 1 EL.CØDES?1

EL 1
CIL?123 25.7

********** WERKSTUK **********

AANTAL ELEMENTEN VAN DE BUITENCØNTØUR?10

10 EL.CØDES?1 4 3 1 3 1 3 1 4 1

EL 1
CIL?184.15 0 -.03 10.2 16 0 .8 0

EL 2
RAD?184.15 185 75 0 -.03 .8 .8 32

EL 3
TAP?204 209.47 0 -.02 3 32 0

EL 4
CIL?209.47 0 -.02 18 16 0 0 0

EL 5
TAP?209.47 210 0 -.02 .25 16 0

EL 6
CIL?210 .1 -.1 30.75 63 0 0 0

EL 7
TAP?210 212.69 .1 -.1 2.6 32 0

EL 8
CIL?212.69 .03 0 8.2 16 0 0 0

EL 9
RAD?212.69 214 -.29 .03 0 .8 .8 63

EL10
CIL?268 0 0 12.7 63 0 .8 0

AANTAL GRØEVEN ØP DE BUITENCØNTØUR?1

GR 1?203.4 22.3 4.8 0.2 0.2 1

AANTAL ELEMENTEN VAN DE BINNENCØNTØUR?2

 2 EL.CØDES?3 1

EL 1
TAP?130.45 124.05 .03 0 3.2 16
?0

EL 2
CIL?124.050 0.03 0 15.8 16 0 0 0

AANTAL GRØEVEN ØP DE BINNENCØNTØUR?1

GR 1?135.89 14.3 9.5 .5 .5 1

DE LENGTE VAN DE BUITENCØNTØUR IS ØNJUIST

PRØGRAM STØP AT 1625

USED     .40 UNITS
```

Fig. 12.

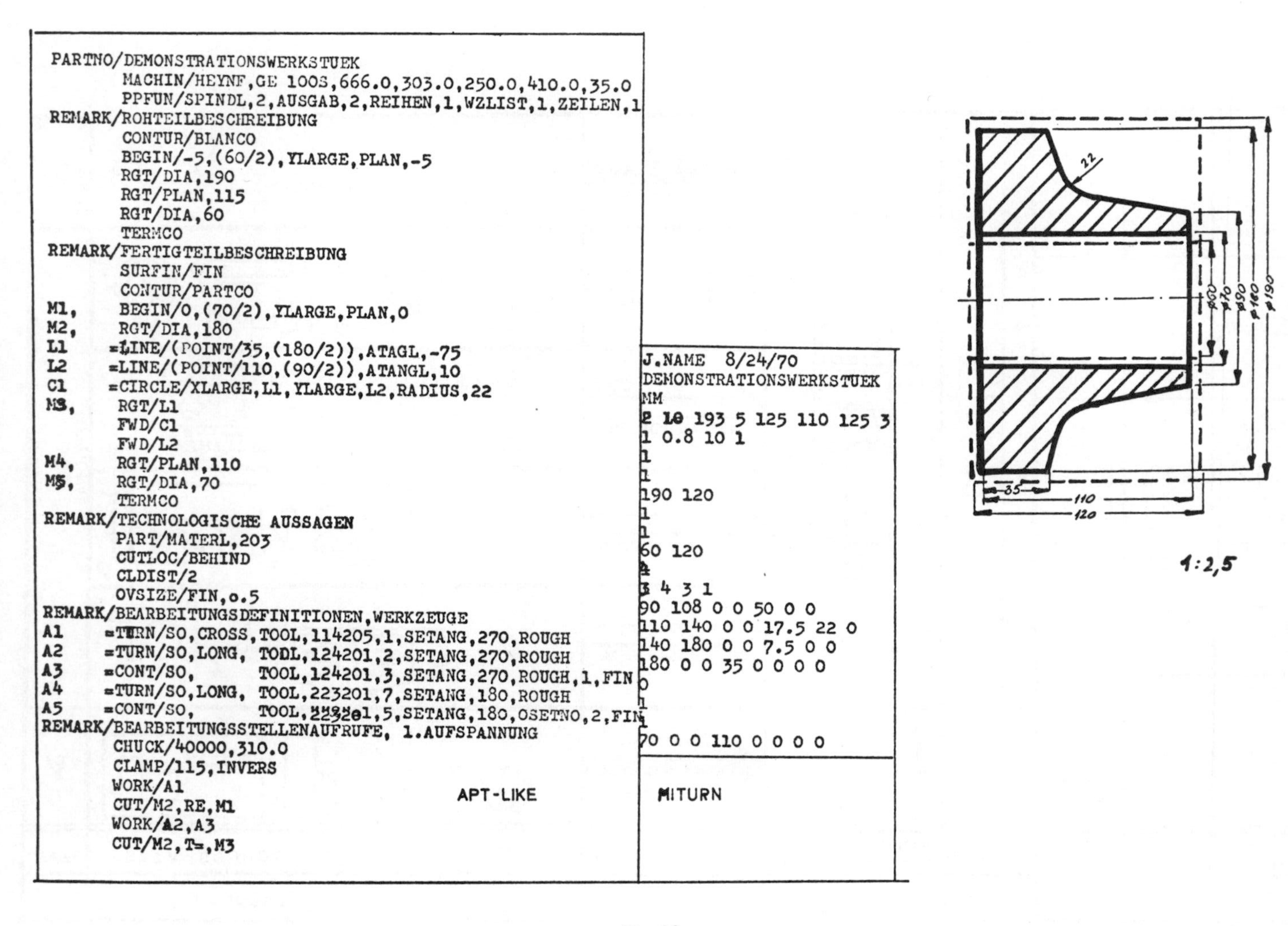
PARTNO/DEMONSTRATIONSWERKSTUEK
MACHIN/HEYNF,GE 100S,666.0,303.0,250.0,410.0,35.0
PPFUN/SPINDL,2,AUSGAB,2,REIHEN,1,WZLIST,1,ZEILEN,1
REMARK/ROHTEILBESCHREIBUNG
CONTUR/BLANCO
BEGIN/-5,(60/2),YLARGE,PLAN,-5
RGT/DIA,190
RGT/PLAN,115
RGT/DIA,60
TERMCO
REMARK/FERTIGTEILBESCHREIBUNG
SURFIN/FIN
CONTUR/PARTCO
M1, BEGIN/0,(70/2),YLARGE,PLAN,0
M2, RGT/DIA,180
L1 =LINE/(POINT/35,(180/2)),ATAGL,-75
L2 =LINE/(POINT/110,(90/2)),ATANGL,10
C1 =CIRCLE/XLARGE,L1,YLARGE,L2,RADIUS,22
M3, RGT/L1
FWD/C1
FWD/L2
M4, RGT/PLAN,110
M5, RGT/DIA,70
TERMCO
REMARK/TECHNOLOGISCHE AUSSAGEN
PART/MATERL,203
CUTLOC/BEHIND
CLDIST/2
OVSIZE/FIN,o.5
REMARK/BEARBEITUNGSDEFINITIONEN,WERKZEUGE
A1 =TURN/SO,CROSS,TOOL,114205,1,SETANG,270,ROUGH
A2 =TURN/SO,LONG, TOOL,124201,2,SETANG,270,ROUGH
A3 =CONT/SO, TOOL,124201,3,SETANG,270,ROUGH,1,FIN
A4 =TURN/SO,LONG, TOOL,223201,7,SETANG,180,ROUGH
A5 =CONT/SO, TOOL,223201,5,SETANG,180,OSETNO,2,FIN
REMARK/BEARBEITUNGSSTELLENAUFRUFE, 1.AUFSPANNUNG
CHUCK/40000,310.0
CLAMP/115,INVERS
WORK/A1
CUT/M2,RE,M1
WORK/A2,A3
CUT/M2,T=,M3
APT-LIKE
J.NAME 8/24/70
DEMONSTRATIONSWERKSTUEK
MM
2 10 193 5 125 110 125 3
1 0.8 10 1
1
1
190 120
1
1
60 120
4
3 4 3 1
90 108 0 0 50 0 0
110 140 0 0 17.5 22 0
140 180 0 0 7.5 0 0
180 0 0 35 0 0 0 0
0
1
70 0 0 110 0 0 0 0
MITURN
22
ø60
ø70
ø90
ø180
ø190
35
110
120
1:2,5

Fig. 13.

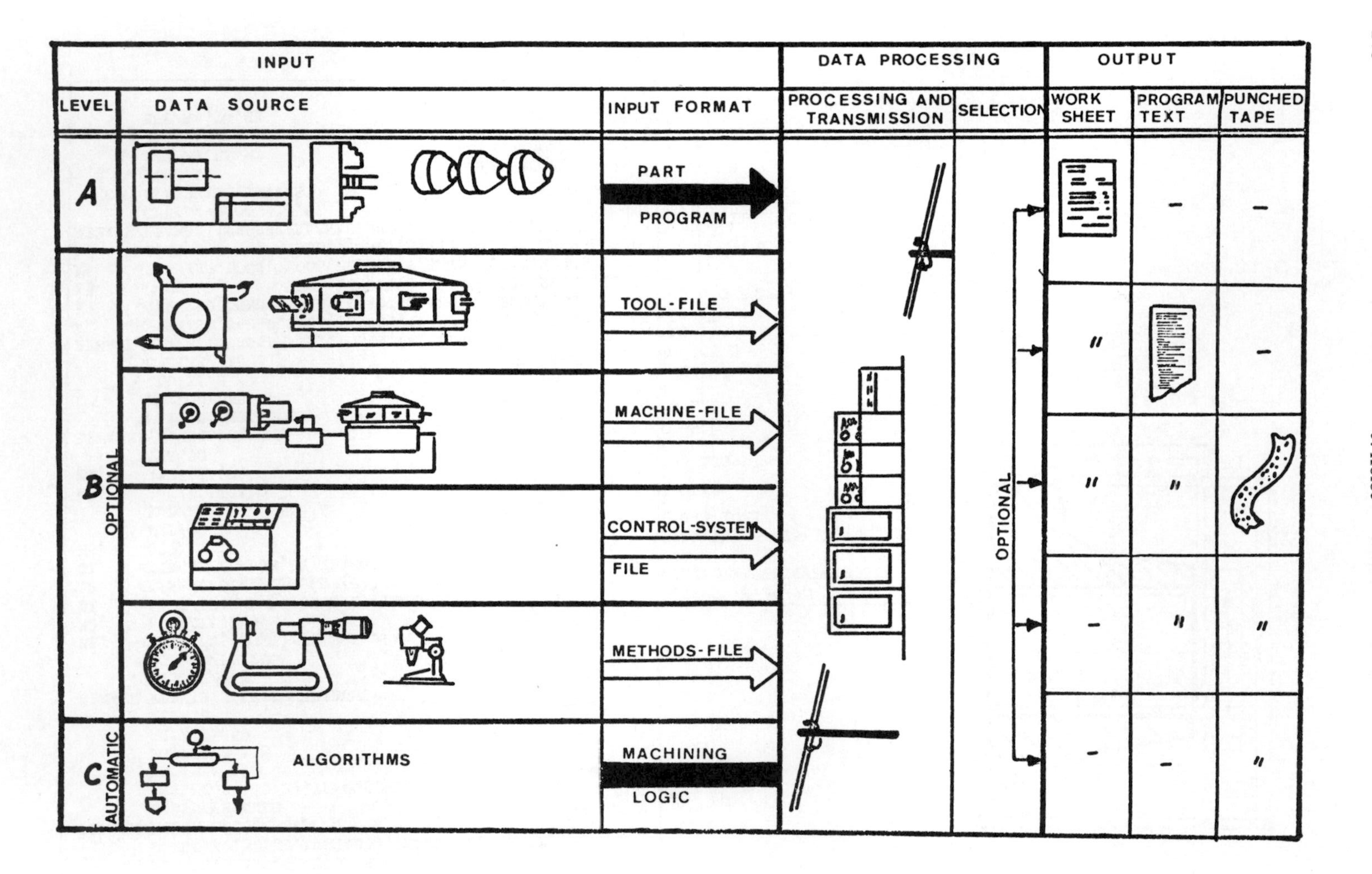
INPUT
DATA PROCESSING
OUTPUT
LEVEL
DATA SOURCE
INPUT FORMAT
PROCESSING AND
TRANSMISSION
SELECTION
WORK
SHEET
PROGRAM
TEXT
PUNCHED
TAPE
A
PART
PROGRAM
B
OPTIONAL
TOOL-FILE
MACHINE-FILE
CONTROL-SYSTEM
FILE
METHODS-FILE
OPTIONAL
C
AUTOMATIC
ALGORITHMS
MACHINING
LOGIC

Fig. 14.

necessarily change for each new workpiece. There are therefore separate inputs for these, which may be used optionally, thus keeping the compulsory part program volume to a minimum.

C Internal system logic, which need not be influenced, as it is based on established physical facts. The system generates its own input for these automated routines.

As stated at the beginning, the approach presented is based on group technology. The system software is modular so that new part families may be easily added at any time. However, great effort has been devoted to developing essential definitions in as general a form as possible. It is difficult to state to what degree this aim has been attained.

A measure of this, perhaps, may be the size of the system software. In its present form MITURN is applicable to more than 50% of all feasible n.c. lathe operations. This estimate is based on very conservative evaluation of workpiece statistics available. The compiler was implemented with no difficulties in a time-sharing system, its inherent limitations in software size. It is hoped that similarly its compactness would be useful in future implementations in programmable direct numerical controls.

The major principles of the technological approach of MITURN are:

- Contouring capabilities of n.c. lathes must be exploited to the full
- The number of necessary tools within one automatic machining cycle has to be kept within the capabilities of most available machines
- The tool-workpiece-machine interference problem during the machining cycle must be solved automatically by the system
- The use of individually selected semi-permanent tooling set-ups (magazines or turrets) must be possible.

A good answer to these problems is to treat an n.c. lathe basically as a tracer-controlled lathe, where the path control does not come from a rigid template but from the programmed input information stored on the punched tape. So the important tools both for external and internal contours must possess full contouring capability (Fig. 15).

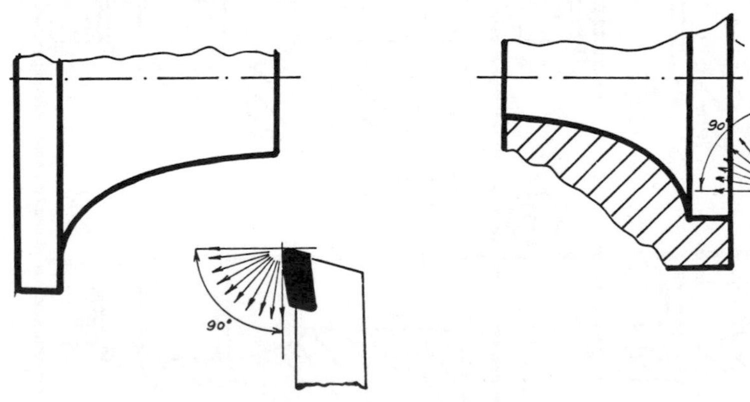

Fig. 15.

There are ten machining operations defined in the system (Fig. 16). These are called by the compiler automatically as the programmed part may require. For each machining operation one or more tools is or are automatically selected. However, the tool file may be established by the

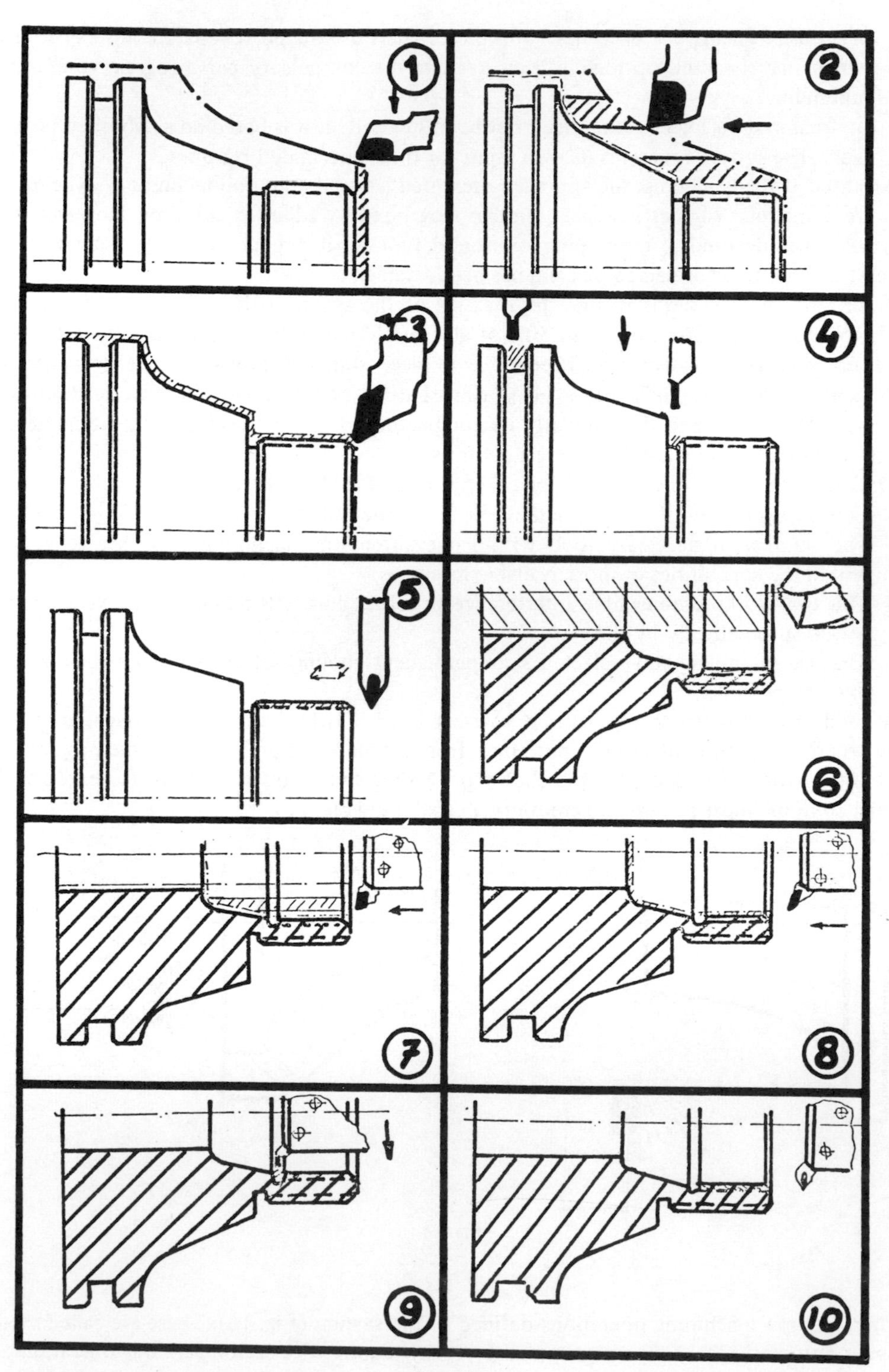

Fig. 16.

user with great freedom provided that the essential requirements of pertinent defined machining operations are met by the tool to be included in the tool file. Thus, operations rather than tools are defined, leaving the selection of the actual 'tool-menu' free for the user.

The MITURN technological processor is based on the decisive importance of the roughing cycle for efficient machining in job lots, where much stock removal is taking place. Not always is a shaped (forged or cast) blank available and one has to work from round bars.

The roughing cycle is defined so that automatic optimisation of the roughing tool path is possible.

There is a functional relationship defined among maximum permissible depth of cut, workpiece overall rigidity and tool rigidity (Fig. 17). This maximum cutting depth is computed before each roughing stroke, as the shape, size, and therefore also the rigidity, of the workpiece changes. The computed value is supplemented by the roughing feed, where the governing factor is the shape of the chip cross-section for the best chip flow. A cutting speed is then computed from a Taylor-like formula, taking care of the cutting efficiency of the tool material (C in the Fig. 17), machinability of the workpiece material (MAC), depth of cut (H^A) and (F^B). A check is then made against power available and if exceeded, recalculation takes place. At this point cutting parameters are available and geometric and kinematical aspects are processed next.

The general form of the roughing cycle as defined in MITURN is given in Fig. 18. Coordinates of points 1, 2 and 3 are computed from the programmed contour geometry. The governing factors of reducing cut between points 1 and 2 are: rigidity and power (H, N), tool-life and chip flow (V, H/F).

Governing factors of contouring cut between points 2 and 3 are: tool-life (V), part geometry (tool path).

Governing factors for back-off and approach between points 3 and 1 are: machine tool characteristics, (rapid traverse rate), part geometry (tool-workpiece interference). Between part 3 and part A a new computation of cutting values is carried out.

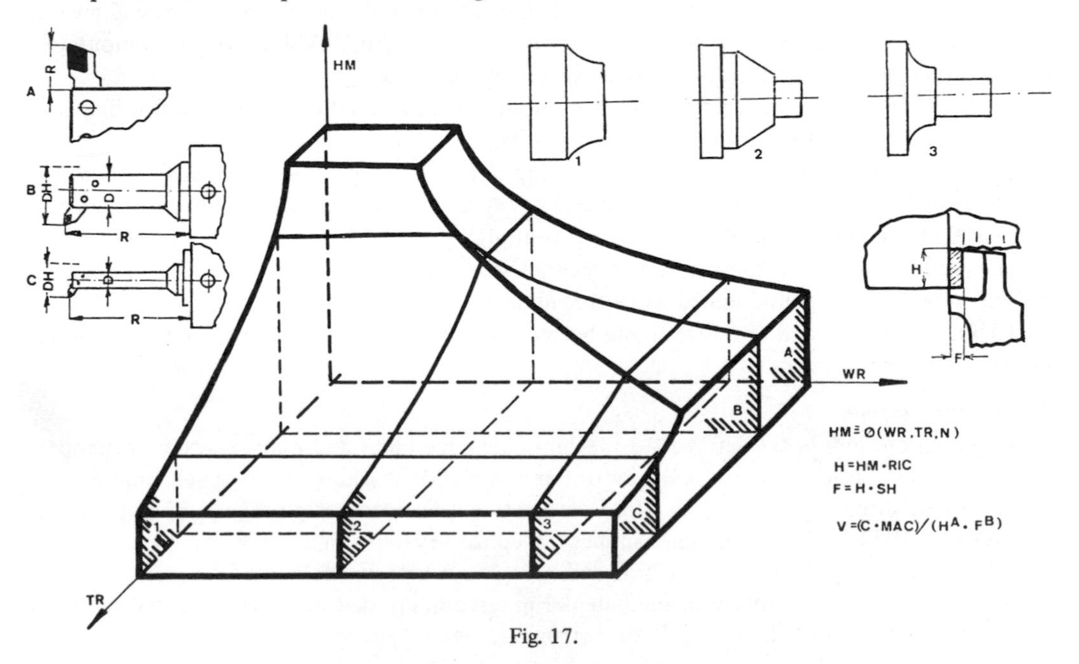

Fig. 17.

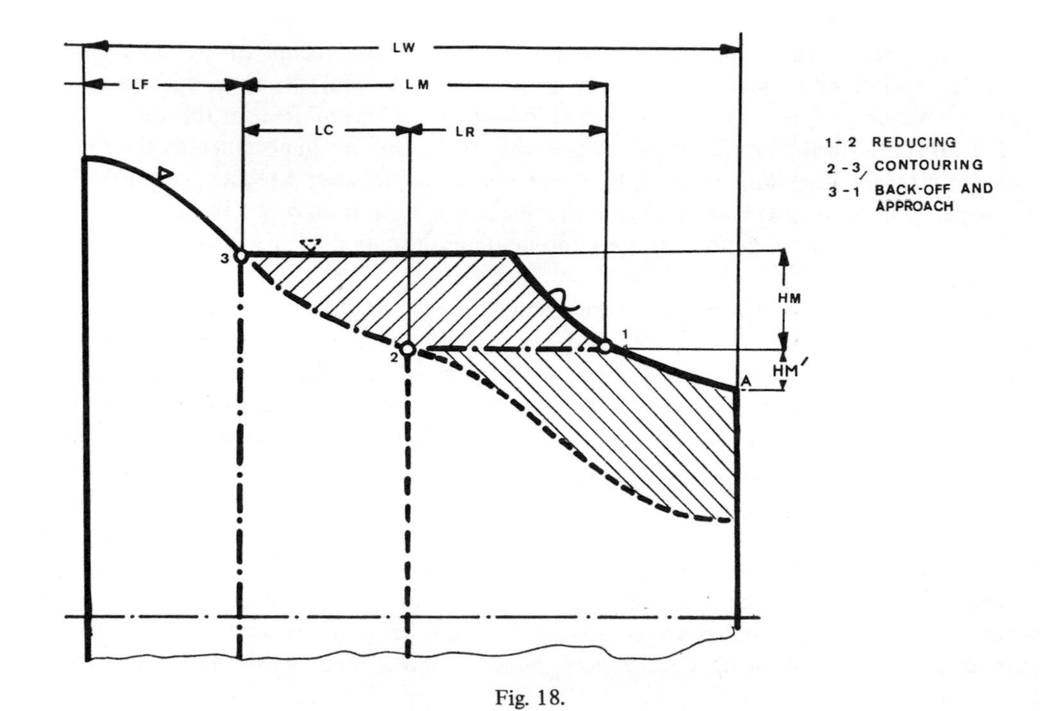

Fig. 18.

Thus a roughing stroke comprises a reducing and contouring part with different governing factors but resulting in a tool path leaving after a single roughing cycle a completely constant finishing allowance irrespective of any contour configuration.

Finishing allowance may be freely selected through thc methods-file, where axial allowance for external and separately for internal contours may be specified. Values for the computation of modified diameters for roughing may be also freely selected.

In the same manner the selection of the finishing feed in relation to surface finish specifications may be influenced through the methods-file.

There are a few 'canned' elements in the input format (Figs 2, 3) such as 45° chamfers, reliefs turned with the finishing tool, standard undercuts, blank cylinders under threads etc. All parameters for these may be also freely chosen through the method-file.

The same applies for some internal technological routines such as woodpecker feed for drilling, repeated compound infeed while threading etc.

These possibilities aim at a favourable balance between a high degree of automation and the necessary flexibility in use.

2.3 *Postprocessing*

Two separate inputs are defined for machine-tool-dependent and control-system-dependent data. Most n.c. lathes with two-axis control and one slide working at any given time may be implemented through the machine file. Conversational postprocessor-programming is planned for version MITURN71. The same applies to control-system characteristics. One may use MITURN for part programming in mm and a machine tool in inches or vice versa. Control systems working with absolute or incremental programming may be accommodated with any punched-tape code (EIA, ISO, 5, 6, 7, or 8-channels, even or uneven parity).

3.0 CONCLUSION

A compact and comprehensive programming system has been presented. This system is implemented on a time-sharing system and possesses a high degree of automation of production planning for n.c. lathes.

BIBLIOGRAPHY

'GF-Drehmaschinen für kleine Serien', *GF information*, Nr. 16 1970; Georg Fischer AG. 8201 Schaffhausen, Schweiz.

KERRY, T. M., 'GETURN – Part Programming Manual', General Electric Company, Schenectady, New York 12305.

KOLOC, J., 'The use of workpiece statistics to develop automatic programming for n.c. machine tools', *Int. Journal for Machine Tool Design and Research*, Vol. 9, pp. 65–80, Pergamon Press, 1969.

'Programmieren, Programmsprache EXAPT, Heiligenstaedt-Postprocessor', *Heyco Mitteilungen*, 6, 67, Heiligenstaedt & Comp., Werkzeugmaschinenfabrik, 6300 Giessen, Germany.

RECENT DEVELOPMENTS IN C.N.C./D.N.C. SYSTEMS

W. A. Carter
Plessey Numerical Controls Ltd

SUMMARY

Significant changes are beginning to take shape in the control of machine tools. Conventional hard-wired controllers are being challenged by the mini-computer which is rapidly gaining acceptance in many fields of control. While the mini-computer is replacing the hard-wired control at the machine tool, larger versions of the mini-computer are being used to control several machine tools in many different modes. This development involves several other aspects of management and production control. Its effects have yet to be fully realised and true cost comparisons with conventional numerical control are by no means easy. Emphasis has to be placed on developments concerning the use of data processing techniques in production control and its effects on the uses of numerical control in industry.

1.0 INTRODUCTION

Along with most technological innovations, numerical control undergoes continuous evolution. It is still a comparatively new method of manufacturing control but several changes have taken place since it was first introduced. Today we see the beginnings of a change that has considerable significance in the development of future control systems, not only for machine tools but also for other production processes and the control of these processes.

Historically, numerical control grew out of a requirement of the aircraft industry for more reliable and accurate methods for the manufacture of complex components. The problem in essence was to manufacture components of which the shape had been mathematically designed and defined. The shapes were not simple but consisted of combinations of circles, conics, point-defined space curves and mesh-of-points surfaces. With this problem in mind in the early 1950s, the U.S. Air Force awarded a contract to the Servo-mechanism Laboratory at Massachusetts Institute of Technology. That laboratory, along with others, was putting to use in the control of machine tools the servo-control techniques that had been developed for such applications, as gun control, radar control and aircraft mechanised control. Under the U.S.A.F. contract the M.I.T. Laboratory was to produce a computer-based system which would assist in the preparation of information for the machine-tool-control system under development.

The M.I.T. development, its subsequent integration into the aerospace industry, and similar developments in the United Kingdom by Ferranti and the British aircraft industry, heavily influenced the path that numerical control took in the first ten years.

The emphasis was strongly on the definition of component geometry and the computation of tool paths required to sculpture the shape of a component from blank material. Computer programming languages to assist in the geometric definition and tool path calculation grew out of this environment. Languages such as APT and PROFILEDATA reflect this period: they are based on a concept of milling contoured components.

Following this early development the potential of n.c. was realised in the manufacture of components that had complexities other than in their contours. Where parts have several faces or bores, such as are present on gear boxes, aircraft engine casings, valve bodies etc., n.c. could provide the type of production improvement we now know to be possible.

The change of emphasis from contouring to these other operations required the development of new types of machine tool. Components with many different hole sizes and operations, several different faces to machine, brought about the development of the machining centre, with automatic tool-changing and rotary positioning tables so that a component could have as much machining done on it as possible without having to move to another machine or be reset for subsequent machining.

These developments helped to spread n.c. into a wider application area, but also to highlight the problem of manufacturing control. The machine tools demanded that all the machining be completely pre-planned, that tools be available and preset, that work-holding techniques be defined, and that the blank material be not significantly different in size or constitution from component to component. All this adds up to a control discipline that in many organisations did not exist. So in many cases the introduction of n.c. served to highlight production control problems, and demanded solutions, which in themselves produced higher productivity for the workshop and increased savings far in excess of those directly stemming from the replacement of conventional machine tools with numerically controlled machine tools.

Both aspects of n.c. in contouring and in machining centre type operations have served to make production engineering in batch production workshops more attractive. Thus higher intellectual ability is now being focused on production techniques and control. The ground has been prepared for the growth of many new ideas and techniques in the area of batch production.

2.0 THE MINI-COMPUTER

While such changes have been progressing at a relatively slow pace, a development outside the machine tool and control system area has shot ahead far more quickly than predicted. It is the mini-computer, a development which radically influences the development of n.c. control systems and the use of n.c. machine tools. Cheap compact, reliable and general-purpose, it offers wide possibilities and scope. It will in many cases replace the n.c. controller as we know it today.

Until recently all n.c. controllers were hard-wired and special-purpose. Special hardware had to be designed and manufactured whenever a controller was interfaced with a machine tool. The mini-computer, which has now begun to enter on the scene, offers the possibility of more flexible integration of machine tool and control system because, when it is used, many machine tool characteristics can be handled by software.

The extent to which the mini-computer takes over or extends the facilities of a hard-wired controller varies considerably from design to design. The important fact to consider is that the mini-computer is having a tremendous effect on control system design and will have an increasing effect as time passes.

3.0 HIERARCHICAL CONTROL

While this development, linked directly to the computer, has progressed, other computer applications to machine tool and machine shop control have progressed in parallel.

Consider production shop control. It is found that three hierarchical levels exist at which control is exercised. These are:

- *Process level*–the machine tool or process is under the local control of a mini-computer.
- *Data level*–numbers of processes or machine tools are controlled directly from small computers, which basically store and distribute the data required at the process level
- *Management level* – provides overall management control of the data distributed and collected from the system at the lower levels.

3.1 *Process level*

Conventional control systems are hard-wired, which means that the individual logic functions to control the machine tool are permanently wired into the system. Interfacing a machine tool with the control system requires the logic pattern to be wired into the system for every integration. Introducing a mini-computer into the system allows the logic function to be performed by software instead of hardware. The system now becomes very flexible for it allows the control system to be integrated with special machine tool requirements. Logic modifications and additions can be made by a simple program change These systems are termed c.n.c. (computer numerical control).

Several other tasks currently performed by hardware can be performed equally well and with greater flexibility using stored programs on the mini-computer. Interpolation is such a function.

Currently a machine tool control system is provided with information to move the machine tool along a certain path at a given rate. Usually this information calls for a large, incremental move from the current position to the desired position, probably a two- or three-axis simultaneous move, and sometimes four- or five-axis in five-axis machining centres. From the information the control system interpolates pulses for each axis at the frequency required for the machine tool cutter to move along the path requested. In two-axis machining the form of interpolation can be linear, circular or parabolic. Above two-axis linear interpolation is used almost invariably.

Interpolation is a repeated, computed function and therefore could be performed by the mini-computer. However, once set up, the system to perform the interpolation does not vary, and therefore can equally well consist of hardware. The advantage of the computer, its flexibility, is of small consequence in this case. Indeed, because of the serial mode of operation of the computed instructions, the computer could be at a disadvantage in timing: the hardware interpolator can work in a parallel mode. Whether to perform the interpolation functions with hardware or software still remains unresolved in the control system field. Although mini-computers are constantly being reduced in cost and increased in speed, equally well it is possible to reduce costs and increase the reliability of hardware interpolators so that they remain strongly competitive.

Another primary function of a control system which the mini-computer may take over is control of the servo-loop. Software control of the servo-loop gives a shorter system-response time and consequently the ability to reduce the following error signal. As with interpolation, controlling the servo-loop by computer requires fast computing and the ability to handle high rates of data transmission.

C.N.C. systems, either being designed or in production, range from systems that perform every function through software, including interpolation, closing the servo position loop and machine function, while others control only the machine logic functions through software, leaving interpolation and the servo-loop control to be serviced as in the current manner by hardware.

Systems under development or marketed by my company range from one that controls the machine functions and control of the interpolators to one that offers complete computer control. The systems operate like conventional controllers in that they are stand alone and use conventional control paper tapes as their input.

How a system handles this information is a function of the individual design. Unlike current systems c.n.c. controller can have a relatively large buffer store in which several control blocks can be stored at one time. This reduces the demands on the paper tape reader and eliminates the possibility of having machine dwells between successive blocks of information. Some control systems read a complete part program into buffer store and have editing facilities so that, for example, the part program can be modified, speed or feed changes can be made at the machine tool, and the modified part program can be immediately used. Such systems require output paper tape facilities to allow modified control information to be punched out in a permanent form.

Putting a mini-computer into the nerve-centre of a control system gives it 'intelligence'. The computer can control functions additional to those performed by conventional controllers. It is possible for the mini-computer to retain information on repeated and frequently used cutting sequences for tapping, drilling etc. Such sequences currently follow hard-wired 'canned' cycles, or are handled by software in a post processor if a computer is used to prepare control information. The computer can be used to monitor machine tool performance and its own control system performance. Routine maintenance and diagnostic checking can be automatic, as can compensation for machine tool slide errors. When sensors and techniques have been developed to make necessary measurements, then the mini-computer will become ideal for adaptive control.

In general, replacing the hardware of a current n.c. system with a mini-computer in all but the most simple systems means that the computer can perform the control functions with equal capability and greater flexibility, and still leave scope to perform even more tasks on the same system. The more functions the machine tool has built in, such as tool changing, pallet loading, etc., and the larger the number of control axes, the more attractive a c.n.c. system becomes.

3.2 Data Level

Whereas c.n.c. handles the data directly at the machine tool, and the system is completely self-contained, other systems are being developed and used which employ a small to medium computer to control several machine tools. This is usually termed *direct numerical control* or d.n.c. Here the computer acts in a supervisory role, storing the control information for all the machine tools in a program library, usually on disc. The computer uses this program library to pass control information to the machine tools as and when they require it. In its supervisory role the computer collects information from the machine tools, recording job or machine status, number of parts produced, machine tool utilisation and performance, cutting tool performance and other data which could be classed as management information.

It is the prime function of a d.n.c. system to exercise stricter control of the manufacturing process and to ensure that realistic and reliable management information is being processed, either at this level or at a higher control level.

Several control system manufacturers, some machine tool builders and a few end-users have developed and installed d.n.c. systems. Control system builders have tended to be conservative in their approach and have used the d.n.c. computer to control a number of machine tools with conventional controllers, but passing the information to the system *behind the tape reader* (b.t.r.). Such a system needs only a minimum of modification to the control at the machine tool. Also it allows the user to interface all of several different machine tools and their various control systems with one d.n.c. system.

Notable among the machine tool builders who have introduced a d.n.c. system are Kearney and Trecker. Their system Gemini uses an IBM 1800 computer to control a number of machine tools with conventional controllers or with c.n.c. controllers. Interfacing the c.n.c. controller with a d.n.c. system is not very different from interfacing with a conventional controller. Control information is passed from the d.n.c. computer to the c.n.c. system in the form of a direct input and does not use the tape reader input. Such a system makes control easier in that the c.n.c. controller, with its buffer store, puts less frequent demands on the d.n.c. computer for information.

Another machine tool firm well established in this field is Sundstrand. Their system differs in approach to that adopted by most control system manufacturers. Sundstrand developed their Omni-control system to act like a d.n.c. computer, but in fact it is a hardware controller. It consists of one control unit and fifteen interpolating buffer units. Each interpolating buffer unit can control one machine tool with up to five axes and 96 machine logic functions. Omni-control has to be driven by a computer, such as an IBM 360, so that stored program information may be passed to the system. Only a simple control unit is required at the machine tool as the one remaining function is to control the servo-loop.

Plessey Numerical Controls' approach is to control machine tools having a possible variety of control systems and controlling them behind the tape reader. It is felt that the real problem is one of communications and information processing. The technical problem of design and control of machine tools from a computer is relatively simple. Information handling and production control pose a far greater problem. This scheme will be discussed in detail later.

3.3 *Management level*

The d.n.c. computer offers control at a local level and can ensure that a given cell of machine tools performs satisfactorily within its control boundaries. In most practical applications it is difficult to divorce and control one cell or group of machine tools effectively without taking the total production and management environment into account. In addition, the size and capacity of a computer usually considered for a d.n.c. application is not capable of processing n.c. information where computer assistance is required for part programming. Generally n.c. processors are large and consequently require a large computer to run them.

Because of these two factors many d.n.c. applications provide for the possibility of linking the d.n.c. computer back to a larger data processing computer, which may be the company's major management computer. It is on this computer that n.c. processing programs such as APT, 2CL or Exapt could be implemented and control information processed. This information could be passed to the d.n.c. computer to form the library of control programs for a day's or a week's production run on the d.n.c. machines. Given good communications and time-sharing facilities, part programs could be up-dated from the shop floor or planning office, not only with respect to speeds and feeds but also to geometry and tool motions. The revised program would be stored or passed to the d.n.c. system for storage and use when required.

With a system where all three levels of control are possible, the ability to hold close production control of the total manufacturing process comes nearer to reality. The d.n.c. system fits well into this picture and offers yet another step to the successful application of automation systems. Through its closer control, d.n.c. will increase the time the machine tool spends in removing metal, in much the same way that n.c. produces savings and greater productivity not entirely through its own merits, but through the increased discipline required to run it.

4.0 ECONOMIC ADVANTAGES OF D.N.C.

The advantages of d.n.c. are probably better understood when viewed in the light of what has happened with the introduction of n.c.

Generally it is accepted that a numerically controlled machine tool does not necessarily improve the return on the unit of capital employed. When an n.c. machine tool is installed in a machine shop where the overwhelming majority of machine tools are conventional, the n.c. machine will in the main be treated like the conventional machine, and no special attention will be given to its support requirements other than preparation of the control tape. Under these conditions one n.c. machine will do the work of approximately four conventional machines, and, as the cost is approximately four times, the return on the unit of capital employed remains the same.

This ratio is supported by evidence which shows that a conventional machine tool in a typical workshop spends approximately 16% of its time removing metal while an n.c. machine tool in the same shop would probably spend about 35% of its time removing metal. The constraints that are imposed by n.c. to introduce more discipline into planning generally increases output per shift to about 2:1 on a conventional machine. This 2:1 ratio, coupled with the increases in metal-removing time, results in about a 4:1 productivity gain.

Obviously there are cases in industry where the n.c. ratio is in excess of 4:1, but almost invariably the excess is due to a better understanding of workshop practice and management control in the presence of an n.c. tool. It has already been stated that, by its very nature, n.c. generates a number of constraints which must be respected if the system is to work at all. It is greater compliance with these restraints which takes the ratio between conventional and numerical control beyond 4:1. Increases of up to 60:1 have been reported, but, when analysed they have been found to be so great because n.c. has shown up the weakness of the previous organisation and the machining methods used. In other works, the n.c. machine has introduced new discipline to the whole workshop technique.

Other indirect savings from n.c. result from down-rating of the skill necessary to operate the machine tool and from more machining operations at one set-up, reduced scrap, and a high order of repeatability.

In general it can be said that the savings of n.c. accrue more from better management, at all stages from design to the finished component, than from the increase in metal removal rate.

It is in this area of improved control over the complete production process that d.n.c. will make its greatest contribution, and, if d.n.c. does contribute to an increase the time spent on removing metal, as it almost certainly will, then further savings will be made.

4.1 *Tape preparation and handling*

From my company's in-house experience in the use of n.c., and from statistics gathered from my company's N.C. Centre, it appears that the planning-to-machining-time ratio is in the order 30:1 when complete processing is used.

Breaking the 30:1 ratio down into time units shows where the planning time is spent:

- *Pre-planning.* Assimilating knowledge of the shape of the component, special features etc., of the operations to be performed, of machine tools required, jigs, fixtures and special tooling.
- *Planning.* Statement or line-by-line planning to define the geometry and tool motions in the language required by the particular n.c. processor employed–i.e. the part programming stage.
- *Prove out.* Proving of the control tape on the machine tool. Initial run with tool backed off from work and subsequent trial runs.
- *Modifications.* Changes in the planning instructions to the computer to produce the required component within the minimum process time, coupled with attempts to obtain high overall cutting efficiency.

Generally the last two phases are not optimised, and this is largely due to the inaccessibility of the computer and the long train of events required to produce and accept a new tape. In almost all cases in British industry the computer is a centralised processor some distance away from the planning department and workshops. In the main, the production managers of British workshops are so grateful to get an n.c. tape that produces the desired shape to the required tolerance that they will avoid any attempt to optimise: they fear the introduction of errors, coupled with the delay inevitable on referring back to a centralised processor.

The presence of a computer, with d.n.c. on a number of machine tools, will contribute to decreasing this 30:1 ratio by giving immediate access within the capability of the computer employed. If, however, the workshop is provided with on-line connection to the central processor on a time-shared and immediate-on-call basis, a d.n.c. installation will not significantly contribute.

But, apart from the fact that British industry is not likely in the foreseeable future to provide each workshop with an on-line computer link d.n.c. still offers advantages in tape preparation and proving, notably the ability to remove one tape reader from each machine tool – it is accepted that the tape reader is probably the greatest single source of unreliability in n.c. A single input to the d.n.c. system provides data for each individual machine, thereby decreasing the number of times that the paper tape has to be made, handled, and checked. There are also savings from the direct, and virtually instantaneous, access to each block of information relating to the movements of each axis of all machines under d.n.c. Compare this facility with the laborious process of running through each tape on each machine to locate the block required for modification. So even with on-line connections to the workshop, d.n.c. contributes to reducing the 30:1 ratio.

4.2 *Direct labour*

Current n.c. machines have contributed to the reduction in the use of skilled men in the machine shop. In most cases they have been transferred to the planning office after training as part programmers. Now they prepare the input instructions for the computer to produce the n.c. tape. Their place at the machine tool has been taken by machine operators.

Whilst machine operators do not necessarily have to be time-served millers or turners, in most cases they must be more than machine minders. There are still decisions that have to be made and actions taken on the control system and machine tool for each component: establishment of datum; manual selection of speeds, feeds and tools unless a sophisticated tool-changer is employed; setting of cutter length and diameter, etc. Thus, in industry today, there tends to be one semi-skilled operator per n.c.

The ability of d.n.c. to store proven programs on disc file, to an extent sufficient at least to cover one shift, and probably more, decreases the action required of the machine operator. The work of the machine tool operator is also reduced by the ability to present discrete instructions, details of tools, jigs and materials per component, time allowed and batch quantity information on a c.r.t. display, and by significant decreases in tape handling.

Thus, with d.n.c. it is now possible for one machine operator to handle more than one n.c. machine. In a row of six machines it is thought likely that only two operators would be required, but clearly this depends largely on the machining cycle and batch size of each component.

4.3 *Reporting system*

With current n.c. there is little or no automatic monitoring of the process except that the n.c. tape defines the shape of the component to be cut. Some control systems monitor the performance of the control system to a limited degree, and diagnostic indicators of a limited nature are sometimes provided. With its reporting system, and its capacity to sample the state of the complete n.c. machine and report on cases of down time, d.n.c. contributes to increasing the availability percentage per shift upon which production can be planned.

W. A. Hubbard[1], of I.B.M. Systems Planning Division, claimed in a paper in 1970 that this reporting facility, coupled with the reduction in scrap, give an improvement amounting to some 10%.

4.4 *Management*

The difficult task of converting the order book into a planned production schedule – from the placing of an order for materials, through production planning of each phase of the manufacture, to the loading of the machine tools – is being considerably assisted by the use of general-purpose computers and the adoption of e.d.p. techniques. Nevertheless, when an unscheduled requirement arises because of a change in the order book, or shortages, or scrap, or failure of equipment, tools, etc., the problem of predicting the effects of corrective action and taking it is very severe.

With its reporting system capability to advise management of the planned and actual production in time, cost and utilisation etc., d.n.c. allows the effect on changes to be assessed quickly. Furthermore, the associated disc file confers the ability to store a large number of programs relating to the current production requirements, so one of the directed machines can perform unscheduled tasks without disturbing the remainder. At the same time the materials, jigs, tools and fixtures relating to the new requirement can be reported.

We have already seen the advantages of current n.c. compared with conventional techniques. They arise as production planning and methods are improved to abide by the constraints imposed. More constraints accompany d.n c., which, in turn, will lead to greater productivity. Probably an effective and continuous system reporting directly from each machine tool will enable management to make rational decisions and see the full effects of any proposed action before production is allowed to continue at the demanded level.

If the full advantages of d.n.c. are to be achieved communication facilities must be extended beyond those between the planning office and the machine tool. Tool and material stores, tool room etc. must be included.

4.5 *Typical costs*

It is assumed that, on average in British industry, machine cycle time is approximately 30 min, the time spent on metal removal is around 35%, and average batch sizes are in the order of

20. Using these values per shift and assuming that the cycle time and the metal removal time are about the same (although in practice it is not likely to be more than 70%), one calculates that an existing n.c. machine tool will produce 5·6 of these components per 8 hour shift in a 45 hour week, which is typical of industry. Just over 6 batches of similar components would be made per month. This agrees reasonably well with the number of tapes that Plessey Numerical Controls supplies to industry through its tape centre.

The cost per hour of an n.c. machine, including machine operator, depreciation, maintenance, replacements, tools jigs and fixtures on a single-shift basis, lies between £10 and £20, depending upon the cost of the machine. An average of £15 might be used. On a double shift this probably falls to £10.

The additional cost of a d.n.c. installation, say controlling a row of 6 machines, the computer, disc file, keyboard input and some form of c.r.t. display at the computer and the machine interface unit, including display boards in communicating devices between machine tool stations and the production control planning office, is probably £60 000.

5.0 QUANTIFYING SAVINGS

As can be seen, some of the savings with d.n.c. can be quantified by analysis of the various tasks being performed with existing n.c. and how these change under d.n.c. These are in the category of direct savings. Other, more indirect, savings, which increase overall productivity and make better use of the capital employed, will only be quantified from practical experience in production work. Nevertheless, an attempt can be made to quantify anticipated savings.

5.1 *Tape handling*

On existing n.c. machines time must be allowed for loading and unloading a tape for each batch of components. Time taken to handle the tape will probably be in the order of 10 or 12 min. While it is possible to load a tape much more quickly than this, in practice the mere fact that the operator has to perform an action will inevitably cause delays.

The time saved by d.n.c. in prove-out will depend largely on the batch size, the size and complexity of the component, the available tape-editing facilities and the turn-round time on the computer (if n.c. computer processing is required).

On a batch size of, say, 20, with a cycle time of 30 min, and allowing for two or three corrections before the tape is finally proved, savings in the order of 30% could be achieved.

5.2 *Reporting system and reliability*

It is not possible at this point in time to realistically quantify the advantages of having an efficient reporting system, but several investigators have put these savings in excess of 10%.

5.3 *Direct labour*

While it is claimed that one operator could handle three d.n.c. machine where with existing n.c. he could handle one, this depends upon cycle time and batch quantity. Generally it could be expected that one operator would handle two machines.

6.0 CONCLUSION

It is clear that all system manufacturers recognise that it is possible, or more probably necessary, to consider information flow in a three-level hierarchy for the successful application of d.n.c. The machine tool is the process or functional level being connected to the control or d.n.c. level. At the highest level of the hierarchy comes the large time-sharing data-processing computer. All systems incorporate all levels, or make provision for interfacing with another level. Similarly, all manufacturers, while they may still pursue different paths at the moment, recognise the need to incorporate n.c., or stand-alone c.n.c., systems by interfacing behind the tape reader.

All manufacturers emphasise information collection and processing. Many systems have display terminals at the machine tool. All have some form of communication with the machine tool. Obviously all manufacturers believe that the d.n.c. system will form part of an overall production control system. It is clear that American manufacturers are attempting to come to grips with organising and optimising the software to meet the housekeeping requirements of the machine tool, the size of the interface computer and the demands of the reporting and communication system.

If the U.K. is to challenge America, and equally well Germany and Japan, considerable emphasis has to be placed on finding a satisfactory solution to the software and communication problems.

REFERENCES

1 HUBBARD, W. A., 'Economic Considerations in Computer Control of N.C. Metalcutting Equipment', in 'management's Key to the Seventies', Edit. M. A. Devries, pp. 251–263, Goodway Inc., U.S.A. (1970).

SESSION VIII

Computer Control of Manufacture

Chairman: Dr I. D. Nussey
Numerical Control Co-ordinator,
IBM (UK) Ltd,
Birmingham

COMPUTERISING THE PRODUCTION OF A NUMERICALLY CONTROLLED MACHINE SHOP

P. Zachar, J. R. Arrowsmith, and M. C. de Malherbe
Canadian Institute of Metalworking,
McMaster University, Ontario

SUMMARY

A description is given of a research programme undertaken at the Canadian Institute of Metalworking, McMaster University, to study the problems associated with the economic utilization of a number of numerically controlled and conventional machine tools, and provide a vehicle both to analyse a given production situation and to ensure the correctness of the decisions made. A computer program is developed simulating the overall operation of an N.C. shop. Equations are derived showing the total cost of manufacture of particular items by numerically controlled and conventional machines. Factors are included allowing for utilization and efficiency to arrive at the true cost of manufacture. The simulation model developed incorporates characteristics of machines, including size, feeds, speeds etc. Part characteristics are coded and the program is devised, allowing for these factors, to make the most economical selection of machine tool.

NOTATION

K	total cost in dollars
A	the once-only cost
B	cost occurring with each batch
C	cost occurring with each piece
m	number of batches
n_i	number of pieces in the ith batch so that the case of unequal batch sizes is allowed
A_T	the economically allowable programming cost
n	total number of pieces

1.0 INTRODUCTION

The program describes computerized methods of the operation of a numerically controlled machine shop, taking various parameters into consideration. This effort should be considered of a preliminary nature. The results of the work described are promising; so much so, that further development of the program will be undertaken at the Canadian Institute of Metalworking with the object in view of producing useful and realistic techniques for optimal use of numerically controlled and conventional machine tools within a given workshop.

2.0 PROGRAM DESCRIPTION

The program provides methods of analysing and selecting machining operations, their sequences, tolerances and other parameters, and selects the most suitable and economic machine for manufacture. A computer simulation of the operation of a numerically controlled machine shop is developed. Basic characteristics of the parts are coded and are compared with characteristics of the N. C. machines available. If the characteristics cannot be matched, the part is rejected as being unsuitable for N. C. manufacture, but if the characteristics can be matched, the program evaluates the economics of manufacture to determine and select the least-cost method, either conventional or numerical control. The program is also arranged to single out any component as being unsuitable for N. C. if manufacture by conventional machines proves to be of lower cost. When one of the N. C. machines offers the lowest cost, the name of the machine and the cost of manufacture are printed out.

As the program is intended to study the economic parameters, it provides an order generator simulating a work load. The study of the factors which contribute to the cost of manufacture by numerical control in relation to the characteristics of the components, and in particular the contribution of programming or tape preparation are studied. The following sections contain descriptions of the program and introduce a concept of complexity factor. This factor may be used to relate cost to part characteristics in which many parameters are involved.

An overall flow diagram for the complete program is shown in Fig. 1. Both simulation and analysis is modes of operations are shown. The first block contains an order generator subroutine which produces a list of orders containing the characteristics of the parts to be made. Subsequent to generation, each order is processed by a subroutine called N. C. Machine Selector. This compares the workpiece characteristics of the order with the characteristics of the N. C. machines, already stored within the program. The output of the subroutine produces a list of numerically controlled machines suitable for making the part and a list of rejected orders when the part characteristics cannot be matched with those of any machine tool. The order is then processed by a routine which generates a time (man hours) for each operation and prints out a listing of orders including the machining operations and their associated times. At this point in the program, the study mode is separated from the analysis mode. In the analysis mode, the orders are processed by a subroutine which calculates the total cost of the order. This total cost is obtained for each N. C. machine on the list of suitable machines. It is also calculated for manufacture by conventional non-numerically controlled machines. The total cost figure for the different machines (including the conventional ones) is compared and the least-cost machine selected. Those orders for which the cost of manufacture by conventional machines is the lowest are separated.

For the rest of the orders, the name of the selected machine and the total cost are printed out. These data are applied to a sub-routine called Total Cost Analyser. This sub-routine is used to study the relation of total cost to each of the part characteristics and will be discussed in detail in a later section of the paper.

It was recognized even in the early days of numerical control that the key to economically successful operation is programming or tape preparation. However, there are still very few data generally available to guide the user of the cost of preparing the tape. A subsection of the program has been developed to investigate this important aspect of the subject. In this section, the operation times are processed through a second cost calculator subroutine. The assumption is made that there is a limiting cost for N. C. production which occurs when the N. C. cost is equal to the cost of conventional non-N. C. production. It follows that there is a limiting or allowable programming cost for each order. It is also clear that this factor is linearly dependent on the quantity of pieces to be made. The subroutine is therefore designed to calculate allow-

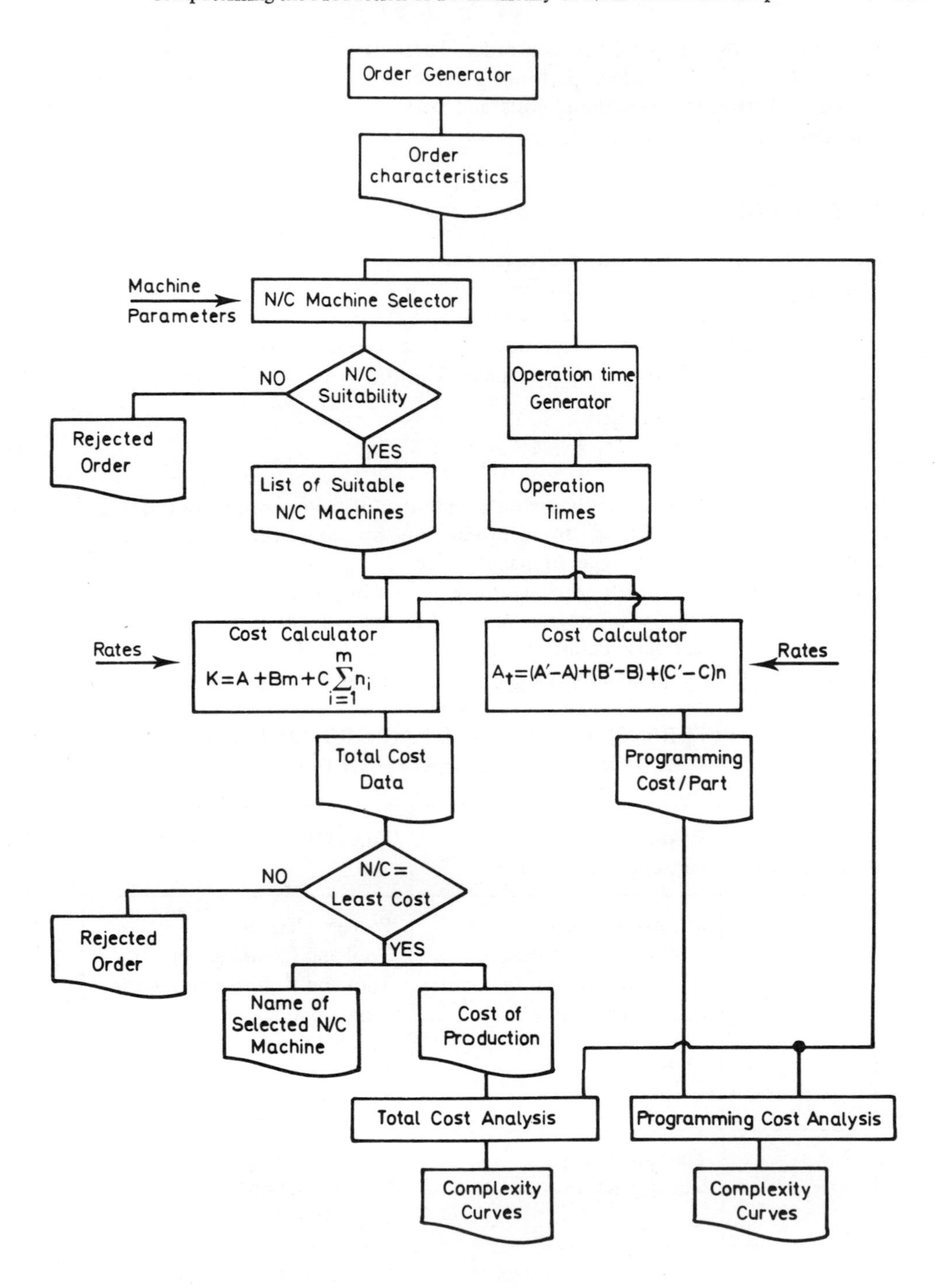

Fig. 1. Overall flow diagram of complete computer program simulating an N. C. shop.

able programming cost per piece for each order. These data are applied to a subroutine called Programming Cost Analyser, which performs a similar task to the Total Cost Analyser. A study of the relation between programming costs and part characteristics was undertaken and the results are discussed later in the paper.

3.0 COST CALCULATIONS

The most significant part of the program is the subroutine for calculating the cost data. The total cost of manufacture of an order for a number of components produced by machining contains three elements: those costs which occur only once and are independent of the number of pieces to be made; those costs which occur with each batch of parts that is made; and those costs which occur with each piece that is made. Thus the total cost is given by the following equation:

$$K = A + B\,m + C\sum_{i=1}^{m} n_i \tag{1}$$

Each of the constants A, B, and C contains a number of factors and these differ in the cases of numerical control production from production by conventional machines. Table 1 enumerates the factors for both types of manufacture.

There are several points worth noting with regard to Table 1. The process planning for N. C. machines is often combined with programming but in this context it is considered as only the preliminary planning necessary before the decision to use N. C. or not is made. The three factors, programming, computer processing and tape proving comprise the overall tape prepar-

TABLE 1. DIFFERENCE BETWEEN THE FACTORS CONTAINED IN THE CONSTANTS IN EQUATION 1

Constants in equation 1	Production by conventional machines	Production by N.C. machines
A	Process planning Tool and fixture design Tool and fixture manufacture Fixture proving	Process planning Tool and fixture design Tool and fixture manufacture Programming Computer processing Tape proving Fixture proving
B	Tool presetting Machine set-up	Tool presetting Machine set-up Inspection
C	Loading/Unloading parts Machining Inspection	Loading/Unloading parts Machining

ation factor previously mentioned. A difference in location of inspection costs should be noted. It is considered that in numerically controlled production only the first and possibly the last of a batch need be inspected so that this factor is related to the batch rather than the piece in the case of conventional production.

Equation (1) is used directly for total cost determination and forms the basis for determining allowable tape preparation costs. However, in order to simplify the analysis in the initial study, it is assumed that all batches are equal in size, so that

$$\sum_{i=1}^{m} n_i = mn$$

and the equation becomes

$$K = A + Bm + Cmn \tag{2}$$

A further assumption has been made which again simplifies Equation (2), that

$$m = 1$$

then

$$K = A + B + Cn \tag{3}$$

This particular aspect of work provides a study for all cases since this factor is by nature independent of the number of batches but decidedly dependent on the total number of pieces. If primed values are used for the case of conventional machines and unprimed values for N. C. machines, Equations (4) and (5) are obtained:

$$K = A + B + Cn \tag{4}$$

and

$$K' = A' + B' + C'n \tag{5}$$

The tape preparation cost is extracted from A of Equation (4) and designated A_T, a new value of A being obtained, giving

$$K = A + A_T + B + Cn \tag{6}$$

If $K = K'$, Equations (5) and (6) may be combined so that

$$A_T = [(A' - A) + (B' - B)] + (C' - C)n \tag{7}$$

The factor A_T is the economically allowable programming cost for an order of n parts and is clearly linearly dependent on n. Since the objective of the study is to relate allowable programming cost to part characteristic, the factor A_T/n is more useful.

4.0 ORDER GENERATOR

Some mention must be made of the subroutine for generating the orders and the part characteristics since the validity of the output of this section affects the entire simulation mode. An analysis of parts showed that there are several factors that are common to all parts and which have a major effect on both the cost of production and the type of machines involved. These factors are enumerated below:

- X, Y, Z overall dimensions.
- Material.
- Tolerances.
- Weight.

- Type of operation to be performed.
- Total number of operations.
- Number of pieces to be produced.

The initial study has been limited to machining centre application, and this places certain restrictions on the above factors. The following limitations have been set on the above parameters to provide a work spectrum within the capabilities of a small shop:

- Overall dimensions $X = 30$ in; $Y = 20$ in; $Z = 20$ in.
- Material was restricted to cast iron and aluminium.
- Tolerances – four possibilities – 0·005 in; 0·0025 in; 0·001 in; 0·0002 in.
- Weight – maximum, 3000 lb.
- Total number of operations – maximum, 100.
- Type of operations – drilling, tapping, boring, reaming, milling.
- Number of pieces – maximum, 100.

Fig. 2 shows the flow diagram for the subroutine. Parameters are assumed to have either

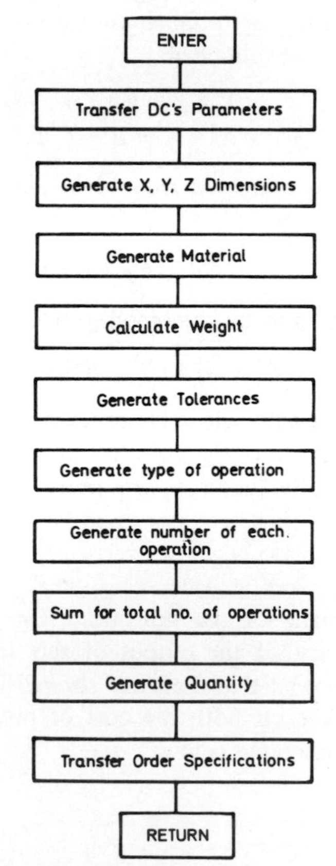

Fig. 2. Flow diagram for the subroutine.

uniform or normal distribution and their values are generated by means of a random number generator and the Monte Carlo rejection method. The complete set of orders, together with their characteristics, are then printed out.

5.0 MACHINE SELECTOR

For this program four machining centres were chosen and their characteristics fed to one side of a decision table. The four machines chosen were: a MOOG-83-1000; an ExCell 108; a Marwin Modulamatic; and a Sunstrand OM1. The order characteristics are applied to the second axis of the decision table and the appropriate comparisons made.

6.0 OPERATION TIME GENERATOR

This subroutine applies a manufacturing time to each operation. It will be appreciated that the variation of possibilities is enormous. For example, a component may require five different hole sizes to be drilled and these may conceivably range from ¼ in. diameter, ½ in. deep to ⅝ in. diameter, 1½ in. deep. Because of this difficulty, the times have been assigned manually in a somewhat arbitrary but practical way. However, for future study the possibility and practicability of establishing a relationship between operations and part size is being considered.

7.0 THE COST ANALYSIS PROGRAM

The prime purpose of the program is to investigate the relation of both total cost and allowable programming cost. These were determined for all the parts and the values obtained plotted. The philosopy underlying this approach is that all machined parts can be categorised according to the parameters that contribute to the cost. It is suggested that a single factor of complexity which embodies the various component parameters can be established. A factor of this nature would be of enormous value to industry since curves of this factor against both total cost and allowable programming cost per part could be provided from which almost immediate cost standards would become available.

In the initial study, a set of typical engine lathe parts, such as headstock covers, quadrants, gear boxes, feedrod and leadscrew brackets and apron frame castings were selected. Fifty-eight parts were analysed and their part characteristics identified. The total production cost and allowable programming cost were determined for all the parts and the values obtained plotted against the quantified value of each part characteristic. The results are illustrated in Figs. 3 – 5. These curves show present limitations of the program. For example, only two materials have been considered, so the curve is as yet not clearly established. In further work it is proposed to use the machinability rating to quantify this parameter. The effect of tolerances needs further study, as the assumption has been made that only one value of tolerance is present in each part. This was done to establish the basic program, which can be used for analysing a particular machine shop and selecting the appropriate machine for production.

As a result of investigating the parameters considered, it is felt that the approach described is a feasible one. Further work is, however, necessary both to refine the procedures and expand the program to consider other factors such as shape in relation to holding methods, type of tooling, the effect of load on the heat generated during the machining cycle, amount of material to be removed, and others that have been omitted initially.

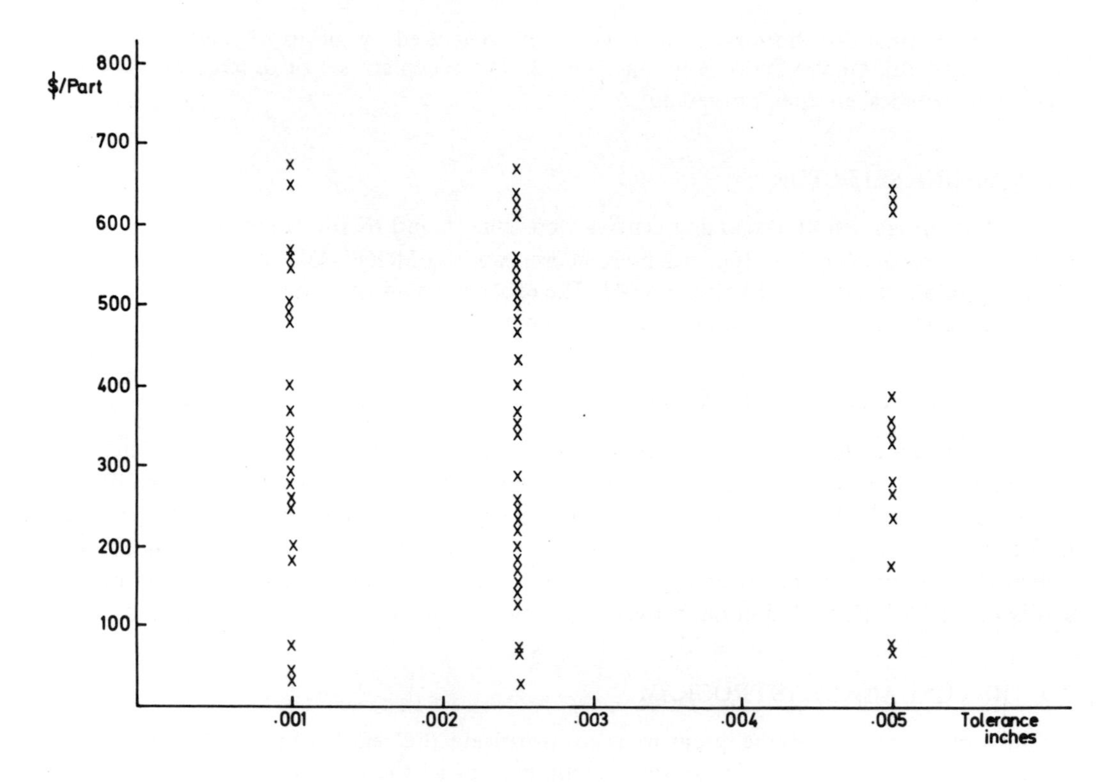

Fig. 3. Total costs as a function of the quantified values of each part characteristic.

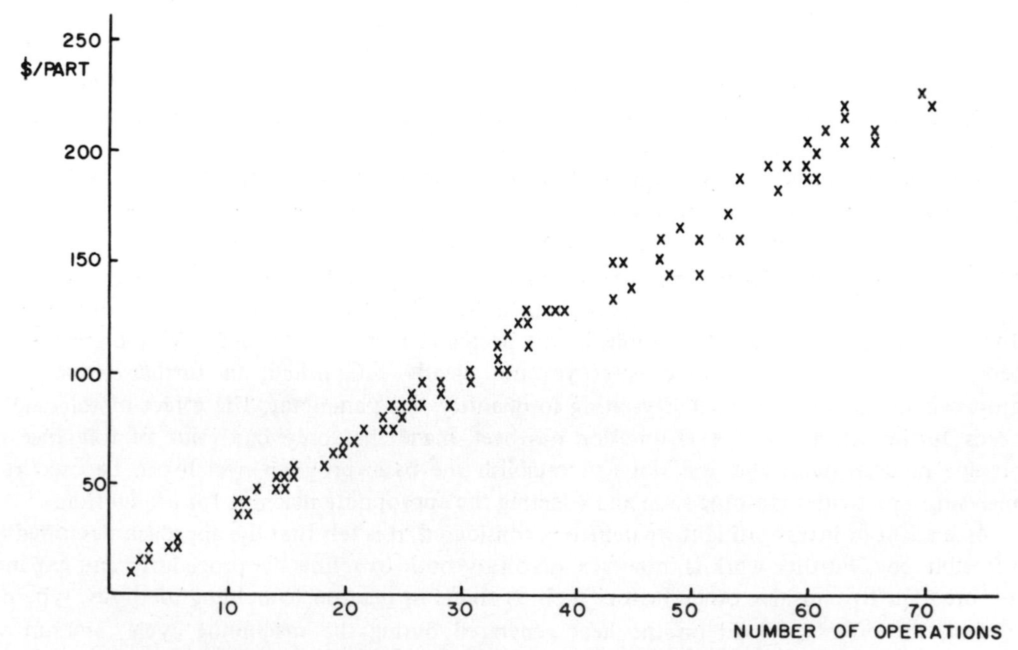

Fig. 4. Total costs as a function of the quantified values of each part characteristic.

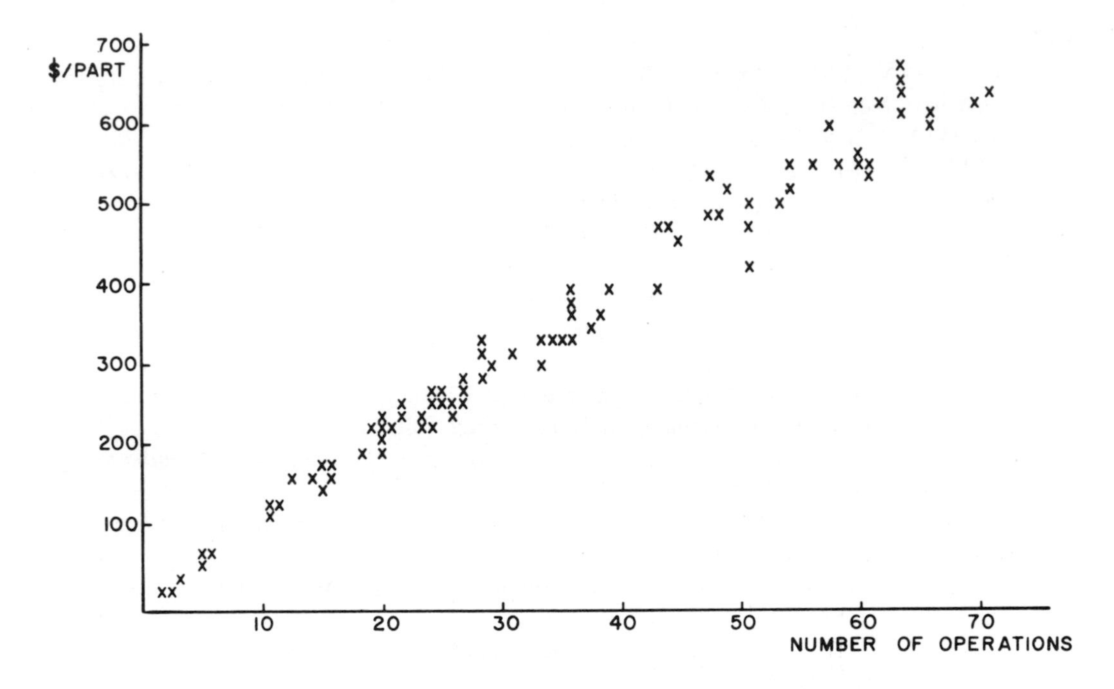

Fig. 5. Total costs as a function of the quantified values of each part characteristic.

8.0 CONCLUSION

An initial computer program has been described capable of analysing the operation of an existing or proposed machine shop and setting up a simulation of a numerically controlled shop operation. Because of their high capital cost, numerically controlled machines must not be idle but kept supplied with suitable work. In most companies, both large and small, the selection of work for the N. C. machines is made rather by 'feel' than by analysis and this can have the result that the company is not realizing the full value from its investment. The analysis mode of the program which is the subject of this paper allows the suitability and economic viability of each proposed part to be determined at the preliminary planning stage and even when quoting for work.

The first application of the simulation has been in the nature of an economic study of the parameters of an N. C. operation and as a result the concept of a complexity factor is suggested. It has been shown that this concept of a single numerical factor relating to cost of production is a practical one. There is, however, a great deal of work to be done to develop the concept into a tool that is useful at the practical machine shop level, and work is proceeding at the Canadian Institute of Metalworking, McMaster University, to pursue this course.

Further work is also proposed tying this program with a machine loading program and allowing consideration of the case when the lowest-cost machine is fully loaded and the next most desirable machine has to be used.

REFERENCES

1. HEIDEMAN, M. V., de MALHERBE, M. C., and HOUSE, G. J., 'Integrated Manufacturing Cycle with Numerical Control Machine Tools', *Proc IEE,* **116**, (11), Nov. 1969.
2. SIDDALL, J. N., *Theory of Engineering Design,* McMaster University, 1967.
3. NIEBEL, B. W., 'Mechanized Process Selection for Planning New Designs', *American Society of Tool & Manufacturing Engineers,* Collected Papers, 1965, Paper 737.
4. BROSHEER, B. C., 'NC Lathe Cuts Small – Part Cost', *American Machinist,* Jun. 15, 1970, 87.
5. 'The Computer is a Manufacturing Tool', *American Machinist,* Jun. 29, 1970, 68.
6. CRAWFORD, R. E., et al. 'How to Improve Production Facilities', *Canadian Machinery & Metalworking,* Jul. 1969, 61.
7. ARROWSMITH, J. R., 'What NC is doing to Production Organization', *Production Conference, 69, Canadian Machinery & Metalworking,* Jul. 1969.
8. NELSON, J. L., 'NC Utilization', *Aeronautic & Space Equipment & Manufacturing Meeting,* Los Angeles, California, Oct. 1968.
9. CROOKALL, J. R., 'How to Look at the Justification Factors', *Canadian Machinery & Metalworking,* Mar. 1969, 82.
10. NEISE, K. A., 'Complete Production Control with Aid of Computers', *Canadian Machinery and Metalworking,* Nov. 1969, 73.
11. HAAS, P. R., 'An Approach to the Rationalization of NC Machine Tools', *SME Technical Paper,* MS70-150.
12. HADDEN, A. A., and GENGER, V. K., *Handbook of Standard Time Data,* Ronald Press Co., 1954.
13. PUCKLE, O. S., and ARROWSMITH, J. R., *An Introduction to Numerically Controlled Machine Tools,* Chapman and Hall, 1964.

THE REDUCTION OF PART PROGRAMMING TIME TO A MINIMUM IN THE PRODUCTION OF DROP FORGING DIES BY SPECIAL APT MACRO

D. French
and
Mrs S. Nada
University of Waterloo, Ontario

SUMMARY

When a numerically controlled machine tool is to be used to produce intricate shapes, it has been found that the part programs may take many times as long to write as the actual cutting time of the numerically controlled machine tool. To overcome this difficulty, it is possible in the APT digital computer program for NC machine tools to build up special subroutines called macros, which contain a program for a particular component, and to store these macros within the APT program. A macro can then be recalled by the name allocated to it. This approach is applied in the manufacturing of dies in the drop forging industry, where this programming problem has been acute. It is shown how the macro method of APT programming is a means of creating a library of special programs. A forecast is made as to future development using group technology.

1.0 INTRODUCTION

When a numerically controlled machine tool is to be used to produce intricate shapes or even relatively simple three dimensional shapes the programming of the part becomes a problem. In many instances, it has been found that the part programmer may take up to 50 hours to produce a program which will give about 1 hour actual cutting time on the numerically controlled machine tool. Generally, if a program has been written for a particular component, and a second part has to be produced similar in configuration to the original part, but with dimensional changes, for example the length of the component or a change in one of its radii, the part programmer must write a new part program. In producing work of a 3 dimensional form the part programmer must have a thorough understanding of the NC language of the digital computer program used to produce the output tape.

The manufacture of dies in the drop forging industry has always been a problem as to the method and cost of manufacture. Copying machines are used extensively. However, a 'Master' must be produced in the first instance for the copying machine. EDM has also been used in the production of dies in the drop forging industry but in this case a male electrode must be produced in order that the die cavity can be produced on the EDM machine. Numerically controlled machine tools are being used for the production of dies for drop forging, but in some instances are proving to be uneconomical, due in the main to the time taken to program the die.

It has been determined by a study of the drop forging industry that the dies used can be classified into a number of geometrical shapes. These shapes would cover about 70% of all dies produced. The geometrical shapes can be classified as:

(1) Cylinder to cylinder
(2) Cone to rectangular shape
(3) Cylinder to spherical shape
(4) Cylinder to hexagonal shape
(5) Cylinder to elliptical shape
(6) Cone to cylindrical shape
(7) Cylinder to rectangular shape

2.0 THE APT MACRO APPROACH

It is possible in the APT digital computer program for numerically controlled machine tools to build up special subroutines called macros which contain a program for a particular component, store the macro within the APT program. This macro can then be 'called' by the name allocated to it. In addition, it is possible by using variables within the macro to alter dimensions and orientation of the part.

When designing a macro for the production of drop forging dies the specification of the macro must be determined. Such requirements were:

(1) The macro program must accommodate all variations in dimensions as necessary.
(2) Filleting between the geometrical parts must be generated.
(3) The shape must be capable of being rotated and/or translated.
(4) The major shape must be capable of having its minor shape, at one or both ends.
(5) The size and shape of the cutter must be dictated by the requirements of the part programmer.
(6) The amount of displacement of the cutter for machining the components must be calculated so that the allowable cusp height between machining passes is obtained.
(7) The program must take into account that when a length of the part is specified, and the cusp height is stated, the number of passes may not be a whole number.
(8) Post processor commands must be included within the macro and changed by the part programmer as desired. For example, to turn a spindle on or off or to reset a new spindle speed or feedrate.
(9) The part programmer should be able to produce the program in the minimum time and with a minimum of effort.
(10) To be capable of producing a male and/or female shape as required.

This paper illustrates a macro written for a cone to rectangular configuration (Figs. 1 & 2). Provision has been made for either one cone and one rectangular shape or two cones and one rectangular shape.

It is not possible to write a macro which will cover every contingency, and certain restrictions must be imposed. In this particular case the restrictions are (referring to Figs. 1 & 2):

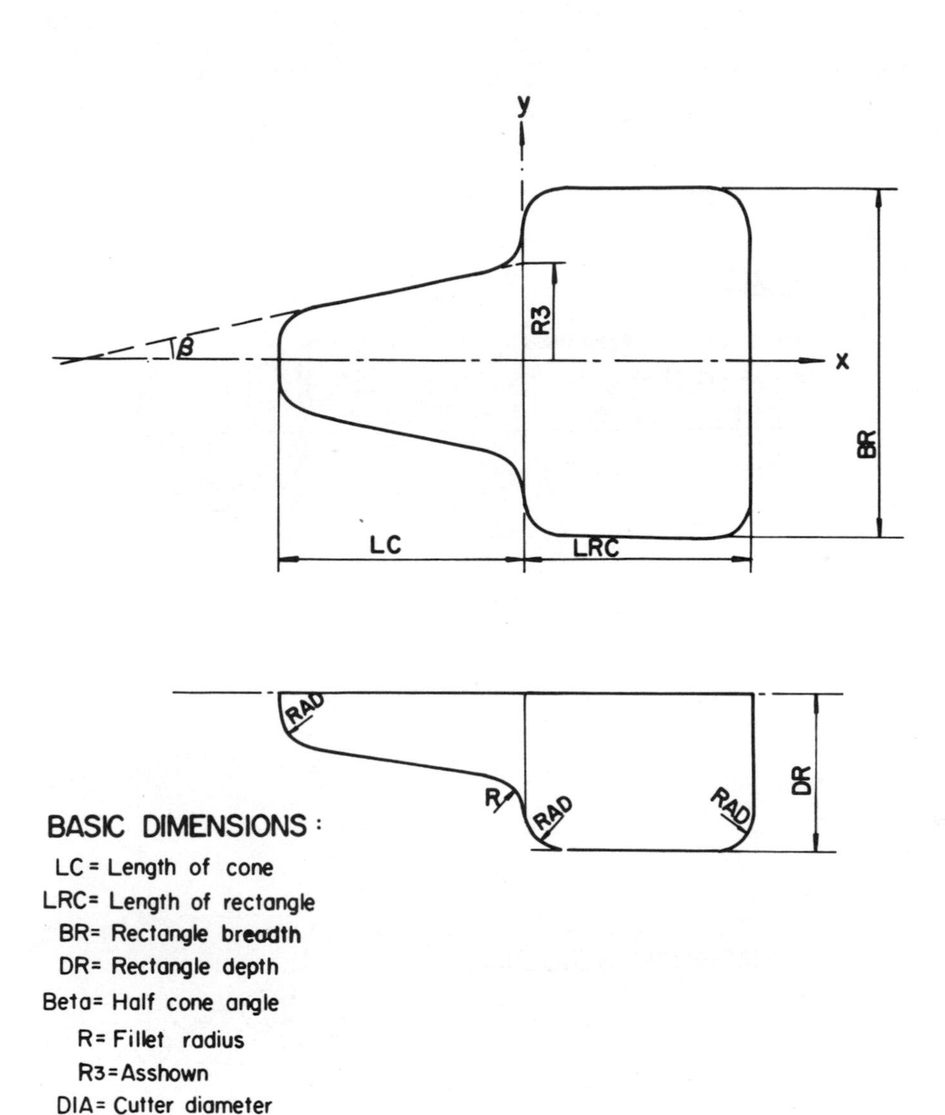

BASIC DIMENSIONS :

LC = Length of cone
LRC= Length of rectangle
BR= Rectangle breadth
DR= Rectangle depth
Beta= Half cone angle
R= Fillet radius
R3 = Asshown
DIA= Cutter diameter
RAD= Cutter radius
HYT= Cutter height
CP= Surface finish cusp
GAM= Matrix angle of rotation
XTI = Translation x– coordinate
YTI = Translation y– coordinate
ZT = Translation z- coordinate

Fig. 1. Cone to rectangle configuration

CONE

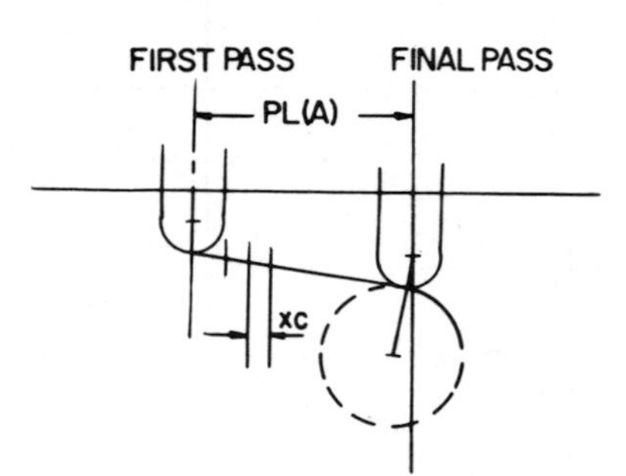

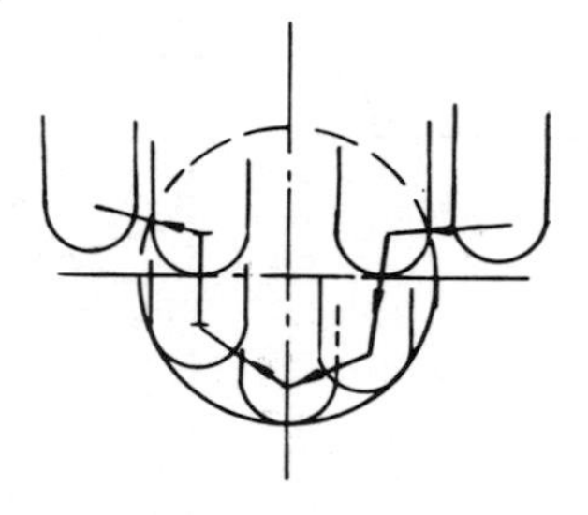

MACHINING SEQUENCE

(a)

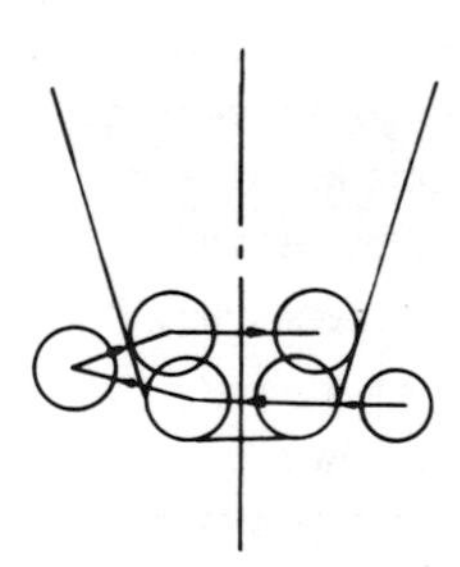

Fig. 2. Machining sequence.

INTERSECTION:

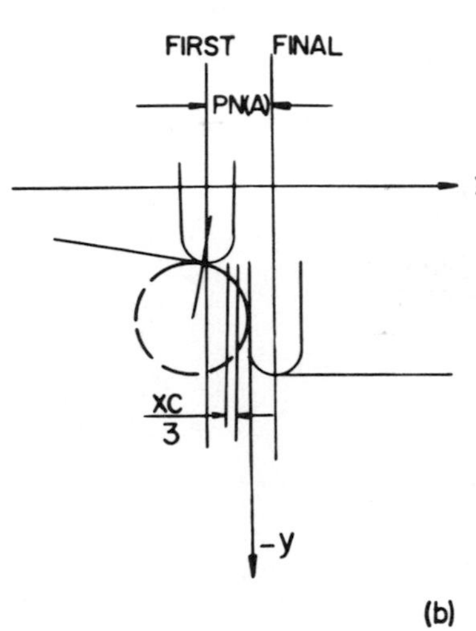

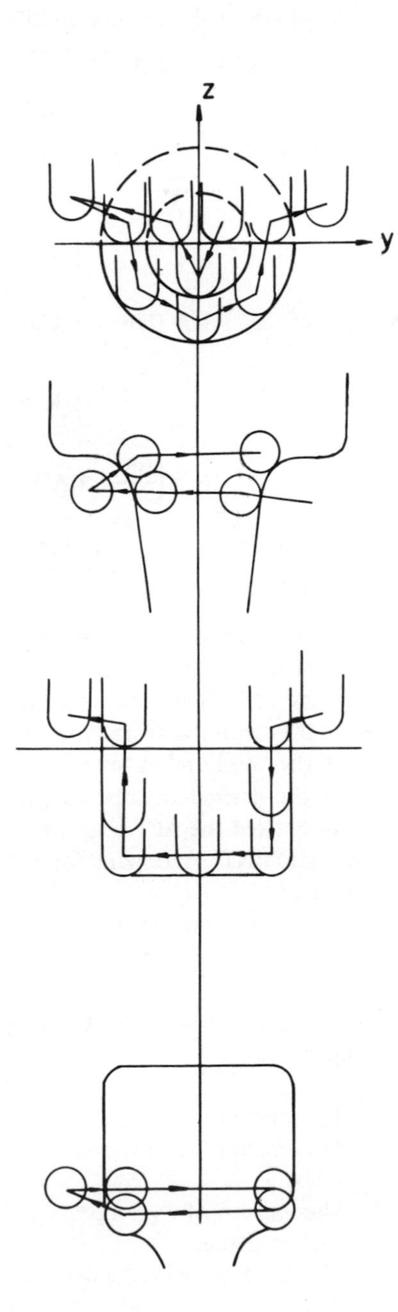

RECTANGLE:

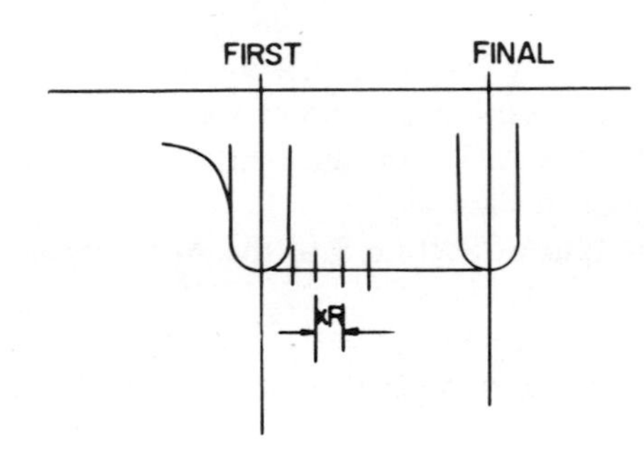

(c)

Fig. 2. (continued)

(1) For machining the male form, the length of the cone (LC) must be such that:

$$LC > R + RAD - (R - RAD)\ \mathrm{Sin}\,\beta + 2\sqrt{(RAD)^2 - (RAD - CP)^2}\ \mathrm{Cos}\,\beta$$

and the length of the rectangle (LRC) such that:

$$LRC > DIA + 2\sqrt{(RAD)^2 - (RAD) - CP)^2}$$

(2)

$$DR > R3$$

$$\frac{BR}{2} > R3$$

(3) If

$$DR < \frac{BR}{2}$$

Put

$$R = (DR - RAD - R3)\tan\frac{90 - \beta}{2}$$

(4) If

$$DR > \frac{BR}{2}$$

Put

$$R = \left(\frac{BR}{2} - RAD - R3\right)\tan\frac{90 + \beta}{2}$$

The flow diagram is shown in Appendix I and the program in Appendix 2.

3.0 USING THE MACRO

The macro will be available in the form of punched cards together with the necessary instructions on how the macro is to be used.

The part programmer will allow a set procedure:

(1) Write the part number statement for the job and any additional relevant information.
(2) Write the 'Machine' statement to call up the appropriate post processor.
(3) State the desired calculation tolerances for the APT program. (N.B. If tolerances are not stated the APT program works on basic tolerances).
(4) The 'CLPRNT' statement for a printed output of Section 3 of the APT program if desired.
(5) Initializing statements as necessary.
(6) The macro call statement and the values of the variables contained in the macro statement.
(7) Completion statements of the program.

The Macro variables are:
(See Fig. 1)

LC	–	The length of the cone			
LRC	–	The length of the rectangular shape	RAD	–	The radius of the 'nose' of the cutter
BR	–	The breadth of the rectangular shape	HYT	–	The cutter height
			CP	–	The allowable height of the cusp between cutter passes
DR	–	The depth of the rectangular shape	GAM	–	Matrix angle for rotation of the shape
B	–	The half angle of the cone			
DIA	–	The cutter diameter	SHAPE	–	For a 2 or 3 figure configuration

R — The fillet radius
R3 — The radius of the cone at the intersection of the cone and rectangular shape
XTI — Translational value in the X axis
YTI — Translational value in the Y axis
SIGN — For cutting of male or female configuration

In addition post processor variables are allowed to enable the machine spindle to be turned or off, the coolant to be actuated, the feedrate to be set.

Note: If additional post processor variables are necessary for a particular numerical control system these may be added to the macro with a minimum of difficulty.

Example of the use of the macro

It is assumed that the dies will be roughed out prior to programming the macro for a finishing cut, that machining takes place along planes perpendicular to the axis of the cone and rectangular shape, and that the shape is symmetrical about the parting lines of the die. The draft angle will be obtained by the cutter.

Example

```
PART NO        CONE TO RECTANGLE CONFIGURATION UNIVERSITY OF
WATERLOO
MACHINE/HILLYER, 1, 1, . . .
CLPRNT
RESERV/ PL, 50, PN, 50, PLN, 50, RD, 50, CYL, 50
CALL/CNREC, LR = 2.5, BR = 4.5, DR = 2.25, LC = 2, R3 = 1.25, $
      BETA = 15, R = 0.4, DIA = 0.3, RAD = 0.15, HYT = 0.15, CP = 0.015, $
      LRC = 4.0, SHAPE = 1, SIGN = 1, SPD = 1000, DIRN = CLW, SPN = OFF, $
      CLN1 = OFF, CLN2 = OFF, FD = 10
STOP
END
FINI
```

The example illustrates the amount of part programming necessary to obtain the output for machining a complete finishing cut of the die.

4.0 CONCLUSIONS

The 'macro' method of APT programming is a means of obtaining a library of special programs for a particular industry or company. The overall object of such a library is to reduce part programming time to a minimum and in addition to enable a standard approach to the machining of a part.

Roughing cuts on both milling and turning operations could be programmed by a macro method.

It can be foreseen that as the application or group technology increases a whole basis of macro programming can be designed working on the coding system of the group technology basis. Such an application could enable a designer to specify details by a coding system which could then be used to call up the appropriate 'macro' to enable the part to be machined or 'drawn' by use of a plotting table.

ACKNOWLEDGEMENT

The research for this paper was supported by the Defence Research Board of Canada, Grant Number 9761-03.

APPENDIX I: Cone to rectangle, flow chart.

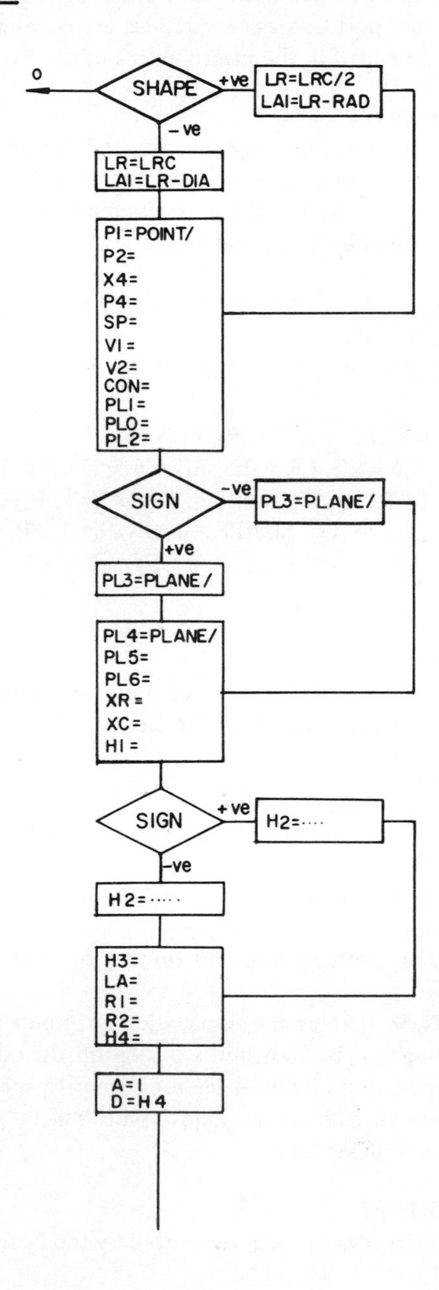

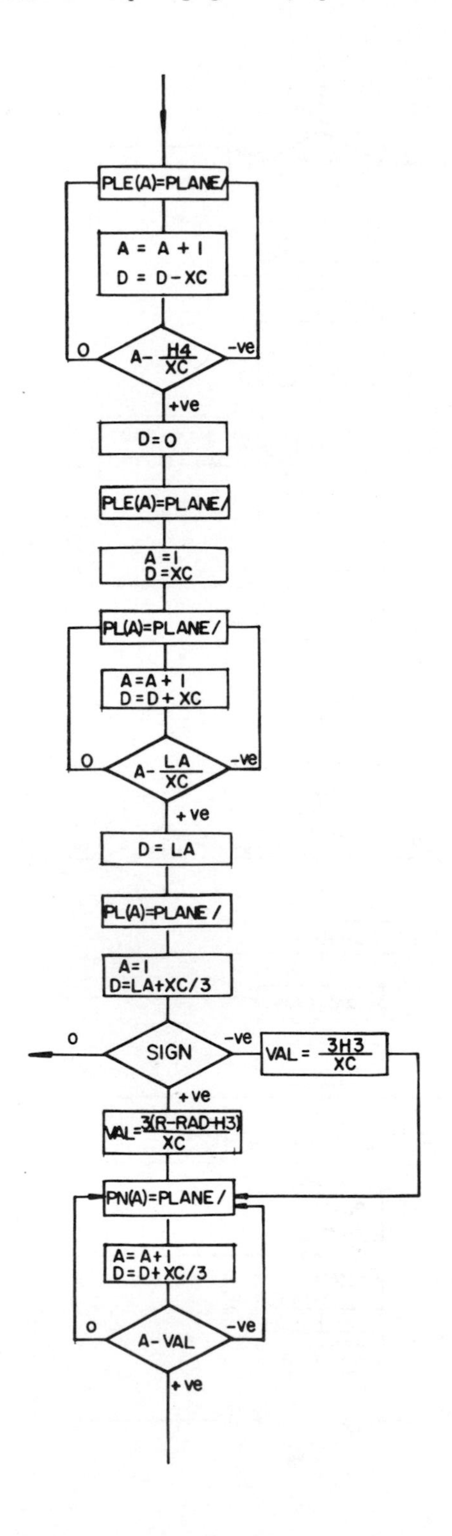
PLE(A)=PLANE/
A = A + 1
D = D - XC
0
A - H4/XC
-ve
+ve
D=0
PLE(A)=PLANE/
A = 1
D = XC
PL(A)=PLANE/
A = A + 1
D = D + XC
0
A - LA/XC
-ve
+ve
D = LA
PL(A)=PLANE /
A = 1
D=LA+XC/3
0
SIGN
-ve
VAL = 3H3/XC
+ve
VAL=3(R-RAD+H3)/XC
PN(A)=PLANE /
A= A+1
D= D+XC/3
0
A- VAL
-ve
+ve

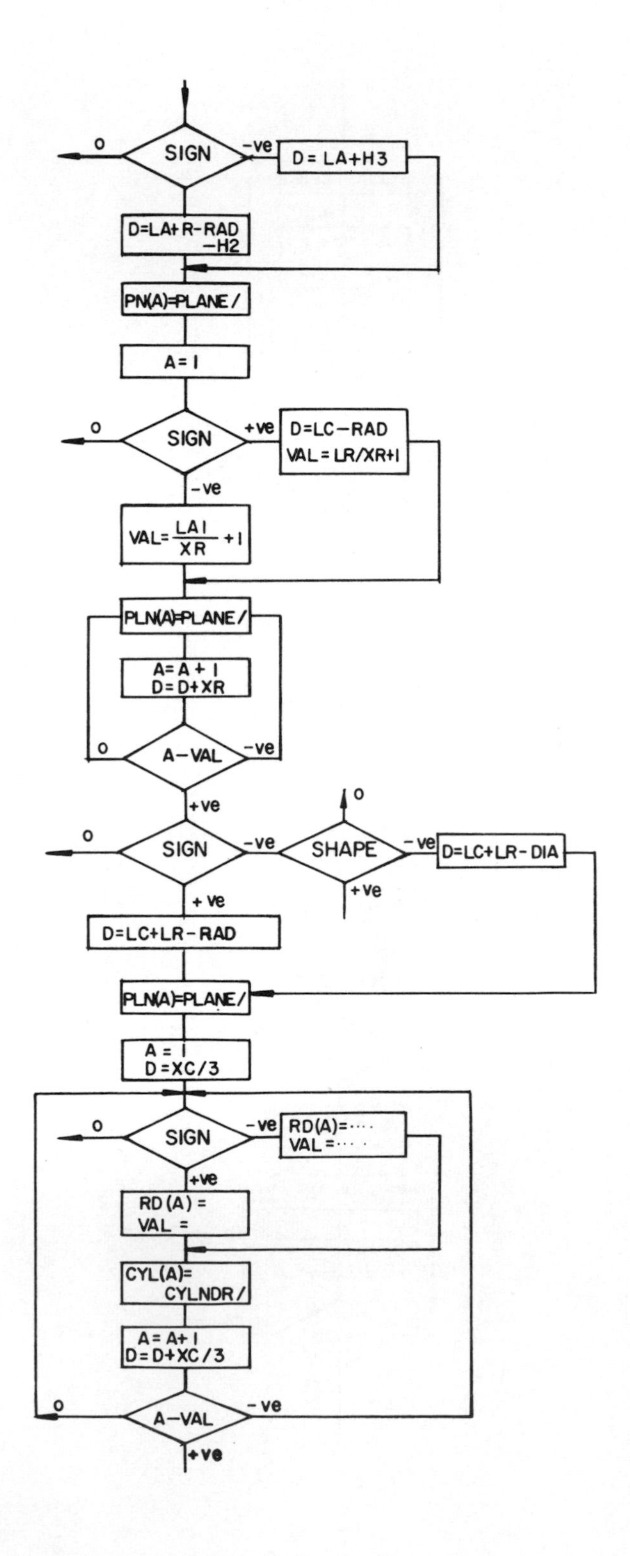
SIGN
0
-ve
D= LA+H3
D=LA+R-RAD -H2
PN(A)=PLANE/
A=1
SIGN
0
+ve
D=LC-RAD
VAL=LR/XR+1
-ve
VAL= LA1/XR +1
PLN(A)=PLANE/
A=A+1
D=D+XR
A-VAL
0
-ve
+ve
SIGN
0
-ve
SHAPE
0
-ve
D=LC+LR-DIA
+ve
+ ve
D=LC+LR-RAD
PLN(A)=PLANE/
A=1
D=XC/3
SIGN
0
-ve
RD(A)=....
VAL=....
+ve
RD(A) =
VAL =
CYL(A)=
CYLNDR/
A=A+1
D=D+XC/3
A-VAL
0
-ve
+ve

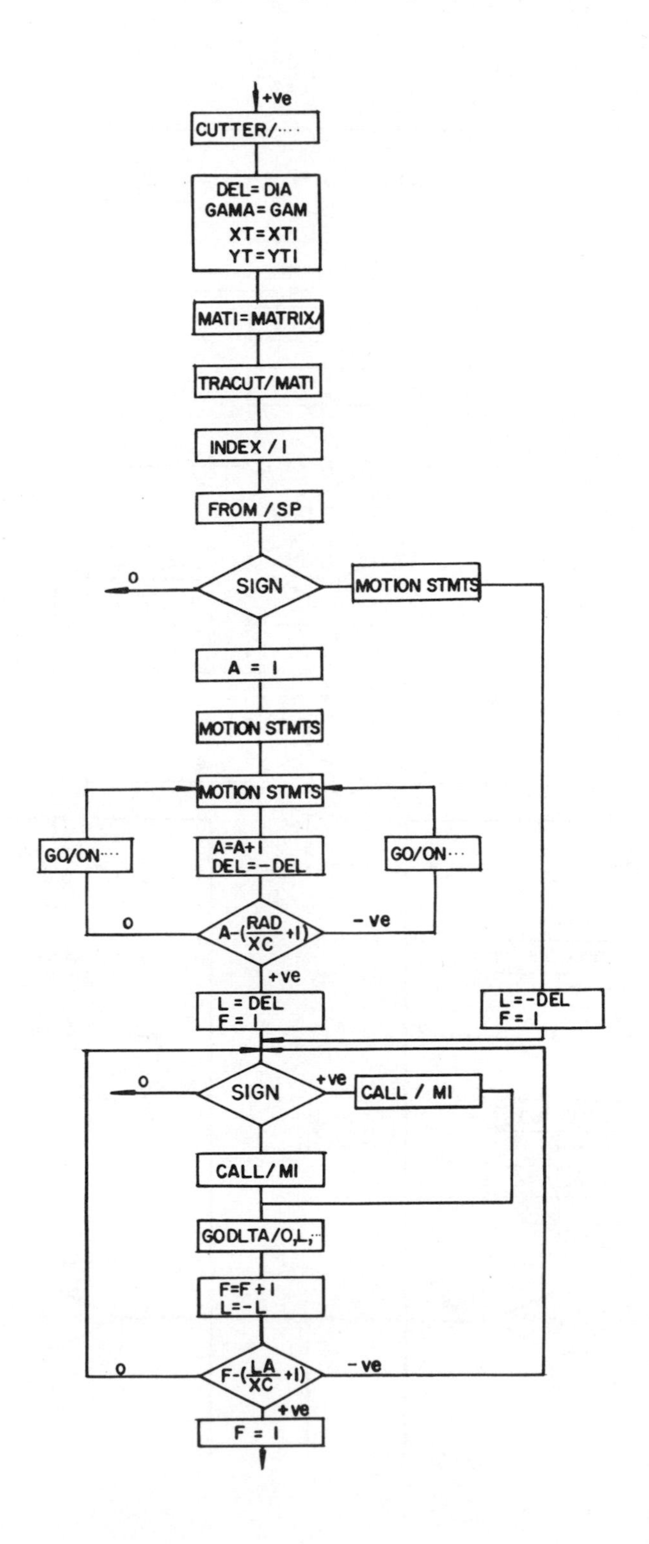
+ve
CUTTER/....
DEL= DIA
GAMA= GAM
XT=XTI
YT=YTI
MATI=MATRIX/
TRACUT/MATI
INDEX / I
FROM / SP
0
SIGN
MOTION STMTS
A = I
MOTION STMTS
MOTION STMTS
GO/ON....
A=A+I
DEL=-DEL
GO/ON...
0
A-(RAD/XC +I)
- ve
+ve
L = DEL
F = I
L=-DEL
F = I
0
SIGN
+ve
CALL / MI
CALL/MI
GODLTA/0,L,..
F=F + I
L=-L
0
F-(LA/XC +I)
- ve
+ve
F = I

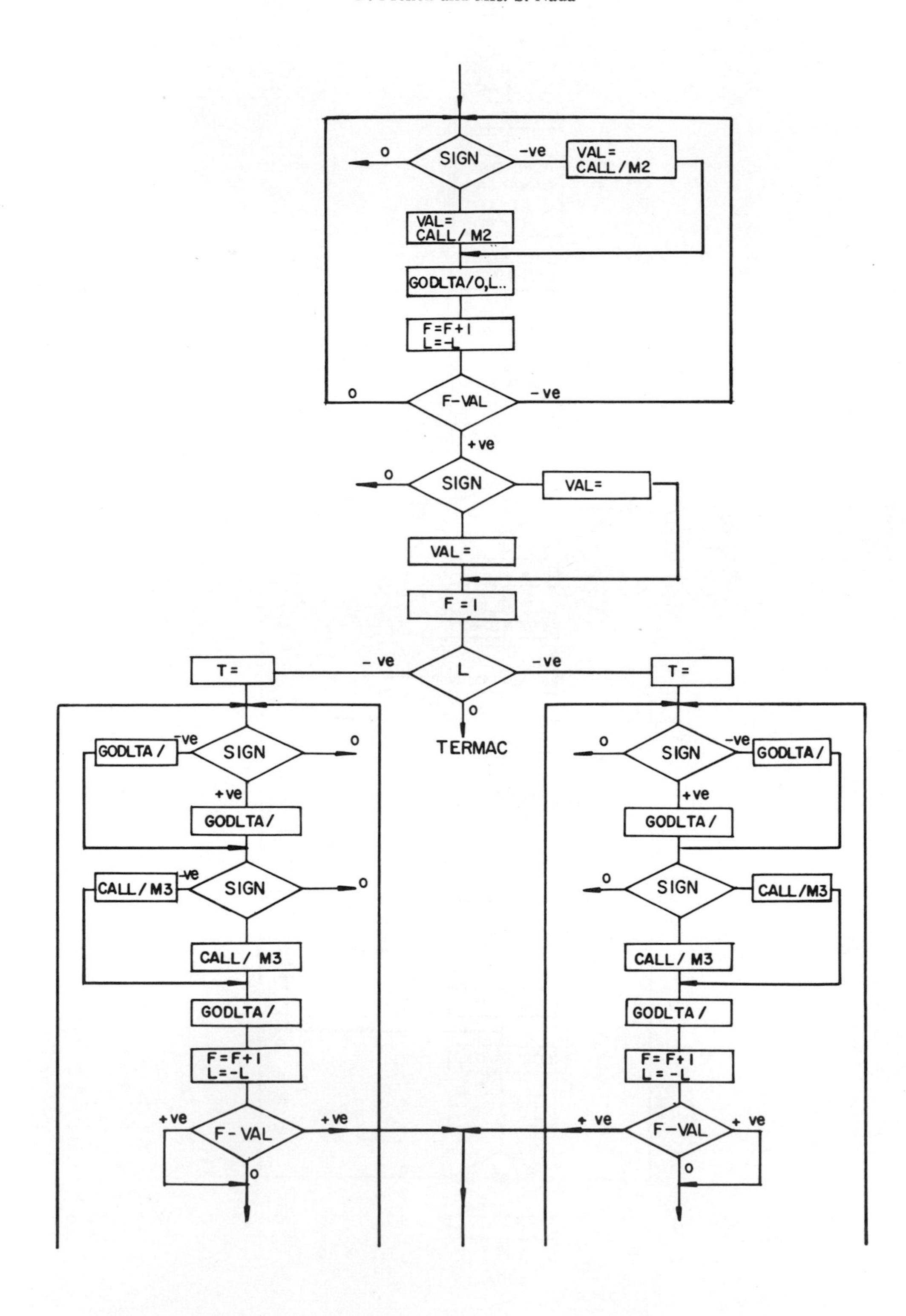
SIGN
-ve
o
VAL=
CALL/M2
VAL=
CALL/M2
GODLTA/O,L..
F=F+1
L=-L
o
F-VAL
-ve
+ve
o
SIGN
VAL=
VAL=
F=1
T=
-ve
L
-ve
T=
o
TERMAC
GODLTA/
-ve
SIGN
o
+ve
GODLTA/
CALL/M3
-ve
SIGN
o
CALL/M3
GODLTA/
F=F+1
L=-L
+ve
F-VAL
+ve
o
o
SIGN
-ve
GODLTA/
+ve
GODLTA/
o
SIGN
CALL/M3
CALL/M3
GODLTA/
F=F+1
L=-L
+ve
F-VAL
+ve
o

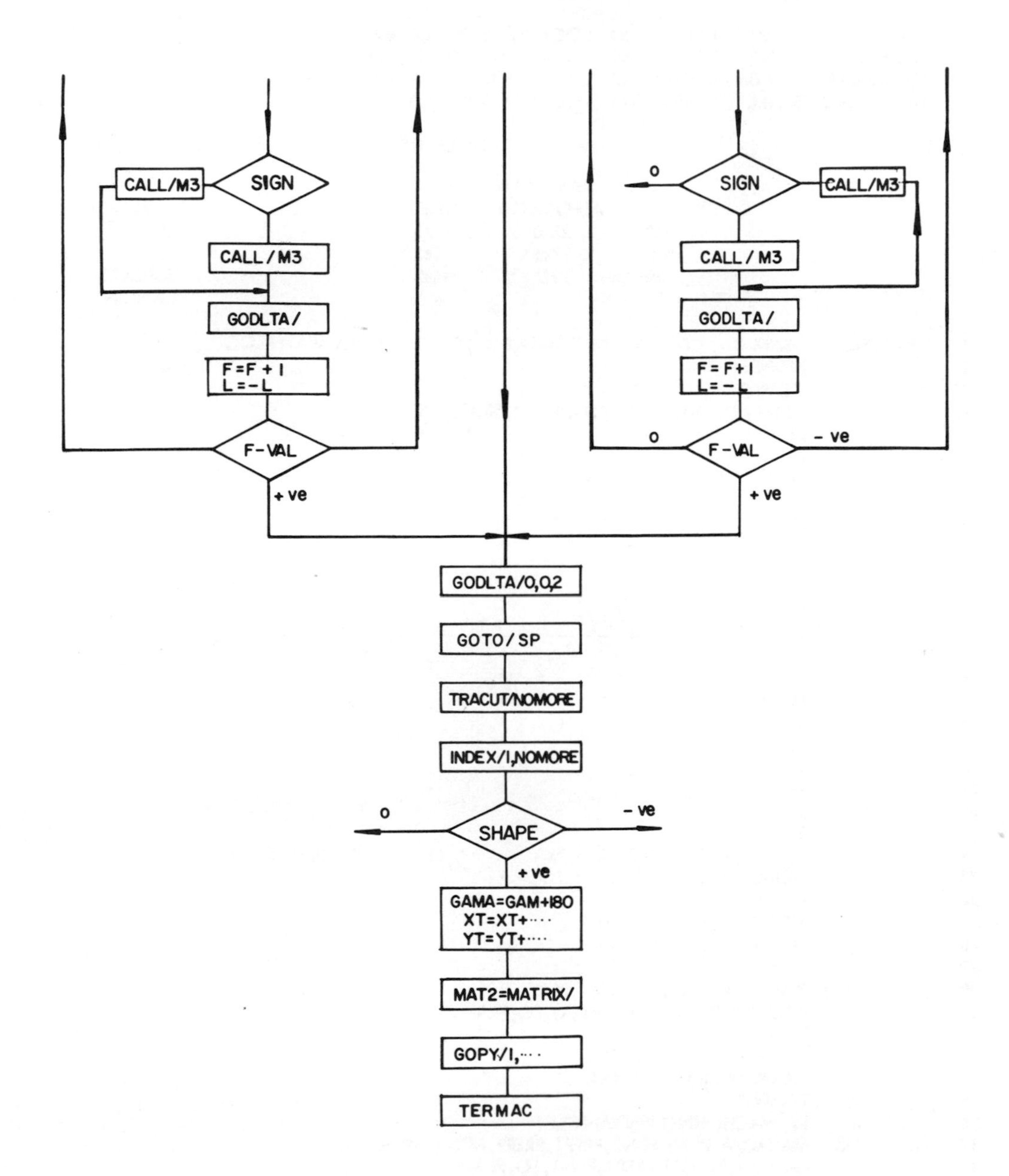

CALL/M3
SIGN
CALL / M3
GODLTA/
F=F+1
L=-L
F-VAL
+ve
0
SIGN
CALL/M3
CALL / M3
GODLTA/
F=F+1
L=-L
0
F-VAL
- ve
+ ve
GODLTA/0,0,2
GOTO/SP
TRACUT/NOMORE
INDEX/1,NOMORE
0
SHAPE
- ve
+ ve
GAMA=GAM+180
XT=XT+....
YT=YT+....
MAT2=MATRIX/
GOPY/1,....
TERMAC

```
N/C 360 APT VERSION 4, MODIFICATION 1        DATE= 70.316 TIME OF DAY IN HRS.

THE MACRO      M1    USES     50  LOCATIONS  IN  CANON
THE MACRO      M2    USES     50  LOCATIONS  IN  CANON
THE MACRO      M3    USES     83  LOCATIONS  IN  CANON
THE MACRO   CNREC    USES   1414  LOCATIONS  IN  CANON

                                              TABLE USAGE DURING INPUT TRANSLATION

                                  PASS ONE
                                 ALLOCATED      USED                        DYNAMIC
                         VST        2750         176                        VST
                         PTPP       2225         262                        PTPP
                         CANON      2225        1597                        SCALARS
                                                                            CANON

 1    PARTNO     SAMIA    CONE TO RECTANGLE    UNIVERSITY OF WATERLOO
 2              NOPOST
 3              CLPRNT
 4              RESERV/PL,50,PN,50,PLN,50,RD,50,CYL,50,PLE,50
 5              $$
 6              $$
 7              $$
 8              $$
 9              $$
10              $$
11              $$
12              $$
13              $$
14              $$
15              $$
16              $$
17              $$
18              $$
19              $$
20              $$
21              $$
22              $$
23              $$
24        M1 = MACRO/A,PREP,MOTION,PLA              $$  MACHINING CONE
25              GO/ON,PL(A),PREP,CON,TO,PLA
26              TLOFPS
27              MOTION/PL(A),ON,PL4
28              GOFWD/PL(A),PREP,PLA
29              TERMAC
30        M2 = MACRO/A,PREP,MOTION,PLA              $$  MACHINING INT.
31              GO/ON,PN(A),PREP,CYL(A),TO,PLA
32              TLOFPS
33              MOTION/PN(A),ON,PL4
34              GOFWD/PN(A),PREP,PLA
35              TERMAC
36              $$ MACHINING RECTANGLE
37        M3 = MACRO/A,PLN1,PLN2,PREP1,PREP2,MTN1,MTN2,PLA
38              GO/ON,PLN(A),PREP1,PLN1,TO,PLA
39              MTN1/PLN(A),PREP2,PL3
40              PSIS/PL3
```

```
41              GOFWD/PLN(A),PREP2,PLN2
42              PSIS/PLN2
43              MTN2/PLN(A),PREP1,PLA
44              TERMAC
45
46
47
48    CNREC=MACRO/LRC,BR,DR,LC,R,DIA,RAD,HYT,CP,BETA,R3,SIGN,GAM,XT1,$
                YT1,ZT,SHAPE                       $$ MACHINING WHOLE PART
49              IF(SHAPE)BB1,IG2,BB2
50    BB2)      LR=LRC/2
51              LA1=LR-RAD
52              JUMPTO/BB3
53    BB1)      LR=LRC
54              LA1=LR-DIA
55    BB3)      P1=POINT/0,0,0
56     P2  =    POINT/(-LC+RAD),0,0
57     P3  =    POINT/(-LC+RAD),1,0
58     X4  = -  R3/TANF(BETA)
59     P4  =    POINT/X4,0,0
60         SP   =  POINT/(-LC-RAD),(-BR/2-(DR+2*HYT)),(DR+HYT)
61     V1  =    VECTOR/3,0,0
62     V2  =    VECTOR/UNIT,V1
63    CON=      CONE/P4,V1,BETA
64    PL11 =    PLANE/P1,P2,P3                     $$  XY-PLANE
65     PL1 =    PLANE/PARLEL,PL11,ZSMALL,(0.25*HYT)
66  PLO    =    PLANE/PARLEL,PL1,Z SMALL,HYT
67     PL2 =    PLANE/PERPTO,PL1,P2,P3
68              IF(SIGN) IK2,IG2,IK3
69  IK2)       PL3 = PLANE/PARLEL,PL11,ZSMALL,DR   $$ RECTANGLE BASE PLANE
70              JUMPTO/IK4
71    IK3)      PL3 = PLANE/PARLEL ,PL11,ZLARGE,DR
72    IK4)      PL4 = PLANE/PERPTO,PL1,P1,P2       $$ XZ-PLANE
73     PL5 =    PLANE/PARLEL,PL4,YSMALL, (BR/2)    $$ RECTANGLE SIDE
74     PL6 =    PLANE/PARLEL,PL5, YLARGE,BR        $$ RECTANGLE SIDE
75      XR =    2*SQRTF (ABSF(RAD**2-(RAD-CP)**2))  $$ INC.  MOTION ON REC.
76      XC =    XR*COSF(BETA)                      $$ INC.  MOTION ON CONE
77      H1 =    R-R*S INF(BETA)
78              IF(SIGN) IL2,IG2,IL1
79    IL1)      H2=(R-RAD)*SINF(BETA)
80              JUMPTO/IL3
81    IL2)      H2=(R+RAD)*SINF(BETA)
82      H3 =    R-H2+RAD
83    IL3)      LA=LC-RAD-R+H2
84      R1 =    ( -X4-H1)*TANF(BETA)
85      R2 =    R1+R*COSF(BETA)
86      H4 =    RAD+RAD*S INF(BETA)
87              A=1
88              D=H4
89    IM1)      PLE(A)=PLANE/PARLEL,PL2,XSMALL,D
90              A=A+1
91              D=D-XC
92              IF(A-(H4/XC+1))IM1, IM1, IM2
93    IM2)      D = 0
94              PLE(A) = PLANE/PARLEL,PL2,XSMALL,D
95              A = 1
96              D = XC
```

```
97        IA1)    PL(A) = PLANE/PARLEL,PL2,XLARGE,D
98                A=A+1
99                D=D+XC
100               IF(A-LA/XC)IA1,IA1,IA2
101       IA2)    D=LA
102               PL(A) = PLANE/PARLEL,PL2,XLARGE,D
103               A = 1
104               D=LA+XC/3
105               IF(SIGN)IW1,IG2,IW2
106       IW1)    VAL=3*H3/XC
107               JUMPTO/IB1
108       IW2)    VAL=(R-RAD-H2) *3/XC
109       IB1)    PN(A) = PLANE/PARLEL,PL2,XLARGE,D
110               A=A+1
111
112
113
114               D=D+XC/3
115               IF(A-VAL)IB1,IB1,IB2
116       IB2)    IF(SIGN) IX1,IG2,IX2
117       IX1)    D=LA+H3
118               JUMPTO/IX3
119       IX2)    D=LA+R-RAD-H2
120       IX3)    PN(A)=PLANE/PARLEL,PL2,XLARGE,D
121               A=1
122               IF(SIGN) IC0 ,IG2,IC3
123       IC3)    D=LC-RAD
124               VAL=LR/XR+1
125               JUMPTO/IC1
126       IC0)    VAL=LA1/XR+1
127       IC1)    PLN(A)=PLANE/PARLEL,PL2,XLARGE,D
128               A=A+1
129               D=D+XR
130               IF(A-VAL) IC1, IC1, IC2
131       IC2)    IF(SIGN)IY1,IG2,IY2
132       IY1)    IF(SHAPE)IA1, IG2, IY2
133       IA1)    D=LC+LR-DIA
134               JUMPTO/IY3
135       IY2)    D=LC+LR-RAD
136       IY3)    PLN(A)=PLANE/PARLEL,PL2,XLARGE,D
137               A=1
138               D=XC/3
139       ID0)    IF(SIGN)ID1,IG2,IN1
140       ID1)    RD(A)=R2+RAD-SQRTF(ABSF((R+RAD)**2-(H2+D)**2))
141               VAL=3*H3/XC+1
142               JUMPTO/IP1
143       IN1)    RD(A)=R2-RAD-SQRTF(ABSF((R-RAD)**2-(H2+D)**2))
144               VAL=(R-RAD-H2)*3/XC+1
145       IP1)    CYL(A)=CYLNDR/P4,V2,RD(A)
146               A=A+1
147               D=D+XC/3
148               IF(A-VAL)ID0,ID0,ID2
149       ID2)    CUTTER/DIA,RAD,0,0,0,0,HYT
150               DEL=DIA
151               GAMA=GAM
152               XT=XT1
153               YT=YT1
154               MAT1 = MATRIX/XYROT,GAMA,TRANSL,XT,YT,ZT
```

```
155          TRACUT/MAT1
156          INDEX/1
157          FROM/SP
158          IF(SIGN)IQ2,IG2,IQ1
159   IQ1)   A=1
160          GO/ON,PLE(A),TO,PLO,TO,PL5
161          GOFWD/PLE(A),TO,CON
162          PSIS/CON
163          JUMPTO/IR3
164   IR1)   GO/ON,PLE(A),TO,CON,TO,PL0
165   IR3)   TLOFPS
166          GOUP/PLE(A),ON,PL4
167          GOFWD/PLE(A),TO,PLO
168          GODLTA/0,DEL,(2*HYT)
169          A=A+1
170          DEL=-DEL
171          IF(A-(H4/XC+2))IR1,IR1,IR2
172   IR2)   L=DEL
173          F=1
174          JUMPTO/IE0
175   IQ2)   GO/ON,PL2,TO,PL11,TO,PL5
176          GOFWD/PL2,TO,CON
177          GODLTA/0,(1.5*RAD),HYT
178
179
180
181          GO/ON,PL2,TO,CON,TO,PL1
182          TLOFPS
183          GODOWN/PL2,ON,PL4
184          GOFWD/PL2,PAST,PL1
185          GODLTA/0,DEL,(2*HYT)
186          L=-DEL
187          F=1
188   IE0)   IF(SIGN)IE3,IG2,IE1
189   IE1)   CALL/M1,A=F,PREP=TO,MOTION=GO UP,PLA=PL0
190          JUMPTO/IE4
191   IE3)   CALL/M1,A=F,PREP=PAST,MOTION=GODOWN,PLA=PL1
192   IE4)   GODLTA/0,L,(2*HYT)
193          F=F+1
194          L=-L
195          IF(F-(LA/XC+1))IE0 ,IE0 ,IE2
196   IE2)   F=1
197   IF0)   IF(SIGN)IF1,IG2,IF2
198   IF1)   VAL=3*H3/XC+1
199          CALL/M2,A=F,PREP=PAST,MOT ION=GODOWN,PLA=PL1
200          JUMPTO/IT3
201   IF2)   VAL=(R-RAD-H2)*3/XC+1
202   IT1)   CALL/M2,A=F,PREP=TO,MOTION=GOUP,PLA=PL0
203   IT3)   GODLTA/0,L,(2*HYT)
204          F=F+1
205          L=-L
206          IF(F-VAL)IF0 ,IF0 ,IT2
207   IT2)   IF(SIGN) IIR,IG2,IIS
208   IIR)   VAL=LA1/XR+2
209          JUMPTO/IIP
210   IIS)   VAL=LR/XR+2
211   IIP)   F=1
```

```
212            IF(L)IG1,IG2,IG3
213   IG1)     T=BR/2-R2-RAD
214            IF(SIGN)AA1,IG2,AA2
215   AA1)     GODLTA/0,T,(2*HYT)
216            JUMPTO/IZ0
217   AA2)     GODLTA/RAD,(4*T),(2*HYT)
218   IZ0)     IF(SIGN)IZ1,IG2,IZ2
219   IZ1)     CALL/M3,A=F,PLN1=PL6,PLN2=PL5,PREP1=PAST,PREP2=TO,$
               MTN1=GODOWN,MTN2=GOUP,PLA=PL1
220            JUMPTO/IZ3
221   IZ2)     CALL/M3,A=F,PLN1=PL6,PLN2=PL5,PREP1=TO,PREP2=PAST,$
222   IZ3)     GODLTA/XR,L,(2*HYT)                    MTN1=GOUP,MTN2=GODOWN,PLA=PL0
223            F=F+1
224            L=-L
225            IF(F-VAL)ZZZ,ZZZ,IH2
226   ZZZ)     IF(SIGN)IIZ,IG2,IIX
227   IIZ)     CALL/M3,A=F,PLN1=PL5,PLN2=PL6,PREP1=PAST,PREP2=TO,$
               MTN1=GODOWN,MTN2=GOUP,PLA=PL1
228            JUMPTO/IIW
229   IIX)     CALL/M3,A=F,PLN1=PL5,PLN2=PL6,PREP1=TO,PREP2=PAST$
               MTN1=GOUP,MTN2=GODOWN,PLA=PL0
230   IIW)     GODLTA/XR,L,(2*HYT)
231            F=F+1
232            L=-L
233            IF(F-VAL)IZ0 ,IZ0,IH2
234   IG3)     T=-(BR/2-R2-RAD)
235            IF(SIGN)AA3,IG2,AA4
236   AA3)     GODLTA/0,T,(2*HYT)
237            JUMPTO/IIA
238   AA4)     GODLTA/RAD,(4*T),(2*HYT)
239   IIA)     IF(SIGN)IIB,IG2,IIC
240
241
242
243   IIB)     CALL/M3,A=F,PLN1=PL5,PLN2=PL6,PREP1=PAST,PREP2=TO,$
               MTN1= GODOWN,MTN2=GOUP,PLA=PL1
244            JUMPTO/IID
245   IIC)     CALL/M3,A=F,PLN1=PL5,PLN2=PL6,PREP1=TO,PREP2=PAST,$
               MTN1=GOUP,MTN2=GODOWN,PLA=PL0
246   IID)     GODLTA/XR,L,(2*HYT)
247            F=F+1
248            L=-L
249            IF(F-VAL)YYY,YYY,IH2
250   YYY)     IF(SIGN)IIE,IG2,IIF
251   IIE)     CALL/M3,A=F,PLN1=PL6,PLN2=PL5,PREP1=PAST,PREP2=TO,$
               MTN1=GODOWN,MTN2=GOUP,PLA=PL1
252            JUMPTO/IIG
253   IIF)     CALL/M3,A=F,PLN1=PL6,PLN2=PL5,PREP1=TO,PREP2=PAST,$
               MTN1=GOUP,MTN2=GODOWN,PLA=PL0
254   IIG)     GODLTA/XR,L,(2*HYT)
255            F=F+1
256            L=-L
257            IF(F-VAL)IIA,IIA,IH2
258   IH2)     GODLTA/0,0,(2*HYT+DR)
259            GOTO/SP
```

```
260            TRACUT/NOMORE
261            INDEX/1,NOMORE
262            IF(SHAPE)IG2,IG2,III
263    III)    GAMA=GAM+180
264            XT=XT+LRC*COSF(GAM)
265            YT=YT+LRC*SINF(GAM)
266    MAT2 =  MATRIX/XYROT,GAMA,TRANSL,XT,YT,ZT
267            COPY/1,MODIFY,MAT2,1
268    IG2)    TERMAC
269            STOP
270            END
271  FINI
```

NOTE: BLANK CARDS INSERTED FOR CLARITY

MACHINING PROCESSES – THEIR THEORY AND CALCULATION OF MINIMUM COST BY COMPUTERS

Jan Bekes
Slovak Technical University, Bratislava

SUMMARY

Machining processes are at the present time designed on the basis of experience. Examples prove that the rules of machining can be expressed by mathematical and logic formulae. These rules build up the basis for the theory of process determination and make it possible to create a mathematical model for the optimisation of machining processes. The main parts of the model for turning are described. The algorithms cover simulation of the cutting process, sequence of operations, scheme of material removal, tool geometry, feeds and cutting speeds for single-and multi-tool machining. The model is in principle available for other machining processes too and is suitable also for optimising the design of new machining units or lines.

NOTATION

For equation 1

N_m – power output of the electric motor (kW)
m – tool-life exponent from formula $T v^m$ = const.
K – constant costs of the machine tool per minute, independent of power output.
e – costs for installed power unit of machine tool
V – volume of material to be removed by machining (cm^3)
k_s – specific cutting resistance (kgf/mm^2)
η – efficiency of machine tool
t_v – time of handling operations (min)

For equation 2

t_{opt} – optimum depth of cut (mm)
t_x – time required to set the next pass, including change of cutting conditions (min)
L – length of tool movement in cut (mm)
d – diameter of work surface (mm)
N_e – effective power on machine spindle (kW)
F_z – permitted tangential cutting force (kgf)
C, x, y – constant and exponents of cutting force equation $F = C . t^x s^y$ where t is depth of cut and s is feed.

For equation 3

γ – rake angle

v_T – cutting speed for minimum costs, expressed as a function of rake angle.

s_{Fz}, s_{Fy}, s_{Ra} – available feeds expressed as a function of rake angle calculated from limiting cutting forces F_z and F_y respectively from limiting surface roughness Ra.

For equation 4

n – number of workpiece elements

k – cross and longitudinal fast feed ratio

L_i – length of the ith element (mm)

DV, DM – maximum and minimum diameter of element (mm)

For equation 5

F – cutting force (kgf)

C, x_F, y_F – constant and exponents of cutting force formula

t – depth of cut (mm)

s – feed (mm/rev)

1.0 INTRODUCTION

Rules for determination of machining processes are expressed mainly as general recommendations. Exact relations, except for the calculation of cutting conditions, are not available. This is being changed by the development of machining processes as a scientific discipline. Though the scientific basis of machining began to form in the middle of the 19th century, the first works in the field of machining process determination were published only in the nineteen-thirties[1,2,3,4,5,6]. These concerned the choice of feeds and cutting speeds.

The remaining problems of machine process determination have not been systematically investigated until the present, and only in recent years have works been published from some other fields.

Methods for determination of machining processes have developed practically only on the basis of experience. The rules for manufacturing processes have been formulated as general instructions and the numerical values that have occasionally been put in are only orientational in character and are valid only for certain average and very roughly determined production conditions (single, series, mass-batch production, etc.). By generalisation of experience, rules are obtained which enable us to propose processes which are suitable from a technological point of view, but which are not optimal for certain production conditions. From some alternative solutions, which are from a technological point of view nearly similar, the most convenient solution is selected by optimum calculation. Thus the chosen process is called an optimal one. For manual processes such a solution is the only one possible, from the point of view of expense as well as of time.

Computers can in a relatively short time process a large amount of information and perform many complicated calculations. Their use eliminates some factors which influence the quality of processes. These include the human influence and the time limitation on large amounts of calculation. This is why processes can be more elaborate in detail if computer-aided. Their quality is variable, depending on the quality and completeness of the input data, and of its accuracy, and this determines the degree of closeness to the optimum solution which can be achieved.

There are in principle three qualitative stages of solution.

1. Fulfilling technological requirements.
2. Choice between possibilities.
3. Optimum solution, calculated from quantitative relations of factors which influence the effectiveness of the solution.

Machining process determination consists in the solution of some sub-problems. The quality of the machining process is affected by the method of their solution.

The sub-problems which have usually to be solved are:

- determination of form and dimensions of workpiece
- determination of machining type
- selection of type and size of machine tool and number of workplaces needed
- choice of clamping, determination of machining operations and positions, including heat treatment and determination of work sequence
- determination of machining allowances for separate machining operations
- determination of machining-allowance sharing
- determination of cutting conditions (feed and cutting speed)
- choice of fixtures and determination of technological requirements in special equipment design
- determination of geometry and characteristics of standard tools, determination of technological requirements in special tool design
- determination of checking procedure, measuring equipment and instruments, determination of technological requirements on special equipment design.

Detailed analysis of the level of present-day sub-problem solutions shows the essential differences in quality. Cutting conditions are optimised satisfactorily only where quantitative relations and exact mathematical methods are used. But the reliability of input data is also questionable here. The other problems of machining process determination are solved mainly by experience, reference to standard processes or general recommendations, and only exceptionally by evaluation of alternative possibilities.

2.0 IMPORTANCE AND FORMULATION OF THE MACHINING PROCESS

The rapid development of the means of production and the change of production conditions in the present technological revolution present a serious problem of harmony between the economic and technological conditions of organisation and between recommendations which are mainly the results of experience.

There is well-founded doubt that experience gained in the past is not fully adequate for production conditions today. The results are that general recommendations as published in literature and in workshop tables cannot build up an exact basis for machining process determination.

All the factors which influence the effectiveness of a process are constantly changing with technological advance. Thus the optimum machining process has only a temporary and local validity, and is affected by the degree of change in separate factors and their influence upon selected criteria of the optimum.

The most important factors influencing the optimum solution are:

- dimension, shape and material of workpieces and details
- cutting tool material and tool design

- the performance, rigidity and accuracy of machine tools and the degree of mechanisation and automation
- the production organisation
- methods used in machining and in production of unmachined pieces
- economic factors such as wages, price of raw-materials, energy, etc.

The influence of changes in these factors upon machining process optimisation depends on the absolute value of change and on the influence of that factor upon the effectiveness of the machining process. Conflicting factors acting in opposite directions can result in a certain stability of machining process data. The demand for maximum effectiveness reacts on the above-mentioned factors.

The influence of changes in factors can be directly investigated only to the extent which present knowledge allows. It is the same with the quantitative determination of the reciprocal influence of separate factors, which the present knowledge of rules for machining processes determination allows. The same can be said also of the qualitative determination of the reciprocal influence of separate factors. Knowledge of qualitative rules enables us not only to adapt a machining process to new production conditions immediately but also to settle the optimum trend to development of other factors. It confirmed, for example, the wisdom of developing cutting materials and tools in recent decades, as directed by effort to realize the most effective cutting conditions.

But the lack of knowledge of quantitative rules for machining process determination slows down technological progress and considerably contributes to the unequal development of research as well as of the means of production in machining. Therefore research on optimisation of machining processes is highly necessary today.

The basic aim of this research must be the quantitative determination of rules expressing the optimum solution of sub-problems of machining process determination. These rules are the exact basis for the algorithm of optimum machining processes, as well as for the solution of many other problems touching the development of machining. Generally known rules for the determination of optimum feed and cutting speed confirm this opinion. Other sub-problems can be derived from quantitative analysis.

Below we show some of our results for the determination of the optimum power output of an electric motor, machining-allowance sharing, calculation of optimal tool geometry and determination of operation sequences. In all cases the minimum machining cost has been chosen as the criterion.

3.0 EXAMPLES OF QUANTITATIVE RULES

3.1 *Optimal power output from the electric motor.*

General experience says that machine tools with higher output are chosen more for series and mass production and for rough machining than for single and small-batch production or finish machining. The quantitative expression of this rule was derived from analysis of the effect of electric motor output on machining costs. Assuming that machining was carried out with tool life for minimum costs and with constant chip section, without any restriction on cutting power, the following equation was derived

$$N_m = \sqrt{\frac{m}{m-1}\,\frac{K}{e}\,\frac{V k_s}{6120\,\eta\,t_\nu}} \qquad (1)$$

Analysis of eqn 1 shows full agreement with the generally acknowledged rules for the choice of machine-tool power output. The optimum power output can be calculated for the set of operations or machining movements with variable chip section and for variable limiting conditions of cutting power as the absolute minimum of the criterial function[7].

3.2 *Machining-allowance sharing*

It is generally known that, with full utilisation of motor output, the layer to be removed can be cut off in several passes only when the passes are sufficiently long for the time required to set the next pass is smaller than the cutting-time saved by using a larger chip section.

Mathematical formulations of this rule have not, to now, allowed the optimum scheme of sharing to be obtained for the stepped workpiece.

Analysis of the interaction of factors has shown that under conditions of full exploitation of motor output and of feed limited by the permitted tangential component of cutting force, F_z the equation for the optimum depth of cut by turning is

$$t_{opt} = \left[\frac{1000\, t_x}{L\pi d} \; \frac{6120\, N_e}{F_z} \left(\frac{F_z}{C_{Fz}} \right)^{1/y_{Fz}} \frac{y_{Fz}}{x_{Fz} - y_{Fz}} \right]^{y_{Fz}/x_{Fz}} \tag{2}$$

Equation 2 shows that the optimum depth of cut and the number of passes is dependent not on the length of passes but on the diameter of the work surface and on conditions limiting the acceptable value of feed and cutting speed. This rule can be quantitatively determined for variable limiting conditions of feed and cutting speed. In this way it is possible to design an algorithm for the optimum scheme for material removal on the stepless or stepped rotational workpiece.

3.3 *Optimal tool geometry*

Optimal tool geometry can be considered to be that which provides maximum tool life or which enables us to meet the technological requirements of surface finish. The analysis of the relationship between tool geometry and machining costs shows that a chosen geometry is optimal only with full tool utilisation, i.e. with restriction of the cutting speed by tool life and when the optimised tool-geometry parameter does not influence the limiting factor of the feed. In other cases the optimum geometry, which depends on further machining conditions (e.g. the optimum rake angle in the area of utilisation of the tool life and by limiting the feed by the radial force F_y), can be calculated as the minimum value of these three equations

$$\min \frac{\gamma\,(v_T\, s_F)}{\gamma}$$
$$s_{Fy} = s_{Fy}$$
$$s_{Fy} = s_{Ra} \tag{3}$$

In this manner the optimal value of other tool geometry parameters for variable feed and cutting speed limiting conditions can be calculated.[9]

3.4 *The pass-sequence*

When the number of workpiece elements exceeds 1 and the plain faces can be turned only from the larger to the small diameter it is necessary to move the tool with rapid feed twice on the face or cylindrical surface, depending on whether the turning starts on the smallest or on the largest diameter. The inequality which decides the operation sequence is:

$$\sum_{i=1}^{n} (2\,k\,L_i + DV_i - DM_i) < \sum_{i=2}^{n} (DM_i - DV_{i-1}) \tag{4}$$

If inequality 4 is true, the movement sequence begins with the length turning on the element with maximum diameter. If, however, the inequality is false the optimum sequence starts from the face turning on the element with the smallest diameter.

The conclusions are in full agreement with practical experience. With disk-form pieces (for which inequality 4 is evidently true), finish-turning starts from the maximum diameter. With shafts, however, the inequality is probably false and finish-turning starts from the minimum diameter. Formula 4 makes a decision possible in doubtful cases also.

4.0 DEDUCTIONS

The examples presented prove that rules found for the optimum solution of sub-problems not only confirm general experience, but also essentially enlarge it. They generalize and exactly limit the area of their validity. It is probable that, for further sub-problems also, the derivation of quantitative rules is possible.

The knowledge of quantitative relations of the factors which influence optimum criteria builds up a basis for theory of machining process determination. The formulation of machining process rules in exact mathematical formulae will, without any doubt, influence the development of machine tools and cutting tools, as well as other production conditions, and will guarantee the further development of machining technology as an exact scientific discipline. The quantitative rules, in their final effect, should give a system which enables us to work out the algorithm for optimum machining processes and thus fully utilise the potential of modern computing techniques for exact solutions in everyday machining practice as well as for future tasks of development.

However, for the present, not all the relations needed for achieving the optimum are known. They are not even known for turning, though this type of machining is most often investigated, e.g. though the mathematical formulac for the obtaining of separate optimum tool parameters were found, formulae are still lacking for simultaneous optimal calculation of all geometrical tool parameters. The quantitative, mathematicallogic formulation of operation sequence rules and their concentration is evidently incomplete and is only beginning.

Instead of using imperfect formulas, an iterative method could be used to consider the procedure for determining some unknown parameters; letting the solution be chosen by the technologist. The technologist can also intuitively choose from the alternative possibilities provided by the computer based on an imperfect selection which satisfies only the known and algorithm built rules.

Such a procedure is generally used in interactive man-computer systems. With this method it seems to be possible at the present moment to create a system for complex computer-aided optimal calculation of machining processes. The mathematical model for determining the optimum turning process will be dealt with further.

5.0 MATHEMATICAL MODEL FOR OPTIMISATION OF THE TURNING PROCESS

The model consists of five main parts. Each part comprises some algorithms with separate input and output, thus the separate parts could be used independently. Such a sharing is accepted in the interactive man-computer system of work. The algorithms are written in FEL-ALGOL language.

5.1 *The complex machinability of workpiece material*

Accurate calculation with inaccurate values has no practical sense. Thus for computer-aided processes one needs to know sufficiently exact quantitative relations between cutting parameters and the characteristics of machinability of the workpiece material (e.g. cutting forces, tool life and the surface roughness). Published empirical formulae are not reliable for use in practical conditions, since these can differ from experimental conditions. In the first part of the model the constants and exponents of empirical equations for given tools and cutting conditions are calculated. The calculation uses empirical formulae for expressing the influence of physical cutting parameters on the machinability characteristics. These formulae are transformed by physical models into the equations in which the technological cutting parameters for special cutting conditions and for the assumed range of cutting condition changes are used. For example by the calculation of characteristics of dynamic machinability the input for the given algorithm is.

- the experimental equations of tangential and normal components of cutting force for orthogonal cutting, expressed as a function of depth and chip-thickness, cutting speed, cutting angle, and material hardness of workpiece;
- the given range of feed, depth of cut and cutting speed;
- the variability of hardness of the workpiece material;
- the geometry of the tool used.

The outputs are the constants and exponents of equations of cutting force components F_x, F_y and F_z in the form

$$F = C_F\, t^{x_F}\, s^{y_F} \tag{5}$$

Some simple algorithms are used for the calculation of technological/economic constants of criterial functions like the costs of tools for one tool life, the costs of one working minute of the machine, the time required to grind one tool etc. These enable us to calculate the required information using data that are normally in evidence in factories, as e.g. amortisations, store stock of turning tools, costs of abrasives etc.

5.2 *Operation sequences and their variants*

The input in the second part of the model is given by a set of workpiece and finished detail drawings, a set of working functions of the machine tools used, of clamping possibilities and the given priority of operations and passes.

The algorithm co-ordinates, step by step, the separate passes to separate machine tools depending on their working functions and with respect to these main principles:

- the minimisation of the number of clampings;
- the possibility of reaching the machined surfaces;
- the priority of passes or operations with respect to productivity;
- the minimisation of tool path with respect to the length of tool movement.

The most accurate surface is chosen for clamping. To merge passes the following criteria are used:

- the technological admissability of merging;
- the technical possibility of merging.

In this part of the model the incidental additional operations are placed in order like the preparation of clamping surfaces for next clamping, bevelling edges, etc. On the output is printed a recommendation of sequence and sharing of passes into operations for all possible variants of given machine tools.

5.3 *Machining allowance sharing*

This part comprises some partial algorithms which solve, for special cases of machining, the problem of optimisation of machining allowance sharing including the optimisation of sharing of tolerance zones into systematic deviations.

The input is given by the set of workpiece and finished detail characteristics, by the set of parameters of the tools to be used, and the workpiece material and machine tools.

The optimal sharing of systematic deviations in the form of tolerance zone (optimal share of tolerance zone on the inaccuracy derived from dimensional tool wear and on the inaccuracy derived from the flexibility of the machine tool – workpiece – tool system) is determined by the minimisation of the sum of cutting time and the time for reset of the tool within one tool life as shown in Fig. 1.

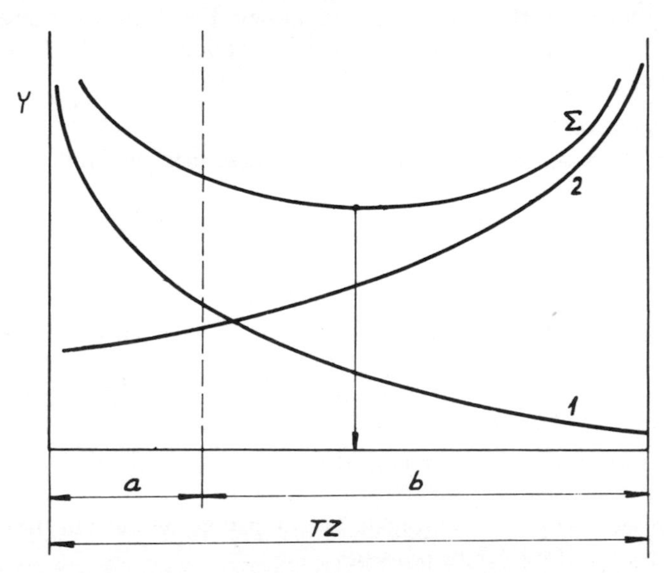

Fig. 1. Diagram for the optimisation of distribution of tolerance on the inaccuracy derived from dimensional toolwear (a) and from flexibility of system machine tool – workpiece – tool (b).
TZ = tolerance zone: 1 = time for reset of tool
2 = cutting time: *Y* = sum of times 1 and 2.

The algorithms for machining allowance sharing are constructed on the basis of machining cost minimisation. Therefore they not only give the optimum depth of cut for separate passes but also optimal feed and cutting speed for single tool machining. The special algorithms solve the machining allowance sharing for rough turning and for finishing, and for the determination of cutting depth for separate rough turning passes for stepless and stepped shafts.

5.4 *Optimisation of tool geometry*

The algorithms solve simultaneously the optimisation of the approach angle and angle of inclination, rake angle and the nose radius of a turning tool for a given depth of cut. The mathematical model of optimisation is based on looking for such geometrical parameters of the cutting edge, where the value of the machined surface with a given tool reaches a maximum.

The inputs are parameters of machine tools, tool material, depth of cut for the pass for which the tool geometry is optimised. The algorithm also gives, with the exception of optimal combination of geometrical parameters, the optimal value of feed and cutting speed for single tool machining.

5.5 *Optimisation of spindle speeds (optimal cutting speed) and the tool changing scheme*

This part comprises some algorithms determining, for different typical machining cases, the optimal cutting speed. For multi-tool machining the special algorithm determines also the optimal tool changing scheme.

The algorithms permit the solution of these cases:

- machining with one tool, with different feeds, depth of cut and machined diameters;
- simultaneous multi-tool machining with several tools on one machine; and
- simultaneous machining with several tools on several machines set in different types of lines.

The input are the parameters of tools and passes, the output comprises the optimal revolutions of separate spindles, machining costs, the productivity of the designed line and the tool changing scheme.

5.6 *General form of algorithm*

The machining processes may be symbolically described by following mathematical relations:

Characteristics of machine tools, tools, the workpiece and the detail are given by numeric models

$$Mi\,(x, y, z);\ Tj\,(x, y, z);\ W\,(x, y, z);\ D\,(x, y, z).$$

By operations on the sets *Mi, Ti, W* and *D* we determine *k*, variability of machining processes. Then from formulae

$$fi\,(x, y) = gi\,(x)$$

we calculate the combinations of technological parameters *x, y* for the best value of criterial function *Y*.

The optimum machining process we obtain by comparing the value of the criterial function *Y* for the variants 1 . . . *k* of machining processes.

Symbols used:

fi – function of technological variables
gi – value or function of limiting conditions
x, y, z – the input, calculated and chosen values of technological variables

The Flow Chart for this algorithm is shown in Fig. 2.

6.0 PROCEDURE OF PROCESSES OPTIMISATION

The chosen procedure where the algorithm of mathematical model is used depends on the degree of optimisation which has to be achieved and in what range separate parameters have to be optimised. In the simplest case the optimisation of feed and cutting speed is satisfactory. This case needs only a simple algorithm for optimisation of feed for single tool machining and the fifth part of the model. The depth of cut distribution and the different alternative procedures can be left to technologist's choice.

If the second part of the model is used we are sure that all alternative processes for working possibilities of the given machine tools were invesitgated. The finding of all alternative

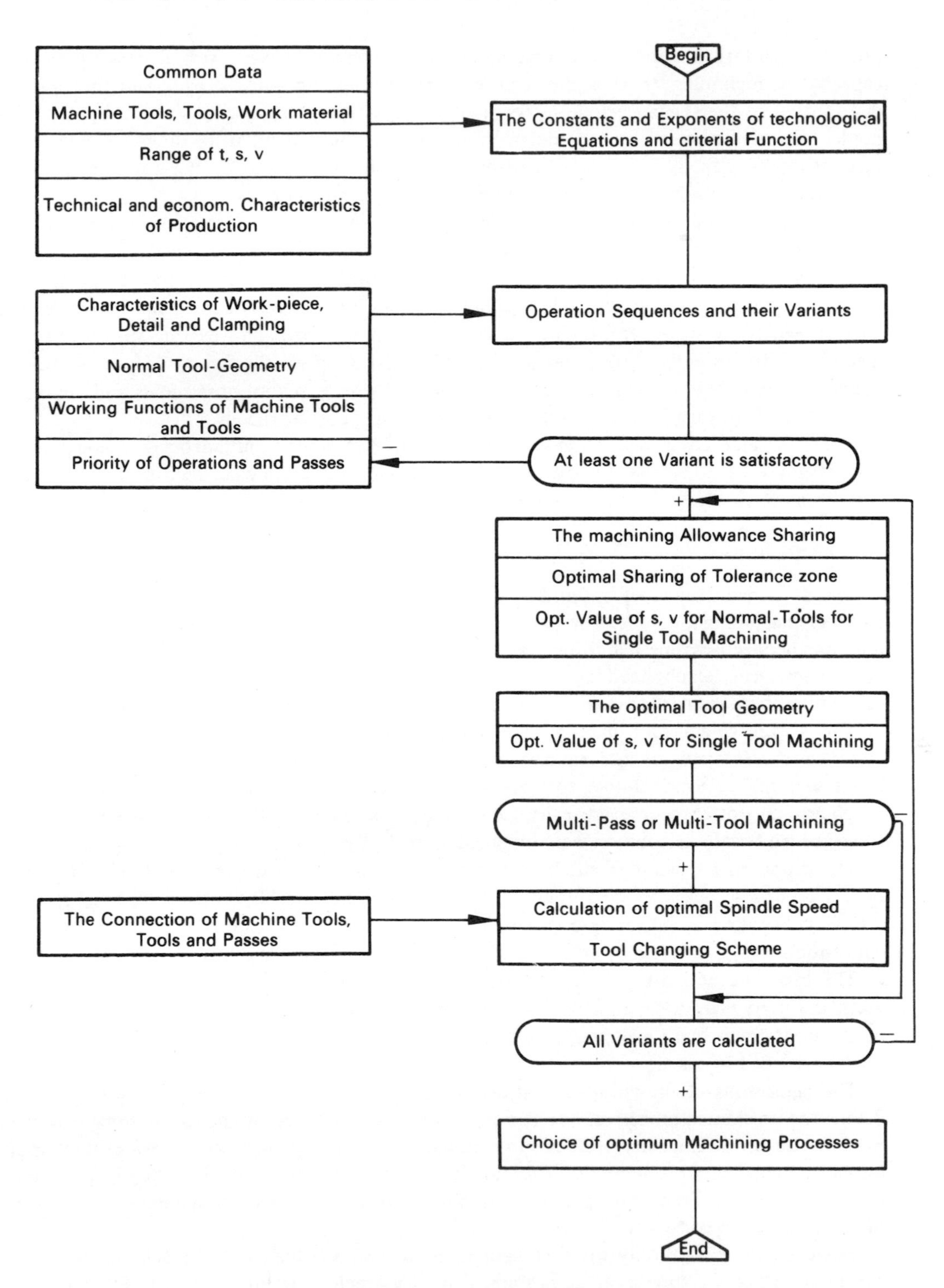

Fig. 2. The flow chart of the algorithm for choosing optimum machining process.

processes, and their detail calculation by the rule, overcomes the practical possibilities and capacity of technologists. It is the same with machining allowance sharing and the optimal distribution of the tolerance-zone in systematical deviations, comprised in the third part of the model. The algorithms of tool geometry optimisation in the fourth part of the model enable us, in connection with its fifth part, to evaluate the economic effect of using non-standard tool geometry and to show all reserves in cutting conditions determination.

The proposed models of optimisation were proved for turning processes of components in stable mass production, where it is certain that the technology used is well tried.

In comparison with actual processes, practically identical results were reached by computing in nearly all cases, dealing with the pass and operation sequence (in some cases several passes have shown to be excessive), rake angle, rough-machining feeds and cutting speeds at single machining. The essential differences were found at the depth of the removed layer in separate operations and their sharing, at the number of tool resets, at tool nose radius and approach angle, at finish turning feeds and at multi-tools machining cutting speeds.

The calculated costs were going down from 18 to 30% while simultaneously raising productivity from 7 to 20%. The quality of products, machined by the calculated process, met the conditions of customer.

7.0 CONCLUSION

The described manner of optimisation for the turning process is in substance applicable also to other machining methods. The basic problem is to know sufficiently the exact quantitative relations between cutting conditions and cutting force components, tool life and surface roughness. These can be obtained by a reasonable number of experiments from the model for calculating the formulae for cutting force components, tool life and surface roughness.

The workpiece and tool material characteristic dispersion and the limits of reliability for the empirical characteristic of machinability also need to be known. In our calculation we used the limits securing 95% probability, so that the real values of cutting forces, tool life and surface roughness do not overcome the calculated values. The data about characteristics of machine tools needed for the calculation were obtained individually.

The proposed procedure is suitable also for the optimisation of the technological conception of single purpose machine tools or lines. The productivity of machines with an optimal technological conception is higher and machining costs are smaller, than of machines where the conception was determined by conventional methods where sense and intuition prevails.

The losses on account of delay, or the low effective performance of machines which arise when the users provide or improve the technology themselves, are smaller when machines and processes are delivered together. Therefore for delivery of machines together with technology a higher price can be asked.

The algorithms of the mathematical model described enable us not only to optimise in all details the machining processes, but also to analyse the influence of change of work material, machine tools, tools, quality of details (e.g. its accuracy), changes of price and wages level upon machining costs and productivity. This enables us, to direct on the basis of trend of development and prognosis, the development of production technology, techniques and equipments in future to the greatest benefit.

However, these were only the first steps taken to discover the rules of machining processes and their complex computer-aided optimisation. The greater part of the task is still before us.

But the modest results obtained have shown that the way we have taken is promising and that from machining science will come already in the near future a very real production force.

REFERENCES

1 LEYENSETTER, W., Die wirtschaftliche Schnittgeschwindigkeit, *AWF Mitteilungen*, Vol. 15 (1933) No. 4.
2 WOXEN, R., A Theory and an equation for the Life of Lathe Tools, Thesis for the Degree of Doctor of Technology, Tekniska Hogskolan (1932).
3 PAASCHE, J., Kennzeichen und Bestimmug der wirtschaftlichen Schnittgeschwindigkeit, *AWF Mitteilungen*, Vol. 15 (1933) No. 7.
4 WALLICHS, A., Schöpke H., Beitrag zur Frage der wirtschaftlichen Schnittgeschwindigkeit, *AWF Mitteilungen*, Vol. 15 (1933) No. 7.
5 DEALE, R. C., Tool Life for Minimum Cost, *Am. Mach.*, Vol. 78 (1935) No. 17.
6 BROWN, R. H., On the selection of Economical Machining Rates, *Int. Jour. of Prod. Research*, Vol. 1 (1962) No. 2.
7 BEKES, J., Optimalizácia výkonu motoru obrábacieho stroja, *Zb. ved. prác SiF SVST Bratislava* Vol. 9 (1969) No. 1.
8 BEKES, J., Optimalizácia rezných podienok pre súbor prvkov – osadený hriadel, *5-th Slovak Conf. of Technology, Tatranska Lomnica 1968,* Zilina CSVTS-Dom Techniky. (1968) p. 427–444.
9 BEKES, J., Jednostupnová optimalizácia geometrických parametrov nástro ja, *Zb. ved. prac SiF SVST Bratislava* Vol. 8 (1968) No. 1.

PROGRAM AUTOPLAN FOR THE DERIVATION OF OPERATIONS AND MACHINING CONDITIONS IN MULTI-OPERATION TURNING

J. R. Crookall
Imperial College of Science and Technology, London
and
J. A. Smith, Glamorgan Polytechnic

SUMMARY

The paper describes the computer program AUTOPLAN which began as a feasibility study into performing the operations planning function for automatic lathes by computer.

AUTOPLAN is written in Fortran IV, and has as input the component geometry in the form of X and Y co-ordinates (for these rotationally symmetrical parts). The tolerances and surface finish associated with each feature of the components are also specified.

The program comprises six subroutines, three of which are concerned with the external features of a component, and the remaining three with the internal geometry. The subroutines perform the functions of recognition of operations required from the component geometry, a division of these among the tooling stations of the machine, and calculation of machining conditions and the associated cycle times.

Some results of computer planning of turned components are given, and the advantages are discussed.

1.0 INTRODUCTION

The object of this work was to examine the feasibility of using a computer for the planning of work for automatic lathes, and thereby reduce its time, potentially to the turn-around of the computer. However, a more fundamental objective was to use the exercise to study the logic which is involved, the breadth of detail and data required, and the recognition and judgment aspects which a human programmer provides. He is working from experience, but the computer must test inexorably and pedantically a wide range of circumstances, which if inadequate of specification, or indeed inadvertently omitted altogether, will produce nonsensical results.

Manual planning involves analysis of the engineering drawing for decisions on the operations and their sequence, and the machining conditions to be used. They are therefore also the requirements of the computer program.

The many different types of automatic lathe (and the semi-automatic capstan and turret variety need not necessarily be ruled out) indicate that a universally applicable program cannot be entertained. Thus an interface has been adopted, which is similar to that existing between a numerical control processor and post-processor, and up to which the program can remain general. At this point, all the operations necessary for manufacture have been specified, together with the machining conditions, and the operations times can be calculated. The operations listing is in 'primary sequence', i.e. necessary precedences (e.g. drilling before tapping) are maintained, but no attempt has been made to derive the 'optimum balance'. The latter involves knowledge of a particular machine-tool layout, and must therefore take place within a 'post-processor' program.

The general program or 'processor' has been named **AUTOPLAN**. Some further details of component specification and recognition have been given in an earlier paper [1]. The present paper describes the general logic and operation of the program, and gives an example of the computed output for a typical test-case component.

2.0 COMPONENT SPECIFICATION

A two-dimensional co-ordinate system of specification is used, comprising the axis of symmetry, and a radial axis, taken at the component parting-off plane. The specification is divided into two separate parts covering the external and internal geometry.

Co-ordinate (XO, YO) would represent the co-ordinates of a point, e.g. a change of section, on the outside surface of the component, and likewise, (XI, YI) would represent a point on an inside surface. Each surface comprises 'sections' which occur between any two adjacent specified points say, (XO(I)), YO(I) and (XO(I+I)), YO(I+I). This particular section is identified as XSECT(I), i.e. taking the lower index of the two adjacent points. It is thus a simple matter to determine a plain section, e.g. cylindrical, facial or taper, from the co-ordinates of its defining points.

Other sections, e.g. threaded, knurled, hexagonal and radiused parts, must be defined directly. Screw threads have been incorporated in the program using a code involving the addition, to the outside radius of the thread, of a number* for each different type thread required, e.g.:

B.S.W.: YO(I) = outside radius + 100
B.S.F.: YO(I) = outside radius + 110

Where a threaded section begins or ends a double specification is used. The co-ordinates of the end of the previous section are quoted together with the co-ordinates of the start of the threaded section. This is done to facilitate recognition, and it means that a 'dummy' section has been specified. A set TPII(I) is used to specify the number of threads per unit length associated with a threaded section I, and this will take the value zero for non-threaded sections.

The coding used for knurled and hexagonal sections is similar to that for threaded sections, e.g:

FINE KNURLING: YO(I) = outside radius + 200
HEXAGON: YO(I) = outside radius + 300

Again, a double specification is used, and so 'dummy' sections exist on either side.

---

* Intervals of 10 in the code numbers were used to ensure that no ambiguity arose due to overlap, and the dummy radius of 100 distinguishes the section as special, i.e. not to be confused with a normal radius: any appropriate numbers could be used.

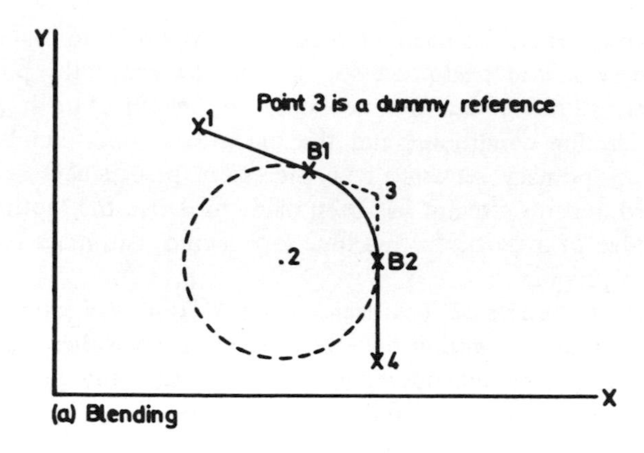

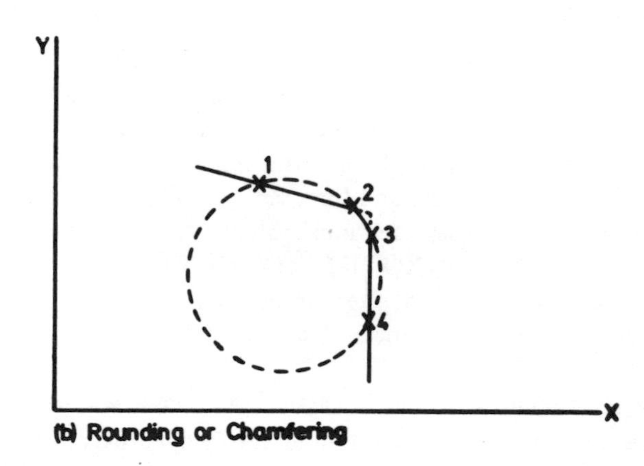

Fig. 1. Specification of blends and rounding or chamfering.

The specification of radiused sections is illustrated in Fig. 1. In Fig.1(a), involving a blending of straight and circular lines, the profile required is 1, B1, B2, 4. The XY co-ordinates of points 1, 2 and 4 are specified, point 3 being specified as a dummy reference. The co-ordinates of the centre of the circular arc are given, XOR(2), YOR(2), and the radius of the circular arc is specified as RO(2). In Fig. 1(b) a blend is not required (e.g. for simply removing a sharp corner) and the operation is much less critical. Code 'ROUND' indicates that a radiused section has no dimensional importance, i.e. merely rounding off a corner, and 'REFORM' indicates that the form of the section is important, e.g. for a pressure-tight sealing.

The system of tolerancing used involves a set SECTO (I, TOLER) associated with the tolerances of each section: I refers to the section number. The tolerance values assigned would depend on individual requirements, and sub-division and extension can be easily incorporated. Similarly, a set SECTO (I, FINISH) is associated with the surface finish of each section of the component, I again referring to the actual section number. Three classes of surface finish have been allowed for, but again, the system can be flexible.

The foregoing applies equally to internal dimensioning, with two exceptions. Knurling is not normally applicable to internal operations. Secondly, for an internal taper, a coding

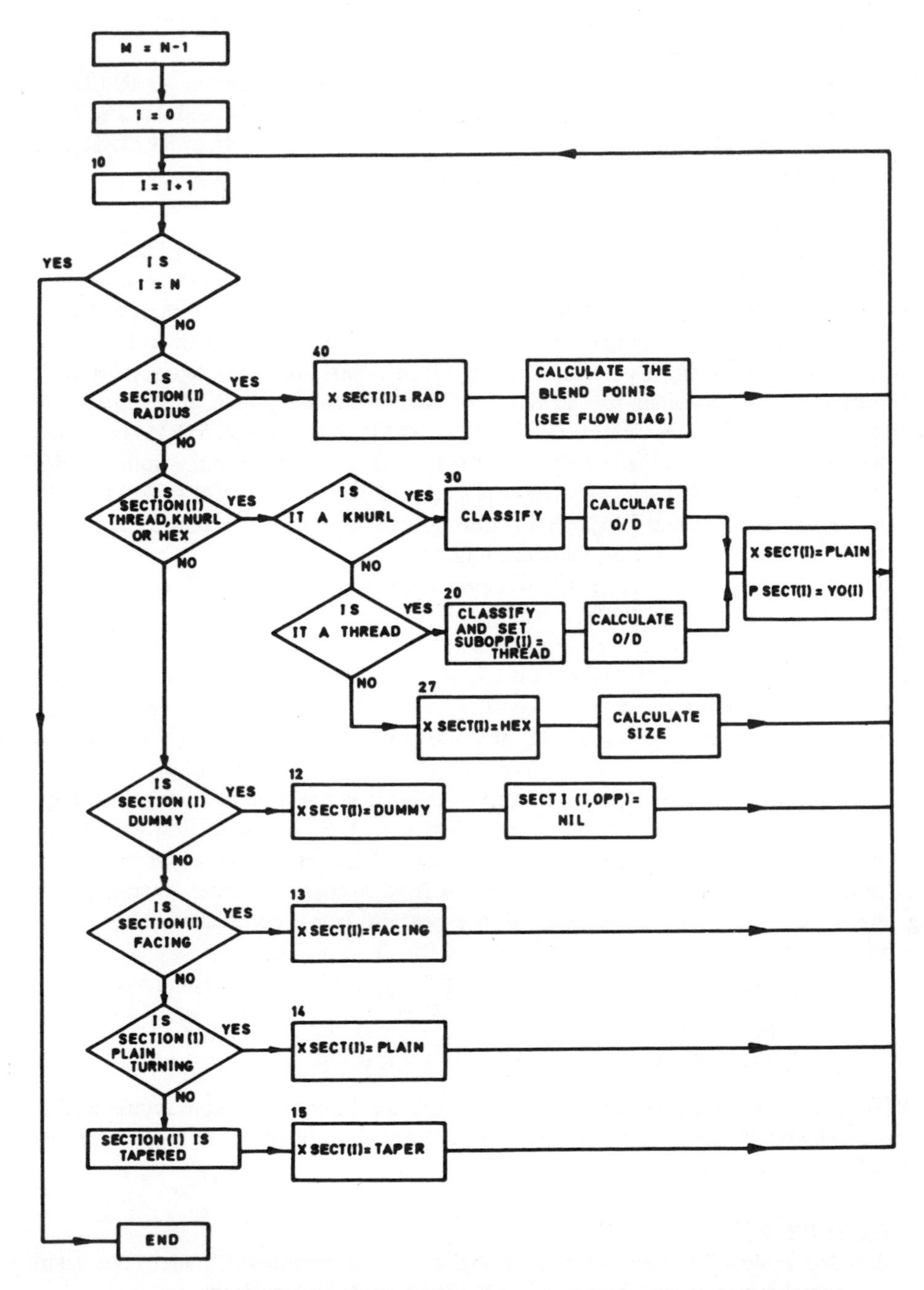

Fig. 2. Flow diagram for component recognition subroutine RECOGN.

'TAPANG(I)' is to be associated with the taper angle of section I. If this is specified by the coding 'STDANG' (referring to a standard angle) the program recognises that this section is formed by the drill point of a previous drilling operation.

3.0 DESCRIPTION OF MAIN PROGRAM

The main program calls upon six subroutines to process the data, in the order RECOGN, SUBREC, MCCOND, for external dimensions, and RECTWO, SUBTWO and MCTWO for internal parts. These subroutines will be described in greater detail; however, as RECOGN and RECTWO are very similar in structure, only the former need be described.

3.1 *Subroutine RECOGN*

The flow diagram is shown in Fig. 2. The subroutine first scans for a dummy specification point to fulfil the conditions XO(I) = 0·0 and YO(I) = 0·0. This is recognised as a radiused section. This test starts from XO(2), YO(2), as XO(1), YO(1), will always fulfil this condition.

The subroutine then tests whether blend points are tangential, and if tolerance is unimportant, and if these conditions are fulfilled the blend points are calculated quite simply. If not, subroutine QUAD is called to calculate two possible blend points, the appropriate solution being decided by the codes IN or OUT. XO(I), YO(I) and XO(I+1), YO(I+1) are then equated to the blend point co-ordinates (where XO(I+1), YO(I+1) is the dummy point specification). For future reference XSECT(I) is made equal to RADIUS, and SECTO(I,OPP) = RAD. I refers to the section number and OPP refers to the operation required on that section. The operation is also added to a list of form-tool operations.

The subroutine then scans for any YO(I) dimension greater than 300, and recognises the section as being hexagonal. SECTO(I,OPP) is made equal to NIL, indicating that no operation is required. The size of the hexagon bar is determined and XSECT(I) made equal to HEX.

The scan proceeds for any YO(I) dimension greater than 200, recognising it as a section which requires knurling, noting the type of knurling required, and the outside radius of the section. Since the section may require plain turning prior to knurling, XSECT(I) is set equal to PLAIN.

The scan then proceeds for any YO(I) > 100, recognising it as a section requiring threading. The type of thread, and its outside radius is then determined. Since the section will almost certainly require a plain turning operation before threading, XSECT(I), is set equal to PLAIN.

Sections remaining after these tests are then classified as parallel, taper or facing, by comparing the co-ordinates on either side of the section being considered. For future recognition purposes, such sections are classified as:

Parallel turning:	XSECT(I) = PLAIN
Taper turning:	XSECT(I) = TAPER
Facing operation:	XSECT(I) = FACING

SUBROUTINE RECOGN then gives a listing of the basic machining operations required (this is evident as 'preliminary recognition' in the AUTOPLAN output given in Fig. 6 for the component in Fig. 5).

3.2 *Subroutine SUBREC*

A more detailed analysis is made of the operations to be performed, under four main sections, plain turning, facing, taper turning and combined form-tool operations.

All sections requiring plain turning are identified, including sections that will require subsequent knurling or threading. The sections are listed under SECTO(I, TYPE) = PLAIN, and the outside radii of these sections are stored in 'LIST(COUNT)'. Comparative tests are then carried out on these outside radii to determine whether or not the operation can be performed from the tool-post, tailstock, or tool-post and tailstock (i.e. for possible overlapping).

These sections are then referenced by the appropriate codes TSTOCK and TPOST.

A search is carried out for facing operations, and these are tested under SECTO(I,OPP) = FACING. A test also identifies the section produced by the parting-off operation, and that formed by the parting-off operation on the previous component, if applicable. The tolerances and surface finish requirements of the remaining facing operations are examined to determine whether a separate facing operation is required, or if the operation is suitable for incorporation in a form tool with some other operation.

Tests determine whether the taper can be produced by a form tool, or whether a taper-turning attachment is necessary, consideration being given to the length of the section, and to the finish required. As a result of these tests SECTO(I,OPP) is set equal to 'TATACH', for a taper turning attachment operation, or 'TFORM', for a form-tool operation. The taper half-angle and 'sense' is calculated, and expressed also as an 'offset' required for a taper-turning attachment.

Sections are then tested sequentially to determine whether they can be combined and incorporated within the same form-tool operation, and if so, they are rewritten, as e.g. SECTN(I) = FORM. These tests take account of surface finish, tolerance, and the maximum form-tool width which can be used.

3.3 *Subroutine MCCOND*

The main program then calls SUBROUTINE MCCOND in which the operations to be performed and the appropriate machining conditions are determined. The operation of this subroutine can be seen from the flow diagram of Fig. 3, and the following brief description.

The first test determines whether or not the bar stock is hexagonal; if not, the largest diameter on the component is read and identified as plain or tapered. If tapered, the appropriate size of bar is selected and the machining conditions for the initial bar reduction calculated. If plain, again the appropriate bar size is selected, depending on the tolerance for the section. The machining conditions for the initial cut are determined, and limits are set for future traverses at the extreme ends of the section. These limits enable the lengths of subsequent cuts to be determined, and prevent overrun of the tool into other sections. A test for a knurling operation is then carried out. The next largest plain section is then identified from the set PSECT(I) and a test performed to determine if the operation can be performed from the tail-stock. If it can, the machining conditions are calculated, and if not, the section will be turned from the toolpost. Tests determine whether a form tool is suitable, and further tests on adjacent sections determine their suitability for inclusion in this forming operation. If a form tool cannot be used, the machining conditions for normal turning are calculated, and further tests determine whether facing operations are required at the ends of the section. The above procedure is repeated for all the plain turning operations on the component.

Machining conditions for the plain turning operations involve the depth of cut, finish, and tolerance requirements for determination of the number of cuts (possibly more than one roughing, and a finishing pass), and the feed rates required. These conditions are evaluated in the 'plain turning' subroutine given in Fig. 4(a).

For sections which require external threading, the length of thread is calculated, and the cutting speed is taken as 0·15 times the speed for plain turning. The machining conditions and manufacturing time are calculated from the major diameter and pitch.

Sections requiring a taper-turning attachment are considered next, and the machining

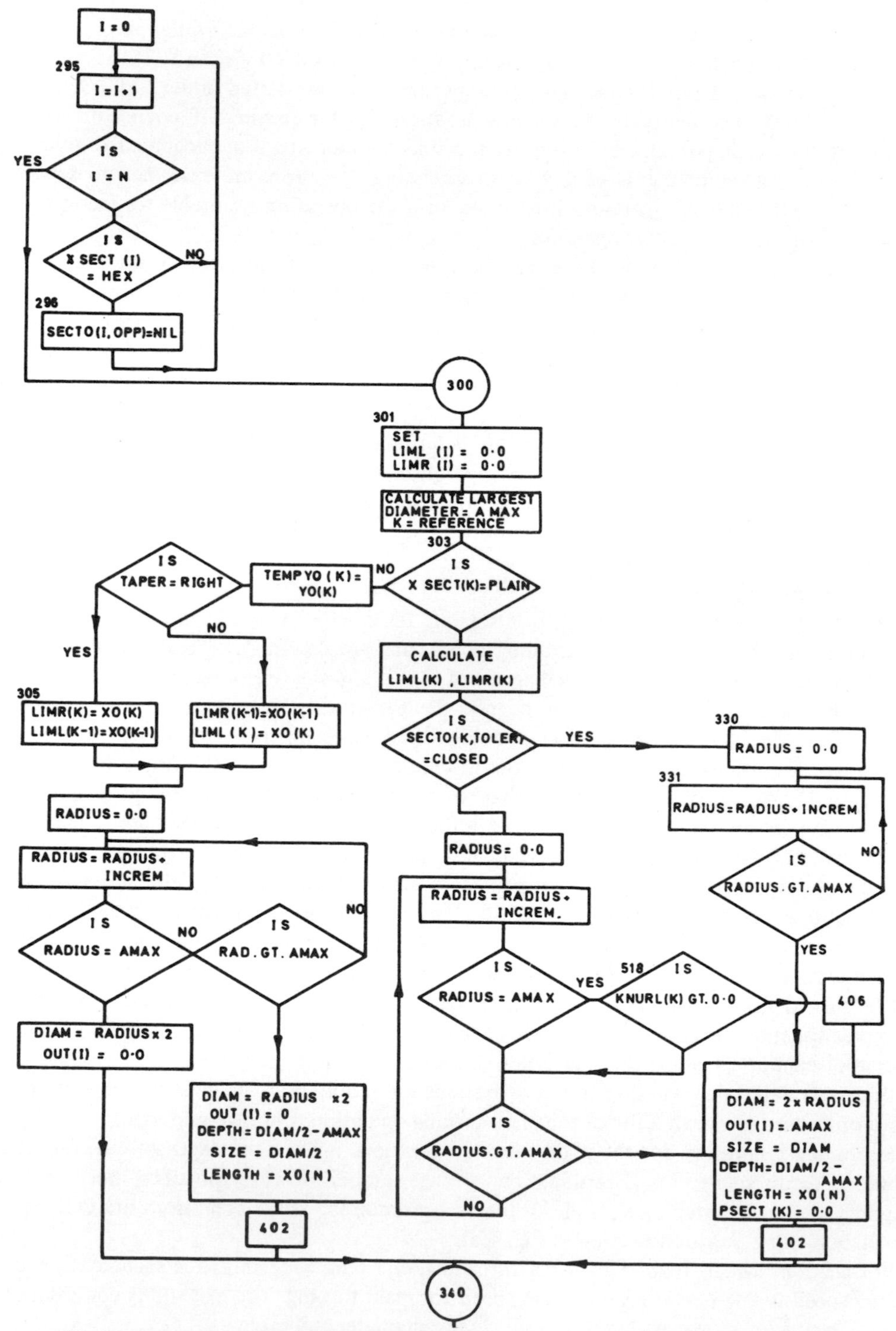

Fig. 3(a). Flow diagram of subroutine MCCOND, for derivation of machining conditions.

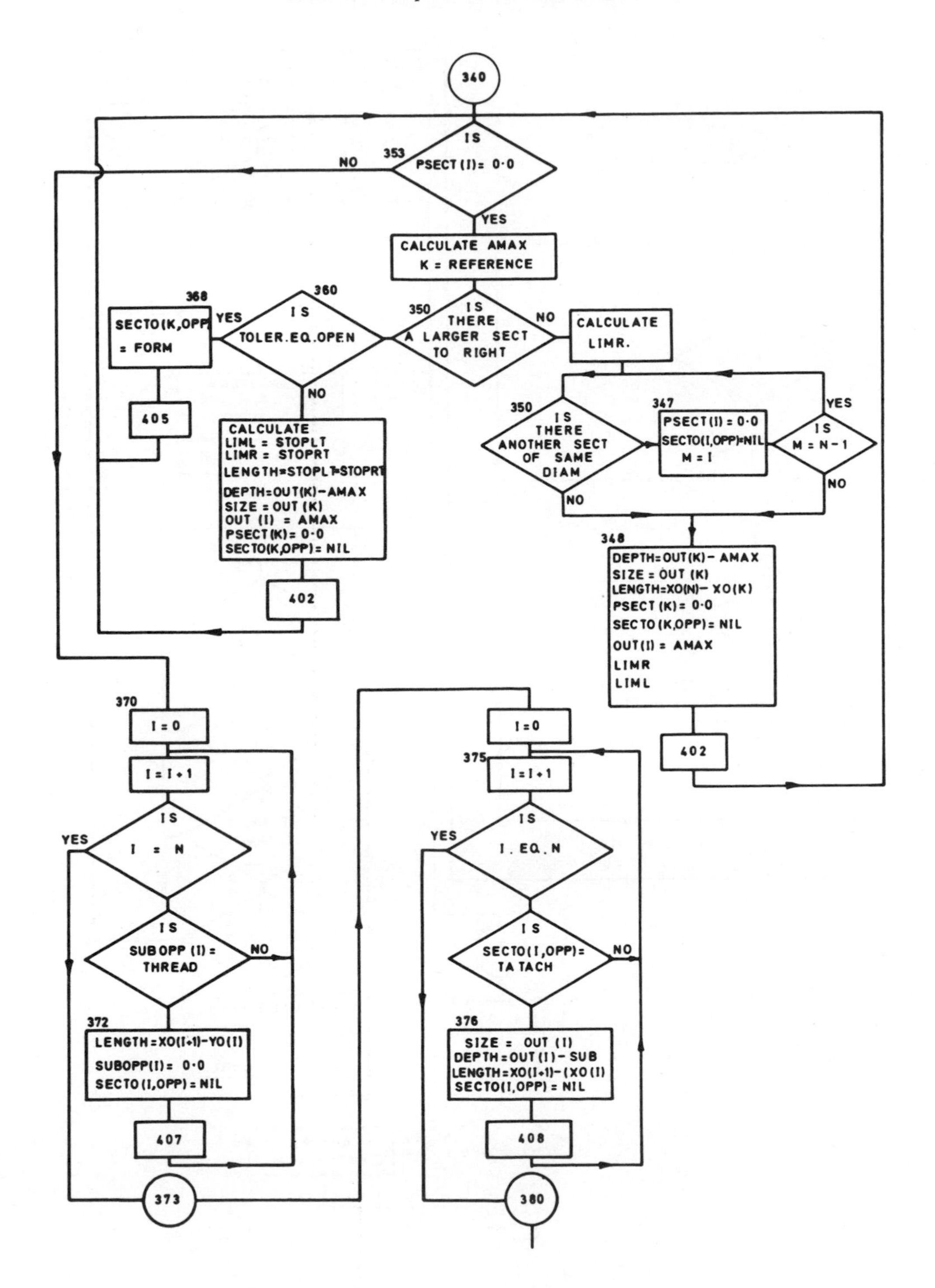
340
353
IS
PSECT (I) = 0·0
NO
YES
CALCULATE AMAX
K = REFERENCE
350
IS
THERE
A LARGER SECT
TO RIGHT
NO
CALCULATE
LIMR.
360
IS
TOLER.EQ.OPEN
YES
368
SECTO(K,OPP
= FORM
405
NO
CALCULATE
LIML = STOPLT
LIMR = STOPRT
LENGTH=STOPLT-STOPRT
DEPTH=OUT(K)-AMAX
SIZE = OUT (K)
OUT (I) = AMAX
PSECT(K)= 0·0
SECTO(K,OPP) = NIL
402
350
IS
THERE
ANOTHER SECT
OF SAME
DIAM
NO
347
PSECT(I) = 0·0
SECTO(I,OPP)=NIL
M = I
IS
M = N - 1
YES
NO
348
DEPTH=OUT(K)- AMAX
SIZE = OUT (K)
LENGTH=XO(N)- XO(K)
PSECT (K)= 0·0
SECTO (K,OPP) = NIL
OUT(I) = AMAX
LIMR
LIML
402
370
I = 0
I = I + 1
IS
I = N
YES
IS
SUBOPP (I) =
THREAD
NO
372
LENGTH=XO(I+1)-YO(I)
SUBOPP(I)= 0·0
SECTO (I,OPP) = NIL
407
373
I = 0
375
I = I + 1
IS
I . EQ . N
YES
IS
SECTO(I,OPP)=
TA TACH
NO
376
SIZE = OUT (I)
DEPTH = OUT (I) - SUB
LENGTH=XO(I+1)-(XO(I)
SECTO(I,OPP) = NIL
408
380

Fig. 3(b)

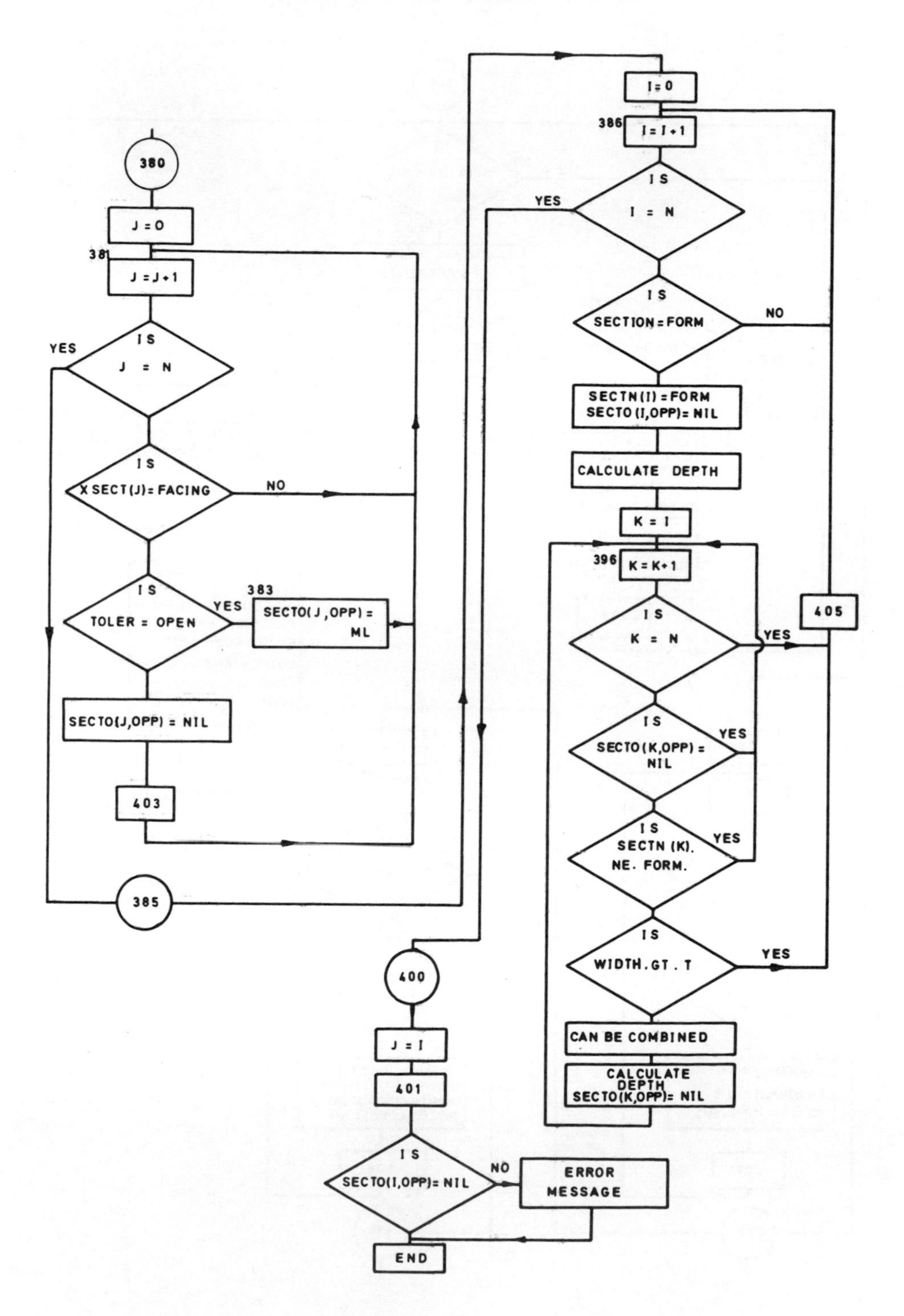
380
J = 0
381
J = J + 1
IS
J = N
YES
IS
X SECT(J) = FACING
NO
IS
TOLER = OPEN
YES
383
SECTO(J, OPP) = ML
SECTO(J, OPP) = NIL
403
385
I = 0
386
I = I + 1
IS
I = N
YES
IS
SECTION = FORM
NO
SECTN(I) = FORM
SECTO(I, OPP) = NIL
CALCULATE DEPTH
K = I
396
K = K + 1
IS
K = N
YES
405
IS
SECTO(K, OPP) = NIL
YES
IS
SECTN (K). NE. FORM.
YES
IS
WIDTH. GT. T
YES
CAN BE COMBINED
CALCULATE DEPTH
SECTO(K, OPP) = NIL
400
J = I
401
IS
SECTO(I, OPP) = NIL
NO
ERROR MESSAGE
END

Fig. 3(c)

conditions established, as for plain turning, in the subroutine of Fig. 4(b).

Facing operations are determined by consideration of the tolerance and surface finish required. If a facing operation is necessary, the spindle speed is determined from the larger dimension of the section to be faced, and the other machining conditions are calculated, together with the operation time.

Remaining form-tool operations, such as radiusing and chamfering, are then considered, together with the possibility of combining operations within the same form tool.

Finally, for the part-off operation, the width of the tool is determined by the depth of cut required. The machining conditions and operation time are then calculated.

As each of the above operations are considered, the reference SECTI(I, OPP) is made equal to zero, so that that section is removed from the operations list, and as each plain turning section is considered, the value of PSECT is also made equal to zero. All the operations required should have been accounted for, but a final test is made, and an error message is printed out if all sections have not been removed from the SECTO(I, OPP) list.

In this subroutine, certain simplifying assumptions have necessarily been made regarding the choice of machining conditions. Thus cutting speeds for various types of operations have been fixed as fractions of that for plain turning. Constraints have also been imposed on feeds, depending on the finish required, and for plain turning, on the power required at the cutting tool. These conditions, based on normal machining practice, are necessary to ensure that sensible choice is made by the program, and no claim is made of completeness, at this stage. However, further consideration is currently being given to a number of these areas.

3.4 *Subroutine RECTWO*

This subroutine is now called to examine similarly the data specifying the internal surface of the component. The only difference is that the tests for knurled sections are not included. This subroutine again 'recognises' the various sections, and lists them under ISECT.

3.5 *Subroutine SUBTWO*

This involves a more detailed analysis of the sections requiring internal facing and cylindrical operations. Tolerances and surface finish indicate whether a separate facing operation is required (i.e. involving a feed motion). For cylindrical surfaces, the associated tolerances determine whether a section can be produced by drilling, or whether drilling will have to be succeeded by reaming or boring, SECTI(I,OPP) being set equal to DRILL, REAM or BORE, whichever operation is required.

3.6 *Subroutine MCTWO*

The main program finally calls subroutine MCTWO to determine the operations and machining conditions required for the internal surfaces. Associated with each plain section is a store IPSECT containing the outside radius of the section. By comparison with succeeding values of IPSECT, a list of recessing operations is drawn up, the width of the recessing tool required being calculated, and the following conditions set:

SECTI(I,OPP) = RECESS
IPSECT(I) = 0·0
ISECTN(I) = FORM

Having removed the plain sections requiring recessing from the IPSECT store, a scan is performed to find the plain section having the largest remaining diameter. A check determines if

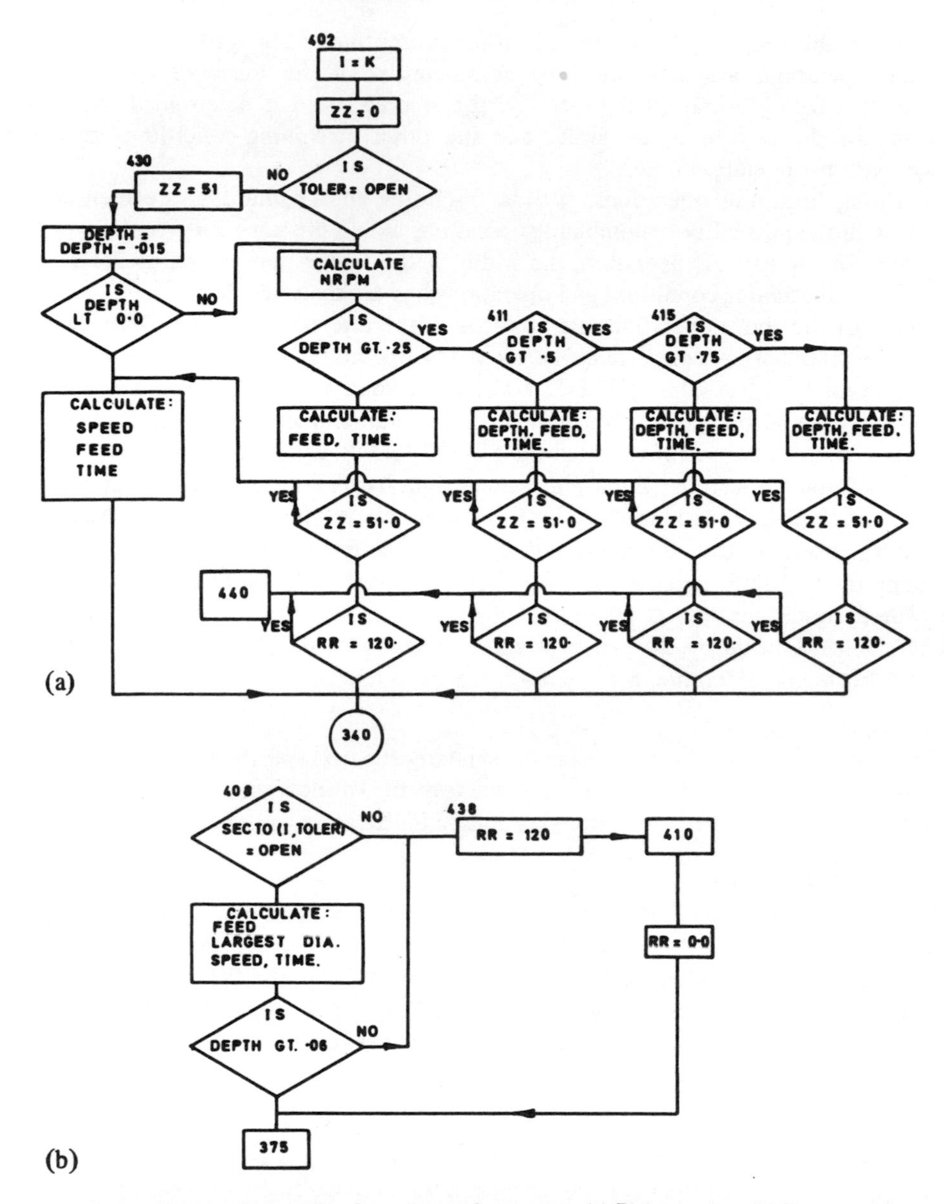

Fig. 4. Subroutine for derivation of conditions for turning. (a) Plain turning. (b) Taper turning.

this section is preceded by an undercutting recess section, and the length of cut is calculated accordingly. SECTI(I,OPP) and IPSECT(I) for this section are removed from the list by equating them to zero. The machining conditions for drilling, and/or boring and reaming, are calculated, together with the machining times. The feeds to be employed are dependent upon the type of operation, and/or the finish required. The speed for reaming is set at one third of that for drilling and boring.

The program then determines whether, prior to tapping, a drilling operation has been carried out. If so the thread length is calculated, cutting speed again being taken as 0·15 of that for plain turning, and the machining conditions and time are calculated. If the section prior to a plain section is a recess, this section is then considered, the feed being decided by the previously calculated recess width, and the machining conditions are calculated. If the prior section required facing, a test on the tolerance determines whether the section can be included with the previous plain turning operation by using a flat-ended drill or form tool. If not, a facing operation is required, and the machining conditions are evaluated appropriately.

If the above two tests are negative, the program tests for taper. Some tapered sections could be formed by the tool point of a previous drilling operation, in which case no separate operation is required. Otherwise, the possibility of using a special tool, such as a tapered reamer, is tried, which if positive, indicates that the section can be included with any other reaming operation by using a specially formed reamer. The remaining possibility is that the section requires a taper-turning attachment, in which case the machining conditions are calculated in the same way as for plain turning.

Final tests determine whether all the plain sections have been considered, and whether there are any sections left unaccounted for. If there are, the possibilities of the sections being tapered, formed, faced or threaded, are considered, in that order. A final check is then made to ensure that the SECTI(I,OPP) list is empty (an error-message being printed out if it is not).

4.0 EXAMPLE

As an example of the use of the program, the computer print-out for the component in Fig. 5 is shown in Fig. 6, the input data specification being given in Table 1. Component recognition (from subroutine RECOGN) in terms of the necessary external operations, appears first, and it can be seen that this is consistent with the component specification. Subroutine SUBREC provides the 'more detailed analysis of operations' which follow. The sequence here is not inviolate, although certain precedences have been observed, such as the obvious necessity to work successively inwards from the shape of the raw stock. The machining conditions follow, given in the operations sequence order, which can be seen to be a technically feasible one.

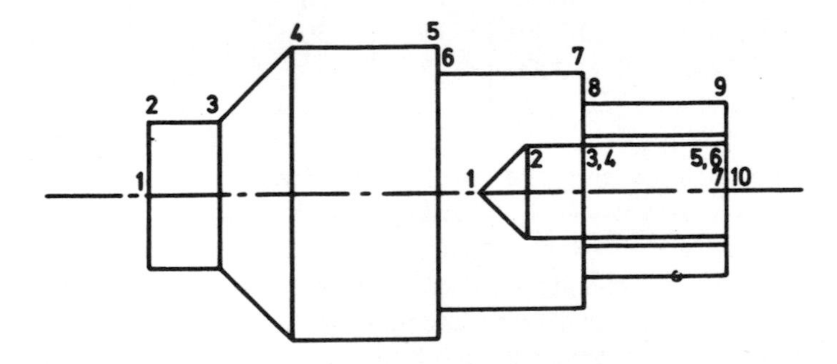

Fig. 5. Typical test component. Input data for this component are given in Table 1 and the computer output in Fig. 6. Numbers refer to co-ordinate data points which must be specified for AUTOPLAN.

In much the same way, the internal operations are next recognised, and then detailed in sequence order, observing similar precedences, such as the necessity for drilling before tapping or reaming, etc. Thus all operations are identified, taken in a technically feasible sequence and their associated machining conditions and cutting times are printed out.

TABLE 1. INPUT DATA FOR COMPONENT OF FIG. 5.

External data N = 10; T = 1·000

POINT	XO (1)	YO (1)	SECTION	TOLER	FINISH	TPIO (1)
1	000·0000	000·0000	1	11	50	–
2	000·0000	000·2500	2	13	51	–
3	000·2500	000·2500	3	13	51	–
4	000·5500	000·5500	4	11	52	–
5	001·0000	000·5500	5	13	50	–
6	001·0000	000·4000	6	13	51	–
7	001·5000	000·4000	7	11	51	–
8	001·5000	000·3000	8	13	50	–
9	002·0000	000·3000	9	13	50	–
10	002·0000	000·0000				

Internal data N = 7

POINT OR SECTION	XI (1)	YI (1)	TOLER	FINISH	TPII (1)	TAPANG (1)
1	001·0000	000·0000	13	50	–	150
2	001·1250	000·1250	13	50	–	–
3	001·2500	000·1250	13	50	–	–
4	001·2500	100·1250	13	50	16	–
5	002·0000	100·1250	13	50	–	–
6	002·0000	000·1250	13	50	–	–
7	002·0000	000·0000	13	50	–	–

However, the operations sequence and cutting conditions may not necessarily be optimal for the component taken as a whole (both internal and external operations) in terms of the minimum production cycle time. Such optimization must be performed by a further 'post processor' program (which is under development) operating on these identified operations in primary sequence, and which would take account of the possibilities of overlapping of operations, depending upon the tooling station layout of a particular machine tool.

5.0 CONCLUSION

Even with a number of simplifications which have been made when working within the framework of the above for the reasons described in the paper, the program AUTOPLAN has reached over 2000 cards. The compilation time for the program on an IBM 7094 is approximately 1·2 minutes, whereas the processing time for an individual component is a few seconds.

```
                    FOR THE EXTERNAL OPERATIONS

          PRELIMINARY RECOGNITION FOLLOWS:

SECTION (1) REQUIRES A FACING OPERATION
SECTION (2) REQUIRES PLAIN TURNING
SECTION (3) REQUIRES TAPER TURNING
SECTION (4) REQUIRES PLAIN TURNING
SECTION (5) REQUIRES A FACING OPERATION
SECTION (6) REQUIRES PLAIN TURNING
SECTION (7) REQUIRES A FACING OPERATION
SECTION (8) REQUIRES PLAIN TURNING
SECTION (9) REQUIRES A FACING OPERATION

          MORE DETAILED ANALYSIS OF OPERATION FOLLOWS:

PARALLEL SECTION (1) CAN ONLY BE MACHINED FROM THE TOOL POST
PARALLEL SECTION (2) CAN BE MACHINED FROM THE TOOL POST OR TAIL STOCK
PARALLEL SECTION (3) CAN BE MACHINED FROM THE TOOL POST OR TAIL STOCK
SECTION (1) REQUIRES A FACING OPERATION

SECTION (1) REQUIRES A PARTING OFF OPERATION
SECTION (5) DOES NOT REQUIRE A FACING OPERATION
SECTION (7) REQUIRES A FACING OPERATION
SECTION (9) DOES NOT REQUIRE A FACING OPERATION

SECTION (3) TAPER CAN BE MANUFACTURED USING A FORM TOOL
SECTION (3) HAS AN INCREASING TAPER LEFT TO RIGHT OF 1 IN 1.000 HALF ANGLE

          THE MACHINING CONDITIONS ARE:

BAR MATERIAL TO BE USED IS1.1250 INCHES ROUND
SECTION (4) ONLY REQUIRES ROUGH TURNING.   MACHINING CONDITIONS ARE:
CUTTING SPEED = 509 R.P.M., FEED RATE = .0500 INS/REV., DEPTH OF CUT =
0.0025 INS., MACHINING TIME = .0611 MINUTES.

SECTION (4) ALSO REQUIRES A FINISH TURNING OPERATION.   MACHINING CONDITIONS
ARE:  CUTTING SPEED = 509 R.P.M., FEED RATE = .0020 INS/REV., DEPTH OF CUT
= .0150 INS., MACHINING TIME = .9632 MINUTES.

SECTION (6) ONLY REQUIRES ROUGH TURNING.   MACHINING CONDITIONS ARE:
CUTTING SPEED = 521 R.P.M.,  FEED RATE = .0500 INS/REV., DEPTH OF CUT =
.1500 INS., MACHINING TIME = .0384 MINUTES.

SECTION (8) ONLY REQUIRES ROUGH TURNING.   MACHINING CONDITIONS ARE:
CUTTING SPEED = 716 R.P.M., FEED RATE = .0500 INS/REV., DEPTH OF CUT =
.1000 INS., MACHINING TIME = .0140 MINUTES.

SECTION (5) DOES NOT REQUIRE A FACING OPERATION.

SECTION (7) REQUIRES A FACING OPERATION, CUTTING SPEED = 716 R.P.M.,
FEED RATE = .0060 INS/REV., MACHINING TIME = .0233 MINUTES, DEPTH OF CUT =
SCIM

SECTION (9) DOES NOT REQUIRE A FACING OPERATION.

.SECTION (2) IS TO BE MANUFACTURE BY A FORM TOOL OPERATION
.SECTION (3) CAN BE COMBINED WITH THE FORM TOOL OF SECTION (2)
SECTION (2) REQUIRES A FORM TOOL OPERATION.   MACHINING CONDITIONS ARE
CUTTING SPEED = 229 R.P.M., FEED RATE = .0015 INS/REV., MACHINING TIME
= .8726 MINUTES.

SECTION (1) REQUIRES PARTING OFF.   THE FOLLOWING ARE THE MACHINING
CONDITIONS:  TOOL WIDTH = 0.0500 INS., CUTTING SPEED = 1153. R.P.M.,
MACHINING TIME = .0723 MINUTES.
```

Fig. 6. [continued at top of page 322]

```
                FOR INTERNAL OPERATIONS

COMPONENT RECOGNITION FOLLOWS:

SECTION (1) REQUIRES INTERNAL TAPER TURNING
SECTION (2) REQUIRES BORING
SECTION (3) IS A DUMMY SECTION
SECTION (4) = .2500 BSW INTERNAL THREAD
SECTION (5) IS A DUMMY SECTION
SECTION (6) REQUIRES INTERNAL FACING
SECTION (6) DOES NOT REQUIRE A FACING OPERATION AS IT IS A DUMMY SECTION
SECTION (2) IS OPEN TOLERANCED AND CAN BE PRODUCED BY DRILLING
SECTION (4) IS OPEN TOLERANCED AND CAN BE PRODUCED BY DRILLING

        THE MACHINING CONDITIONS ARE:

SECTION (2) REQUIRES A DRILLING OPERATION.   MACHINING CONDITIONS ARE:
MACHINING TIME = .0218 MINUTES, CUTTING SPEED = 292 R.P.M., FEED RATE =
.0025 INCHES.

SECTION (4) REQUIRES A TAPPING OPERATION.   MACHINING CONDITIONS ARE:
CUTTING SPEED = 344 R.P.M., FEED RATE = .0025 INS., MACHINING TIME = .0349 MINS.

.TAPER SECTION (1) IS FORMED FROM THE DRILL POINT OF THE PREVIOUS OPERATION.
```

Fig. 6. AUTOPLAN output for the component in Fig. 5, giving preliminary recognition, operations analysis, and machining conditions for external and internal component geometry. See Fig. 6, on page 321

Thus more economical use will be made of computer time if a number of components are processed with each run of the program. This is possible as the program is enclosed in a 'DO LOOP', the number of components to be processed being read in before the component specification data.

Although the scope of the program has been necessarily restricted so far, making the practical usefulness as it now stands somewhat limited, the results obtained from the components on which it has been tried are encouraging. This leads the authors to conclude that computer planning of automatic lathes is a feasible proposition, and that a number of benefits both in the practice of manufacture, and for certain researches into the production technology of this area, can result.

Several advantages in the application of a computer to the planning function can be cited. The use of such a program in practice would result in a considerable saving in planning time. Comparison of the turn-around time of the program (potentially a matter of minutes) with that taken for manual planning, makes this advantage clear. Then there are the normal advantages associated with data processing of relatively large quantities of detailed information, involving component geometry, the technological considerations of machining, and the implied economics.

The work as described here is considered to be at a fairly early stage. Further work is in progress on the use, not only of optimum individual machining conditions, but in optimization of combinations of operations, some of which must be performed individually, and possibly in a certain sequence, whilst others can be overlapped, with a resulting reduction in overall cycle time, and hence the cost of the component produced.

REFERENCE

1 CROOKALL, J. R., and SMITH, J. A., 'A component-recognition program for computer planning of automatic lathes', *Proc. Hungarian Scientific Soc. of Mech. Engnrs. Conf. 'COMPCONTROL '70 on Industrial Application of Computers,* Miskolc, Hungary, 7-11 Jul., 1970, V, 89 – 100.

A COMPUTER SYSTEM FOR OPTIMISING PROCESS CONTROL

J. A. Smith
Glamorgan Polytechnic

SUMMARY

A description is given of a computer system for the optimisation of the inspection function for a process subject to predictable variability of manufacture. Optimum process setting and resetting points are determined on a criterion of minimising unit cost of manufacture. The computer program accepts periodic sample data from the process being controlled. It determines optimum process setting and resetting points, anticipates when resetting will be necessary, and determines the overall production rate. The results of a simulation test, together with the graphical decision rules produced, are presented.

NOTATION

- μ_1 initial process mean setting
- σ_1^2 initial process variance
- μ_2 final process mean setting
- σ_2^2 final process variance
- ϕ constant, related to process mean variations
- ϵ constant, related to process variance variations
- l_1 lower limit of component
- l_2 upper limit of component
- K constant related to process resetting point
- Q rate of production per unit time, during manufacture
- p proportion defective produced by process
- τ process manufacturing time per manufacturing cycle
- τ' process resetting time per manufacturing cycle
- T overall manufacturing cycle time $= \tau + \tau'$
- C_m cost per unit time for the process working (includes operator cost, cost of materials, lighting, power and machine depreciation)
- C_r cost per unit time for the process being reset (includes operator and/or setter costs, fixed costs and machine depreciation)
- C_d cost per reject item
- C manufacturing cost per unit
- P selling price per unit
- Z profit per unit time
- α ratio of productive to resetting time $= \tau/\tau'$
- β ratio of resetting to productive time $= \tau'/\tau = 1/\alpha$
- R process capability ratio $= (l_2 - l_1)/\sigma$

1.0 INTRODUCTION

There is no technical reason why a computer should not completely control the manufacture of products from the production planning stage, through machining, to final inspection and despatch. Manufacture can be considered as comprising the following stages:

(1) Production planning.
(2) Process or machine tool set up.
(3) Manufacture or machining.
(4) Inspection.

Berra and Barash[1] and Crookall and Smith[2] have proposed computer systems for the automatic production planning of lathe work. The stages of set up and machining can be automated using the principles of numerical control. The object of this study is the computerisation and optimisation of the inspection function for a process subject to predictable variability of manufacture. Typical of such a process would be the manufacture of components on an automatic lathe, where inspection is concerned with critical dimensions. Other examples include metal rolling, extrusion and pressing operations, where predictable variability due to tool wear occurs.

2.0 DEFINING THE PROBLEM

Conventional process control is facilitated by means of statistical quality control charts. The statistical control of quality has the objective of minimising scrap. This need not be the best objective to pursue, especially under conditions of tool wear in which the process needs frequent resetting. It is suggested here that, with the facility of high-speed calculation offered by the

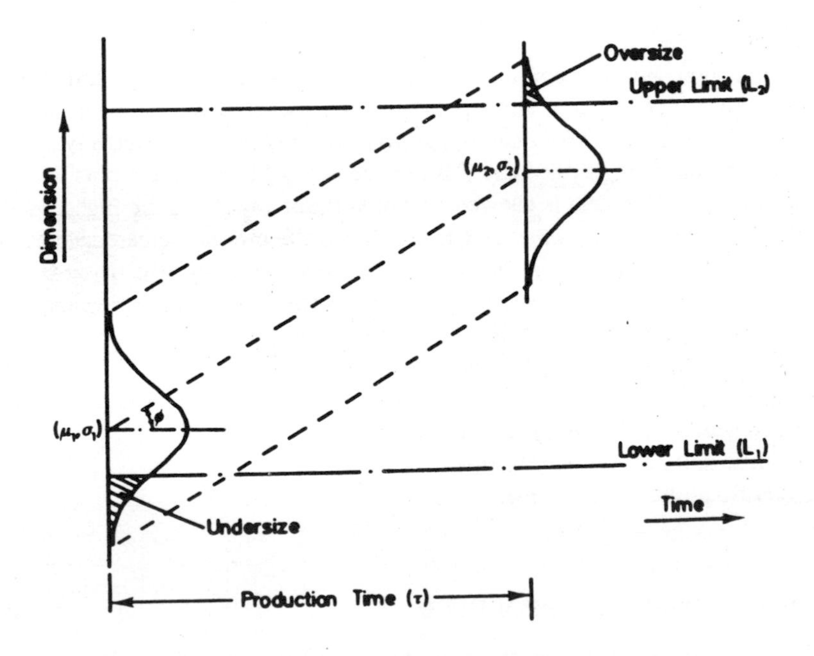

Fig. 1. The effect of process variability and tool wear.

digital computer, the pursuit of other objectives, such as minimising unit cost of manufacture, would be a feasible proposition.

During manufacture, due to the inherent variability of a production process, component dimensions conform approximately to the normal distribution curve. The mean setting of the process μ, and the process variance σ^2, may vary with time. The effects of process variability are shown in Fig. 1. The process commences at time $t = 0$, with initial process setting μ_1, and standard deviation σ_1. After a time t, the mean setting μ_2 and standard deviation σ_2 are given by

$$\mu_2 = \mu_1 + \phi t \tag{1}$$

$$\sigma_2 = \sigma_1 + \epsilon t \tag{2}$$

The component dimensions at time t are assumed to be normally distributed according to the equation

$$y = \frac{1}{(\sigma_1 + \epsilon t)\sqrt{(2\pi)}} \exp\left[-\frac{(x - (\mu_1 + \phi t))^2}{2(\sigma_1 + \phi t)^2}\right] \tag{3}$$

For convenience

$$y = \sim N(\mu_1 + \phi t, \sigma_1 + \epsilon t) \tag{4}$$

To ensure that no rejects (3σ confidence limits) would be produced, the initial process mean setting μ_1 would be at $(l_1 + 3\sigma_1)$, and the process reset when the process mean reached $\mu_2 = (l_2 - 3\sigma_2)$, l_1 and l_2 being the lower and upper part drawing limits, respectively.

The overall manufacturing cycle time T is defined as machining time τ plus process resetting time τ'. The longer the process is allowed to run without resetting, the higher is the ratio of manufacturing time to resetting time, and the higher is the process percentage productive time $100\tau/(\tau + \tau')$. However, the longer the process is allowed to run after $\mu_2 \geqslant (l_2 - 3\sigma_2)$, the more defects are produced. This suggests optimum points for process setting and resetting, the optimum points being determined by the criteria on which the system is to operate.

Three criteria immediately suggest themselves:

- To ensure that the proportion defective produced by the process will not (within prescribed probabilistic limits) exceed a stated level.
- To minimise unit cost of manufacture.
- To maximise profit per unit time.

3.0 PROPORTION DEFECTIVE PRODUCED

From Fig. 1 it can be seen that the proportion defective p produced by the process, for different setting and resetting points, entails the solution of the following equation:

$$p = 1 - \frac{\int_{o}^{T} \int_{l_1}^{l_2} \sim N(\mu_1 + \phi t, \sigma_1 + \epsilon t)\, dx.dt.}{\int_{o}^{T} \int_{o}^{\infty} \sim N(\mu_1 + \phi t, \sigma_1 + \epsilon t)\, dx.dt} \tag{5}$$

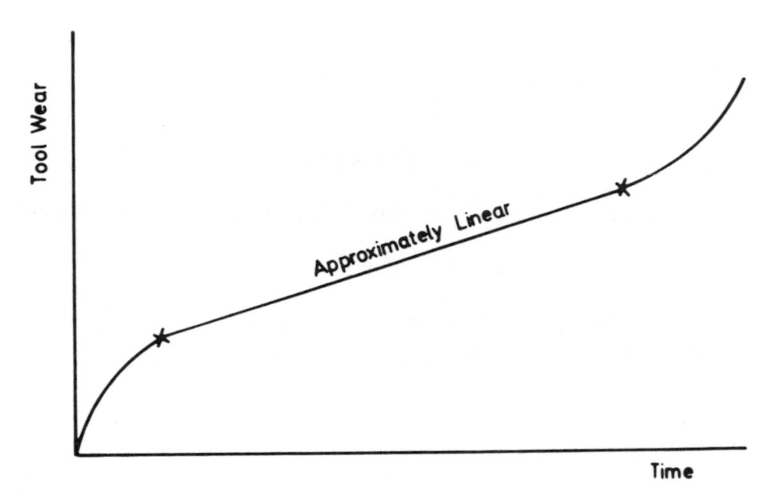

Fig. 2. Relationship between tool wave and time.

In a conventional single-point cutting-tool metal-cutting operation, the rate of tool wear is as shown in Fig. 2. Hence, for a large proportion of the tool life, tool wear is approximately linear with respect to time. Therefore, ϕ becomes a constant which can be predicted for specified machining conditions. A further simplification is introduced: $\epsilon = 0$, i.e. the process variability remains sensibly constant throughout the manufacturing cycle. Therefore,

$$p = 1 - \frac{\int_{o}^{T} \int_{l_1}^{l_2} \sim N(\mu_1 + \phi t, \sigma)\, dx.dt}{\int_{o}^{T} \int_{o}^{\infty} \sim N(\mu_1 + \phi t, \sigma)\, dx.dt} \tag{6}$$

The proportion defective produced by the process is negligible when $l_2 - \mu_2 \geqslant 3\sigma$, and when $\mu_1 - l_1 \geqslant 3\sigma$. Hence, Equation (6) has been evaluated for time intervals of t_1 and τ, where

$$\mu_1 = (l_2 - 3\sigma) \qquad \text{at time } t = t_1$$

and

$$\mu_2 = (l_2 - K_i\sigma) \qquad \text{at time } t = \tau$$

The proportion defective produced during this time interval is denoted by p', where

$$p' = 1 - \frac{\int_{t_1}^{T} \int_{l_1}^{l_2} \sim N(\mu_1 + \phi t, o)\, dx.dt}{\int_{t_1}^{T} \int_{o}^{\infty} \sim N(\mu_1 + \phi t, \sigma)\, dx.dt} \tag{7}$$

The above equation was solved for 30 different resetting levels, i.e. for 30 different values of K_i, taking the following values:

$$K_i = K_1 - [0{\cdot}1 \times (i - 1)]$$

where $K_1 = 3{\cdot}0$

and $i = 1, 2, ...30$

Equation (7) was solved using the Monte Carlo simulation method. A random normal generator was used to simulate the production of components whose critical dimension is

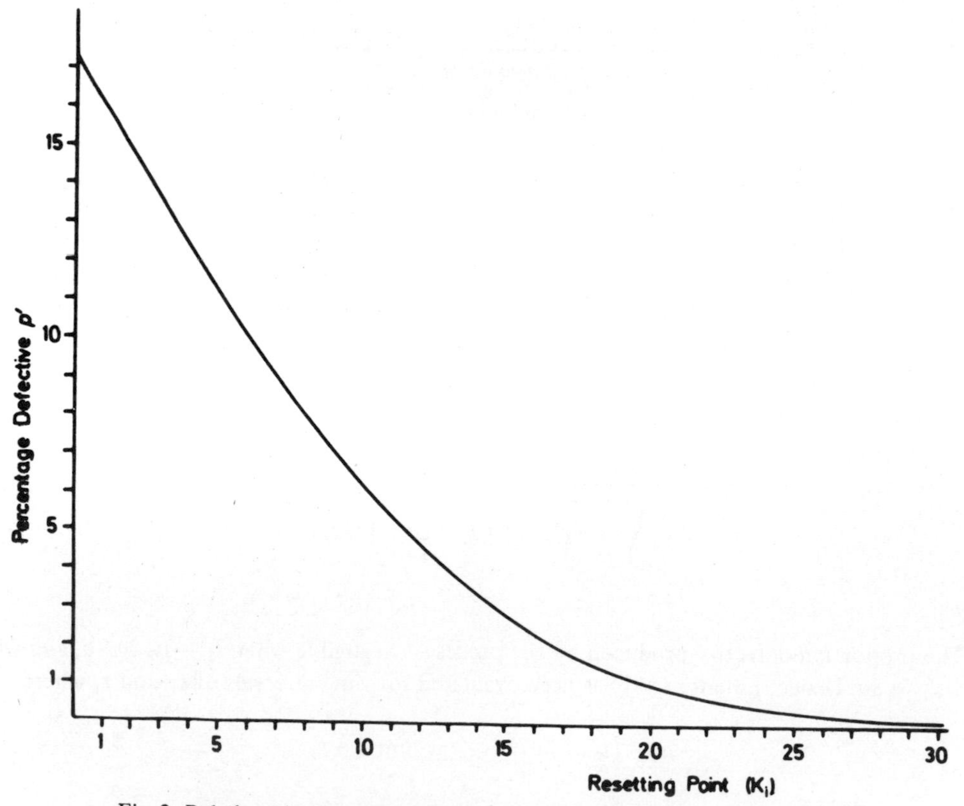

Fig. 3. Relationship between percentage defective and process resetting level.

normally distributed according to Equation (3), when $\epsilon = 0$. Random number streams of 5000 gave the desired confidence limits in the results obtained. The proportion defective produced between the times t_1 and τ, for the different resetting levels, are shown in Fig. 3.

The total proportion defective produced in a manufacturing cycle is therefore given by

$$p = p'\left(\frac{\tau - t}{\tau}\right)$$

4.0 MATHEMATICAL MODELS

4.1 *A cost minimisation model*

$$\text{Total cost per manufacturing cycle} = \tau . C_m + \tau' . C_r + \tau . C_d . Q . p \quad (9)$$

$$\text{Quantity of acceptable products per cycle} = \tau . Q . (1 - p) \quad (10)$$

$$\text{Manufacturing cost per unit } C = (\tau . C_m + \tau' . C_r + \tau . C_d . Q . p)/[\tau . Q . (1 - p)] \quad (11)$$

$$\text{or } C = (\alpha . C_m + C_r + \alpha . C_d . Q . p)/[\alpha . Q (1 - p)] \quad (12)$$

$$\text{where } \alpha = \tau/\tau'$$

This function has to be minimised.

4.2 *A profit maximisation model*

$$\text{Production rate per unit time} = \left(\frac{\tau}{\tau + \tau'}\right) Q \quad (13)$$

$$\text{Number of acceptable components per unit time} = (1 - p)\left(\frac{\tau}{\tau + \tau'}\right) Q \quad (14)$$

$$\text{Revenue per unit time} = P . (1 - p)\left(\frac{\tau}{\tau + \tau'}\right) Q . \quad (15)$$

$$\text{Manufacturing costs per unit time} = C_m\left(\frac{\tau}{\tau + \tau'}\right) + C_r\left(\frac{\tau'}{\tau + \tau'}\right) + p\left(\frac{\tau}{\tau + \tau'}\right) Q . C_d \quad (16)$$

$$\text{Profit per unit time } Z = P(1 - p)\left(\frac{\tau}{\tau + \tau'}\right)Q - [C_m\left(\frac{\tau}{\tau + \tau'}\right) + C_r\left(\frac{\tau'}{\tau + \tau'}\right) + p\left(\frac{\tau}{\tau + \tau'}\right) Q . C_d] \quad (17)$$

$$\text{or } Z = \left(\frac{1}{1 + \beta}\right)[P . (1 - p) . Q - (C_m + C_r . \beta + p . Q . C_d)] \quad (18)$$

$$\left(\text{where } \beta = \frac{\tau'}{\tau}\right)$$

This function has to be maximised.

5.0 COMPUTER SYSTEM

A diagram of the computer system is shown in block form in Fig. 4. There are two sources of data input to the program.

(1) *Initial input data.* Here, data regarding component specification, critical dimensions and tolerances are fed in, together with production rates, machining conditions, cost data and process capability information. From this information, the initial process setting is determined, consistent with the cost minimisation objective.

(2) *Sample.* This is the second source of data input. This consists of periodic sample data from the process being controlled. Although the program is capable of accepting large samples, sample sizes of $n > 6$ are not recommended. The program is capable of accepting sample data at irregular time intervals, but regular sample information is preferred.

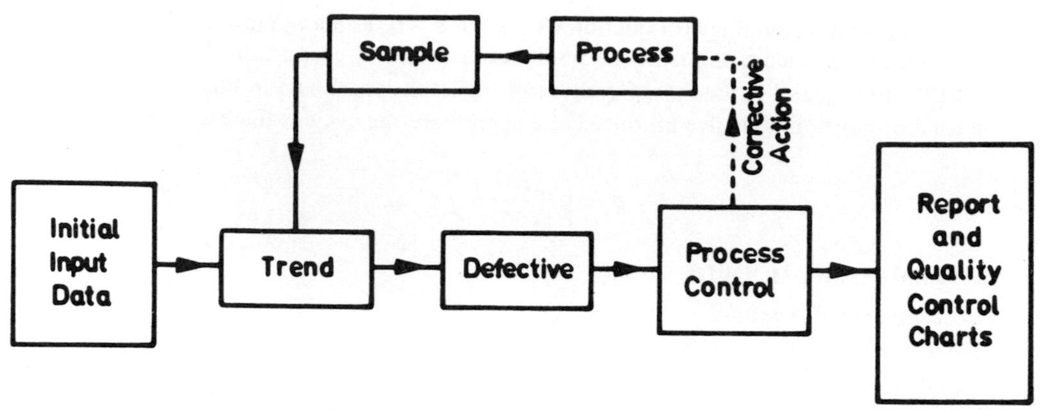

Fig. 4. Computer system.

These sample data are processed through subroutine TREND. Here the process mean value, the process variance, and the trend in mean drift are estimated. In addition, statistical tests of significance are conducted on the variance and the regression line slope.

This information becomes the input to subroutine DEFECTIVE, where the proportion defective produced for various setting and resetting points are calculated. The stored information from the graph shown in Fig. 3 is used for this purpose, the proportion defective being calculated from Equation (8).

The mathematical model that optimises process setting is included in PROCESS CONTROL. This section of the program determines optimum process setting and resetting points, anticipates when resetting will be necessary, predicts the level of rejects that will be produced, and determines the overall production rate. Provision is made to impose a defective constraint, i.e. $p \leqslant w$, where w (the maximum defective allowable) is to be specified. This constraint will override the optimising function.

The program output consists of information to be fed back to the process being controlled to enable corrective action to be taken when necessary. A written report is also produced periodically indicating process trends. This is accompanied by conventional quality control charts.

For the system to function efficiently and effectively, the time lag in the feedback mechanism must be reduced to as small a value as possible, and so, for on-line process control, the sample data must be recorded and transmitted to the computer as quickly as possible. The practical application of the system for on-line process control will have to await developments in in-process gauging equipment with digital read-out transmitting sample information via a data link to an on-line real-time computing unit, where the turn-round time could be measured in seconds.

The validity of the system has been tested by generating simulated component data which has been processed through the computer program.

6.0 SIMULATION RESULTS

The production of components having normally distributed dimensions, and constant variance was simulated. A time base was built into the simulation model to allow conditions of tool wear to be introduced and controlled.

The results of a simulation test, together with the graphical decision rules produced, are now presented.

Test conditions. Five rates of tool wear were investigated; ϕ = 0·0005; 0·001; 0·0015; 0·002; 0·003. Associated with each rate of tool wear seven process capability values R were tested. $R = 6, \ldots 12$ in increments of 1, where $R = (l_2 - l_1)/\sigma$. The above tests were repeated for ten different ratios of cost of machine being reset, to cost of machine running, for C_r/C_m = 0·1,... 1·0 in 0·1 increments. The process variance was taken to be constant at $1{\cdot}0 \times 10^{-6}$ and the process resetting time τ' as 0·5 for each test.

The optimum process setting and resetting levels for each of the above 350 tests were determined. The associated proportion defective was determined, and the ratio of manufacturing to resetting time calculated.

The simulation results obtained from the tests of C_r/C_m = 1·0 are included, and represented graphically in Figs. 5-7.

The optimum resetting level associated with the different values of R, for the five rates of tool wear, can be determined from Fig. 5. This information would enable the process engineer to decide when process resetting should be carried out in order to minimise unit cost of manufacture.

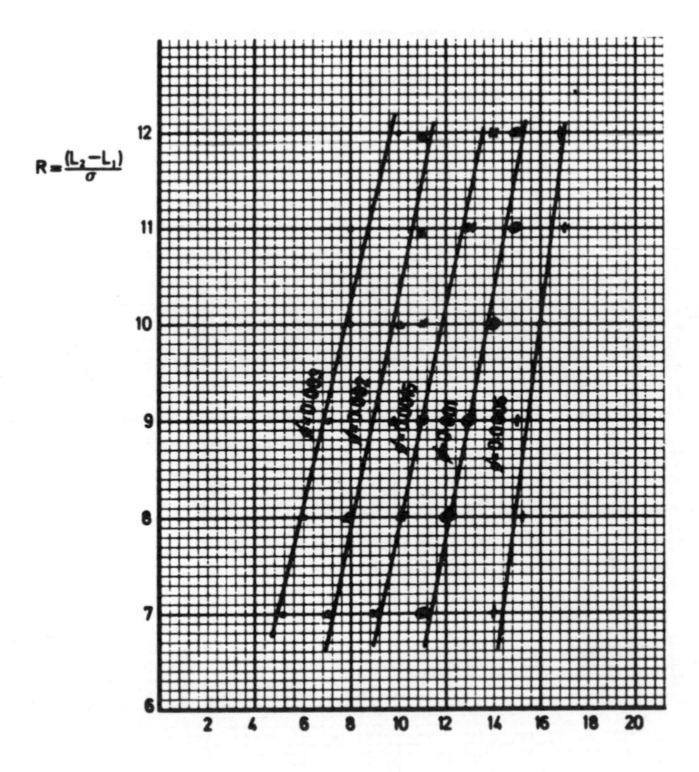

Fig. 5. Relationship between process capability and optimum resetting point for C_r/C_m = 1·0.

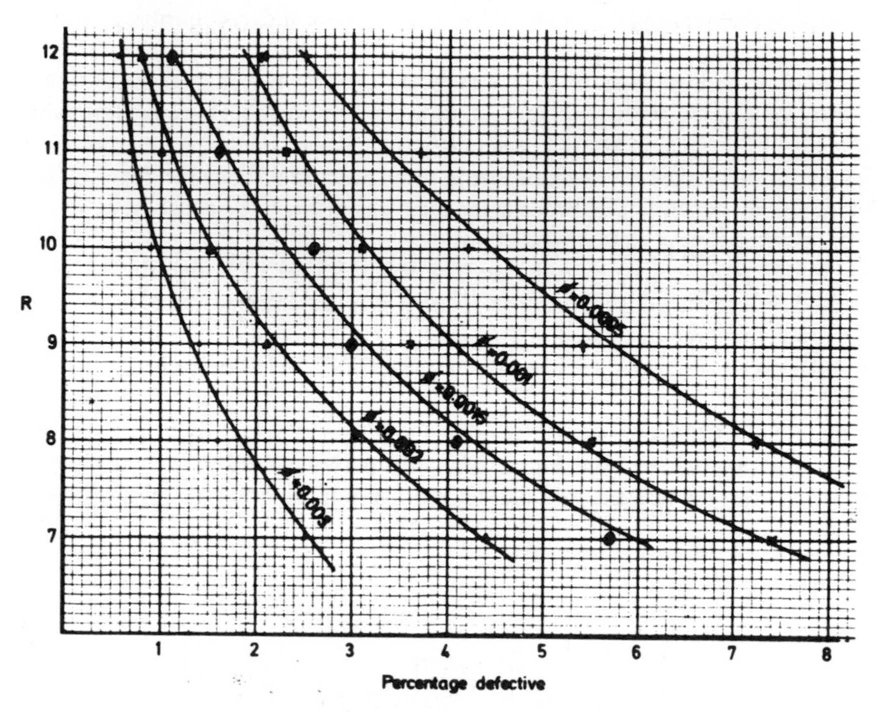

Fig. 6. Relationship between process capability and percentage defective for optimum resetting (C_r/C_m = 1·0).

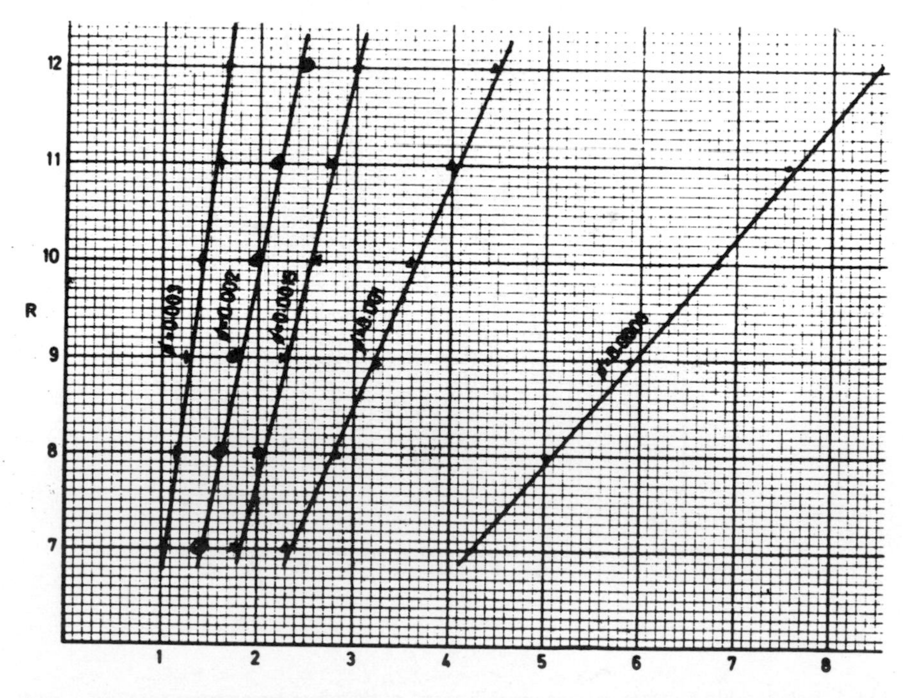

Fig. 7. Relationship between process capability and proportion manufacturing time for C_r/C_m = 1·0.

Associated with these optimum resetting points, the corresponding proportion defective being produced by the process can be determined from Fig. 6. From Fig. 7 the ratio of process manufacturing to process resetting time can be obtained. This means that the production rate can be ascertained.

7.0 CONCLUSION

The system described has been designed primarily for on-line process control. However, its use in this connection will have to await developments in in-process automatic component sizing, and on-line real-time computing systems. The program has, however, current application for process control. For given manufacturing data, costs and component specification, the rate of tool wear and, hence, rate of process drift, can be estimated. Feeding this information through the computer program would allow decision rules to be obtained for process setting and resetting. Valuable information for estimating and planning purposes are also obtained.

REFERENCES

1 BERRA, P. B., and BARASH, H. M., 'Automatic Process Planning and Optimisation for a Turning Operation', *Int. Jnl. Prod. Res.* 7 (2), 93-103, 1968.

2 CROOKALL, J. R., and SMITH, J. A., 'Component – Recognition Programme for Computer Planning of Automatic Lathes', *Conference Compcontrol 70*, Miskolc, Hungary, 1970, *Hungarian Scientific Society of Mechanical Engineers.* 5, 89-99, 1970.

SESSION IX

Design of Production Systems

Chairman: Mr J. Hughes
Corporate Director,
Honeywell Ltd

INTERVARIATIONAL PRODUCTION ANALYSIS

P. Mihalyfi
State Office of Technical Development, Hungary

SUMMARY

Any industrial activity should be performed with maximum economic effectiveness. These activities include factory planning, development, organisation and operation. In this paper the author bases his method of dealing with these problems on 'Intervariational Law'. This should determine the 'weight' with which the economic effectiveness is influenced by each of the production characteristics in consequence of their mutual interaction. Two types of characteristics will be used: well-known traditional ones and special characteristics known as the 'degree of massivity', 'the degree of continuity', etc. The interaction of the production elements is expressed by the intervariation of the production characteristics. This is quantified by the intervariation of the production indicators. The method has two phases. The first comprises the determination of intervariational laws, in the form of an intervariational law equation. This is accomplished on the real basis of the effective values of twenty to thirty production indicators, determined in as many production units (for example factories) as possible, for instance thirty to fifty factories, and processed by computers. The intervariational law equation leads to the realisation of the second phase of the method: its practical application with the help of computers.

NOTATION

a	asymmetry exponent
C_t	degree of completeness
E_t	relation of the extreme values $V_{t\max}$ and $V_{t\min}$
F_r	degree of continuity (or fluency)
K_a	absolute capacity
M_{an}	absolute basic degree of massivity
$\bar{M}_{rc}$	composite average of the relative basic degree of massivity
M_{rn}	relative basic degree of massivity (of an operation n)
$\bar{M}_{rs}$	simple average of the relative basic degree of massivity
$\bar{M}_{rw}$	weighted average of the relative basic degree of massivity
n	any individual operation
p	degree of order, or of order of magnitude, or of development, or ranking or rating.
P_{ki}	degree of magnitude of the indicator i in the production unit (for example factory) $k; i = 0, 1, 2, 3, \ldots k = 1, 2, 3, \ldots$
$\bar{Q}_n$	charge of a workplace in a given year, hours/year
Q	average charge

r_{min} real length of route
r_{real} maximum length of route
t_n time of the operation n per year
t_{in} the same, inside the group
t_{out} the same, outside the group
V_{kn} united indicator value in the production unit (for example factory)
V_{op} original indicator value
V_{tmax} maximum theoretical indicator value
V_{tmin} minimum theoretical indicator value
V_{tp} theoretical indicator value
$\hat{V}_{tT}$ geometrical average in the central point M

1.0 INTRODUCTION

1.1 *Object of intervariational production analysis*

Any industrial activity should be accomplished with optimum results, that is each activity should be optimised. The activities of industrial and production engineering should be optimised. Such activities are, for example, the planning and detailing of projects and the development, the organisation and operation of production units (for example factories).

1.2 *Optimisation of production*

The optimisation of these activities should be achieved by optimising the operation and functioning of the production unit (for example factory) and by optimising within it the production process itself.

Generally, the operation of the production unit may be considered optimum if its economic effectiveness is maximum, since this is criterion and accordingly its most important characteristic.

Intervariational production analysis should promote such optimisation by determining an intervariational law applicable to all production units of a branch of industry, or to the whole industry.

Such an intervariational law should express the intervariation of the indicator of economic effectiveness with all other indicators of the production units in question.

1.3 *Affinities with systems, engineering and cybernetics*

Any production unit, taken as a 'case' for intervariational analysis, should be considered as a system.

At the same time, the totality of these units forming a branch of an industry, should be considered also as a system, but one of greater complexity. These systems are to be considered according to the principles of systems engineering.

It may be supposed that intervariational analysis has some affinity with cybernetics, because according to our supposition intervariational laws regulate the economic effectiveness of the production units.

1.4 *Combination of macro-aspects and micro-aspects*

The analysis is concerned with entire production units and the intervariational law should be expressed in relation to these entire units, on a macro-aspect basis.

Simultaneously, the analysis should be based upon a micro-aspect as well, considering the studies in depth needed for the determination of intervariational laws. For example, for the sake of several indicators, the analysis must take into account the individual operations of the production process. This can be effected only by sampling these individual operations, which means taking into account possibly only a small number of them.

This means that the intervariational analysis should be based upon a macro-aspect in its substance, and at the same time upon a micro-aspect in the practical determination of the real values of some of the production indicators in the production units.

1.5 *Concept of the production unit*

The concept of the production unit can be very widely extended. The production unit may be a shop, factory, a company, a concern, a branch of an industry, or an entire national industry.

Any of these units may be analysed by this method. In any case, we should select sufficient production units to get a good approximation in the determination of the required intervariational law.

1.6 *Degree of order*

A fairly common approach is to analyse the variation of the values of some indicators over a period of several years. In this analysis, we do this only in a second part of the work. Here we are analysing first of all on the one hand the variation of the values of every indicator by itself, in accordance with their order of order, and on the other hand the intervariation of the values of all indicators amongst themselves. As stated earlier, the degree of order could also be called degree of order of magnitude, ranking, rating or development.

1.7 *Basis of method*

The immediate object of this type of analysis is the determination of so called intervariational laws. This should be accomplished by means of a large amount of real data, taken from the practical operation and functioning of the production units in question. In accordance with this the intervariational laws, computed upon the basis of these data, will reflect the practical realities of the operation of the production units.

1.8 *Necessity of computers*

The existence of computers suggested in the author's mind a few years ago the idea of the method of intervariational production analysis.

Without computers, the analysis of a system with so many variables would be too complicated for practical purposes. Without computers a very large group of investigators would require to spend an immense amount of time to establish the intervariational laws.

There are several well-known production functions which determine the relations between the important factors of production. But, generally, the number of these factors taken in account is very small, due to complexity of the task in resolving the matrices of many variables.

This means generally that there is a trend to simplification to reduce the real production process, or the real operation of the production unit to a relatively simple process of some three, four of five variables.

Much has been achieved in exploiting interesting and very important production functions. However, with current computer capability, it would be anachronistic not to try to take into account a much greater number of variables or indicators. Hence, instead of considering three, four, or five variables about twenty to thirty or more might be considered.

1.9 *Some important suppositions*

Intervariational production analysis is based upon several suppositions. Some of these are:

- The number of production elements participating in the operation of a production unit, and in the production itself, is infinite.
- The number of production characteristics which could be elaborated is infinite.
- The number of production indicators which could be formed is also infinite.
- Economic effectiveness is the only characteristic entirely comprehensive in the sense of bearing the effect of all other characteristics of the production units forming the system in question.
- The same applies to the indicator quantifying the economic effectiveness.
- The infinite number of production elements are interacting and the infinite number of production characteristics and indicators are mutually intervarying. This means that the values of each indicator, in their simultaneous variation, are varying with the values of all other indicators.
- This intervariation does not mean a dynamic variation in the course of time, but a simultaneous, (that is, in a given time interval, measurable) static differentiation of the values of given indicators in the production units in question. Theoretically, instead of a time interval, we should have an instant. Practically, we have an interval of a year, because this interval seems to be the most suitable for our purposes.
- It is supposed that the method will be applicable in the field of forecasting prognostics, futurology, and will be used to promote prognosis as well, on the one hand, by the intervariational laws concerning the intervariation in a relatively short period of time, and on the other hand by intervariational laws concerning the intervariation in a relatively long period of time.

In this paper the author concentrates on the first of the types of intervariational analyses just mentioned.

2.0 PRODUCTION-ELEMENTS, CHARACTERISTICS AND INDICATORS

2.1 *Production elements and their interaction*

In the context of this intervariational analysis, we call 'production elements' all those factors participating in the operation of the production unit and, within that, in its production process.

We thus describe all elements whether or not their participation in the operation of the production unit is intentional or accidental advantageous or disadvantageous, useful or harmful.

Accordingly, such production elements are, for example, all the workers, equipment, tools, power supplies, buildings, and all other intentionally and accidentally participating as well. These latter elements are mostly disadvantageous, but there are also advantageous ones amongst them.

2.2 *Production characteristics and their intervariation*

Relative quantities and values of the production elements utilised and the manner of their utilisation serve as bases for forming 'production characteristics', as termed in the context of intervariational production analysis.

We may distinguish between traditional and special production characteristics.

Traditional characteristics are, for example:

- Economic effectiveness
- Productivity
- Capacity utilisation, etc.

Special characteristics are, for example:

- Massivity
- Continuity
- Completeness

It should be emphasised that one of the basic features of intervariational production analysis is the forming and application of special characteristics.

2.3 *Production indicators and their intervariation*

In the context of intervariational production analysis, indicators are those values which serve to quantify, that is assure, the numerical expression of, the production characteristics.

Naturally, the indicators may also be traditional or special ones, in accordance with the corresponding characteristics quantified by them.

Every production characteristic may be quantified by one or by several indicators.

2.4 *Importance of special production characteristics and indicators*

As mentioned, one of the features of intervariational production analysis is the formation and application of special characteristics and indicators, which express and quantify correctly the various implications of the operation of the production units and within them the production process.

For a comprehensive mathematical approach of general validity, giving an acceptable approximation, it is indispensable to form mathematically suitable and correct indicators.

For instance, the 'massivity', that is the property of approximating the character of mass production, has such a great effect upon the economic effectiveness of production that it should be considered indispensable, in a comprehensive mathematical approach, to form mathematically applicable indicators of this 'massivity'. Some of these will be given in Section 3.

2.5 *Method based upon analysis of intervariation of indicators*

The values of each indicator will be used to form an 'intervariational distribution curve' and the values of the indicators in each production unit will be applied to form 'case equations'.

By means of these two logical mathematical tools, the analysis of the intervariation of the values of indicators will be effected, leading to the determination of an intervariational law in relation to the system in question.

2.6 *Individual systems and the integral system*

In every intervariational analysis there are two types of system: the individual system and the whole or integral system. For example, if we are analysing factories of an industrial group, each factory is an individual system and the industrial group is an integral one.

2.7 *Suitable form of indicators as variables*

The indicators should have a form suitable for expressions of general validity. For this purpose it is useful to express them in relative form and to eliminate dimensions as far as possible.

2.8 *Simultaneous advantageous and disadvantageous effects*

It will be supposed that each characteristic and, accordingly, each indicator has simultaneously both advantageous and disadvantageous effects upon economic effectiveness. Thus, the effect of any characteristic and of any indicator is a resultant effect.

3.0 SOME EXAMPLES OF SPECIAL PRODUCTION CHARACTERISTICS AND INDICATORS

3.1 *Examples to be treated*

We treat here only a small number of all the special characteristics and indicators mentioned so far, to indicate their nature.

We discuss firstly the characteristic 'massivity' and some of its indicators, because we suppose that it is the characteristic which has the greatest influence on the economic effectiveness.

It must be emphasised that all the indicators based upon various individual operations may be computed with the aid of operation samples. This is important, because often there are so many operations that it would not be economically acceptable to take them all into account.

3.2 *'Massivity' of production*

We distinguish various indicators to express this characteristic.

Most of these indicators are originally related to an individual operation of the production process.

By forming averages of these values, we get values related to a whole production process.

One of the indicators of 'massivity' is the relative basic degree of massivity of an individual operation n:

$$M_{rn} = \frac{t_n}{Q_n} \tag{1}$$

where: t_n = sum of the times of the individual operation n in a given year
Q_n = sum of all the times during which the equipment or operator performing the operation n is loaded with work during a given year.

In the case of the absolute basic degree of massivity M_{an} of an operation n, we should divide by K_a = 8760 hour/year which is the absolute capacity, instead of by Q_n.

To express the massivity of a whole production process, composed of several individual operations, three types of average may be used:

(i) Simple average of the relative basic degree of massivity:

$$\bar{M}_{rs} = \frac{I}{\bar{Q}} \frac{\sum_{n=I}^{Z} t_n}{Z} \tag{2}$$

where: $\bar{Q}$ = average of the Q values in the production process in question
Z = number of individual operations

(ii) The weighted average of the relative basic degree of massivity:

$$\bar{M}_{rw} = \frac{I}{\bar{Q}} \frac{\sum_{n=I}^{Z} t_n^2}{Z} \tag{3}$$

$n = I \; t_n$

(iii) The composite average of the relative basic degree of massivity:

$$\bar{M}_{rc} = \sqrt{(\bar{M}_{rs} \bar{M}_{rw})} \tag{4}$$

The dimensions are eliminated from each of these indicators, leaving:

$$\frac{\text{hours/year}}{\text{hours/year}}$$

There are several other indicators of massivity, which are not discussed here.

3.3 *Degrees of continuity of production*

One of the indicators of the degree of continuity takes into account the length of the materials handling routes or internal transfer routes:

$$F_r = \frac{\sum_{n=I}^{Z} r_{min}}{\sum_{n=I}^{Z} r_{real}} \quad (5)$$

where: r_{min} = the minimum length of the route of the internal transfer of workpieces between two work stations or items of equipment, according to an imagined ideal lay-out for mass production;

r_{real} = the real or effective length of the route of the internal transfer of workpieces between two work stations or items of equipment, according to the real lay-out.

The values of F_r have no dimension.

The other indicators of continuity are not discussed here.

3.4 *Degree of completness of production*

One of the corresponding indicators is the degree of completeness based upon the times of the individual operations of the production process:

$$C_t = \frac{\sum_{n=I}^{Z} t_{in}}{\sum_{n=I}^{Z} t_{in} + \sum_{n=I}^{Z} t_{on}} \quad (6)$$

where: t_{in} = time of the operation n carried out inside the given group of work stations, for example, a production shop;

t_{on} time of the operation n carried out outside the given group of work stations.

The value of C_t has no dimension.

4.0 SOME QUESTIONS OF THE METHODOLOGY OF INTERVARIATIONAL ANALYSIS

4.1 *Two phases of realisation*

The first phase of intervariational analysis is devoted to the determination of the real values of the indicators in the production units selected, and to the determination of the intervariational law, using computers to resolve the matrices formed by the values of the indicators determined in the production units.

The second phase is the application of the results of the intervariational analysis, which means the overall application of the intervariational law devised for the system in question.

4.2 *Intervariational distributions*

We have mentioned already that one of the bases of intervariational analysis is the intervariational distribution expressed in the form of distribution curves. As an example, Fig. 1 shows two intervariational distribution curves. One is the real and the other the theoretical curve.

The real curve is determined on the basis of the real values of the indicators in the selected production units. The theoretical curve is determined by computation made on the basis of the real values of the original curve.

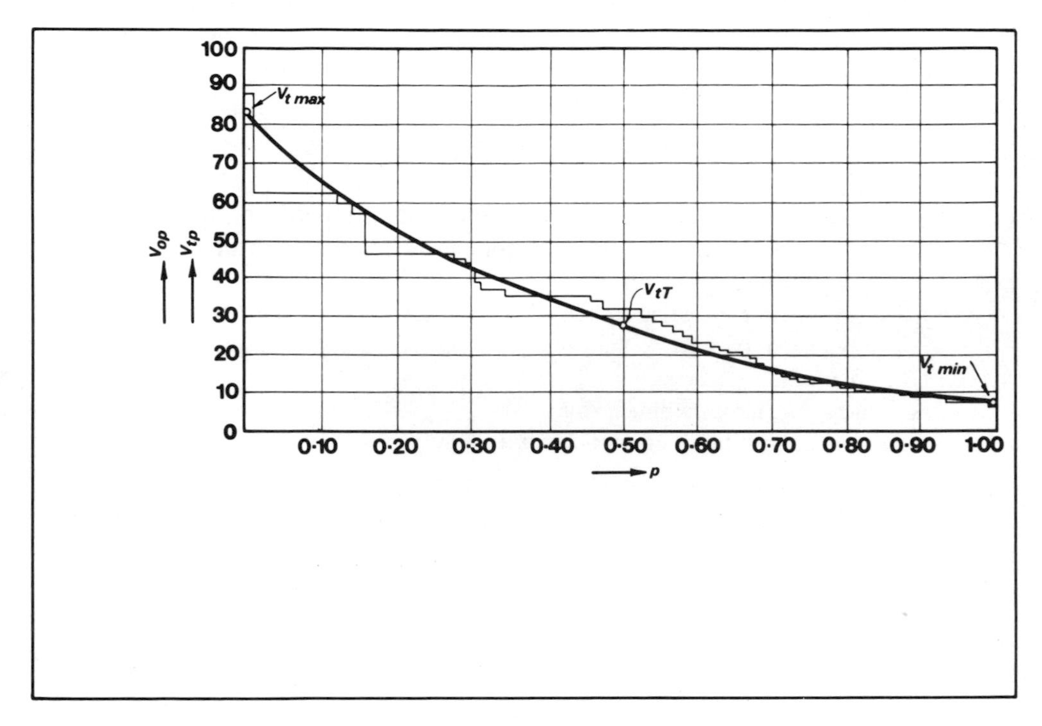

Fig. 1. Intervariational distribution curves. Example of principle. Original curve represents real values (V_{op}) of a given production indicator in production units (e.g. factories) taken into account. Theoretical curve, represents corresponding theoretical values (V_{tp}). p is the 'degree of order'.

On the basis of studies on the distribution of the values of various indicators, the author made the following suppositions:

- By setting the values of any one of the production indicators in a special way, shown by the original stepped curve of Fig. 1, a stepped curve of stochastic (yet apparently regular) character is obtained.
- By means of suitable techniques the original curve may be approximated by a theoretical curve.

The special way of forming these curves involves weighting the numbers of order of magnitude of the indicator values with weighting numbers for each production unit, these latter numbers expressing the importance of each unit. Such weighting numbers may be, for instance, the production values of the selected production units in the given year.

The special way of forming the distribution curve consists first of all in the expression of the p abscissa values. These are relative numbers, from 0 to 1, computed by dividing the cumulated sums of the weighting numbers by their greatest cumulated sum.

These p abscissa values are those which we call degrees of order (see section 1.6) or degrees of magnitude.

The equation of the theoretical curve appears in Fig. 2. In this E_t is the relation of the extreme values V_{tmax} and V_{tmin}.

The a asymmetry exponent has the greatest effect upon the 'thickness' of the curve (see Fig. 2).

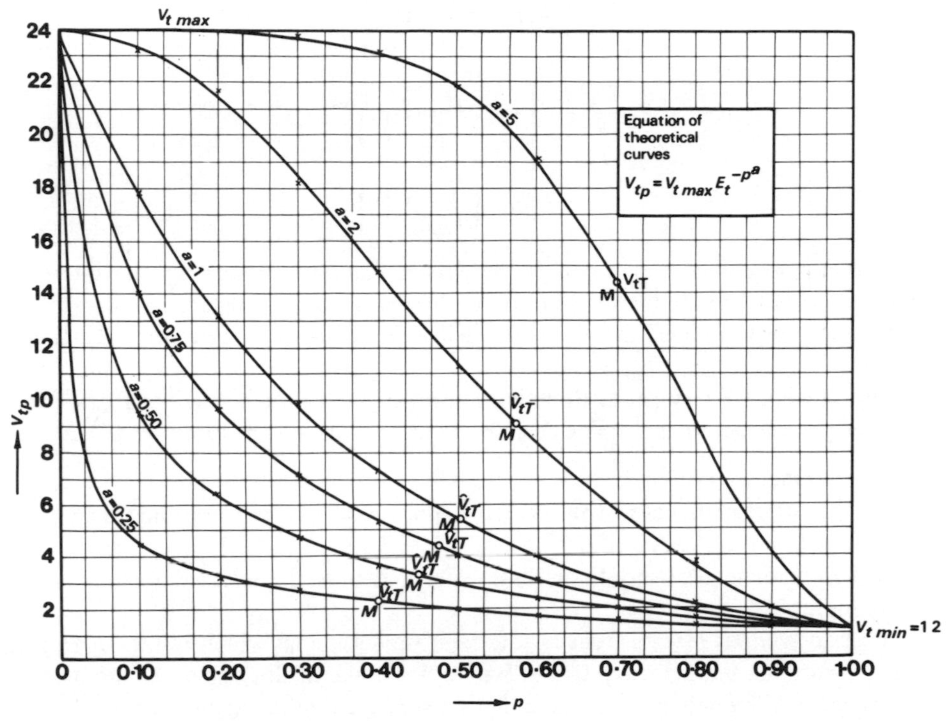

Fig. 2. Example of principle of set of intervariational distribution curves with given value $V_{t\ max}$ = 24. Curves differ only in value of exponent of, asymmetry, a. For $a \neq 1$ the values of 'total geometric average' ($V_{tp} = V_{tT}$) are asymetrically situated ($p \neq 0{,}50$).

It is supposed that through increasing the number of 'cases', that is, the number of selected production units, the theoretical curve approximates more closely the original one. Theoretically, with an infinite number of values the two curves should coincide.

The 'boldest' supposition is that the equation in Fig. 2 may be applied to any one of the production indicators. If this is true, this is a great advantage in the computations.

Under this supposition, we may determine for every indicator its degree of order, p. As mentioned, the value of p varies between 0 and 1. It is an advantage in the computation that the p-values of the theoretical curve of every indicator vary between the same values, 0 and 1. One advantage of this lies in the possibility of directly comparing the various indicators. The other advantage is that the value of the degree of order of the economic effectiveness, that is the value p , may be considered as the weighted arithmetical or eventually geometrical average of the degrees of order, that is the p-values of all the other indicator values.

The value of the degree of order, that is the p-value, may be calculated from the equation of the theoretical curve (see Fig. 2).

$$P = \left(\frac{\ln V_{tmax} - \ln V_{tp}}{\ln E_t} \right)^{1/a} \tag{7}$$

In another form:

$$P = \left(\frac{\ln V_{tmax} - \ln V_{tp}}{\ln V_{tmax} - \ln V_{tmin}} \right)^{1/a} \tag{8}$$

Thus:

$$P = \left[\frac{\ln \left(\frac{V_{tmax}}{V_{tp}} \right)}{\ln \left(\frac{V_{tmax}}{V_{tmin}} \right)} \right]^{1/a} \tag{9}$$

There is an important conclusion to be drawn from the theory of the intervariational distribution treated here. If the distribution of the values of the indicators is formed according to the foregoing assumptions, the values of each indicator approximate to a special geometric progression.

For this reason, we may get arithmetic averages of the values of the production indicators with greater precision if we determine them by means of the geometric averages.

4.3 *Case equations*

Besides the intervariational distribution, the other basis of intervariational analysis is by the case equations.

The logico-mathematical model of these equations must be selected through a 'dialogue' with a computer.

We must try various models. Finally, the right model is that which results in an intervariational law giving the best approximation. We do not make in this paper any study in depth of the case equations. We only mention here that we suppose that the application of the intervariational distribution theory and especially the degree of magnitude or development p will yield valid results for the case equations.

For every 'case', which means for every selected production unit, we should raise a case equation. For example, this could be written in principle as follows:

$$p_{ko} = p_{k1}\, x_1 + p_{k2}\, x_2 = \ldots p_{kn}\, x_n \tag{10}$$

where: k indicates the case, that is one of the selected production units

$1 \ldots n$ are related to the various indicators, variables of the system

x are the weighting numbers, determining the weight of each indicator in the system

We must not think that equation 10 means a linear intervariation! The p-values are exponents in the constructed form of the original V values.

It may be correct to try the following type of model also:

$$p_{ko} = p_{k1}{}^{x_1}\, p_2 k_2{}^{x_2} \ldots p_n k_n{}^{x_n} \tag{11}$$

The x values may involve the asymmetry exponent a and the value E_t or rather in E_t as well.

The x values may be dependent variables, which may imply that the system has a great complexity.

The p_{kn} values belong to the V_{kn} values of the 'united variable' corresponding to an imagined 'united indicator'. We use the term 'united' because such a variable should unit the effects of the infinite number of indicators not formed and not taken into account.

These p_{kn} values may be determined by their constructed intervariational distribution curve. The values of that curve may be calculated by means of the case equations through iteration.

5.0 *Applications of the method*

There are many possibilities of application. Here we mention only two of these. Let us mention first a diagnostic application. After having elaborated the 'intervariogram' (Fig. 3), without the arrows we should make it available to industry. Thus any production unit could buy such an intervariogram and insert first its own real values represented by the continuous-line arrows and afterwards the values of the tolerance calculated for its own indicator values, represented by the dotted-line arrows. When a continuous line arrow for a real value falls below the two dotted-line arrows for the corresponding tolerance, the possibility of the occurrence of some disturbing situation is indicated.

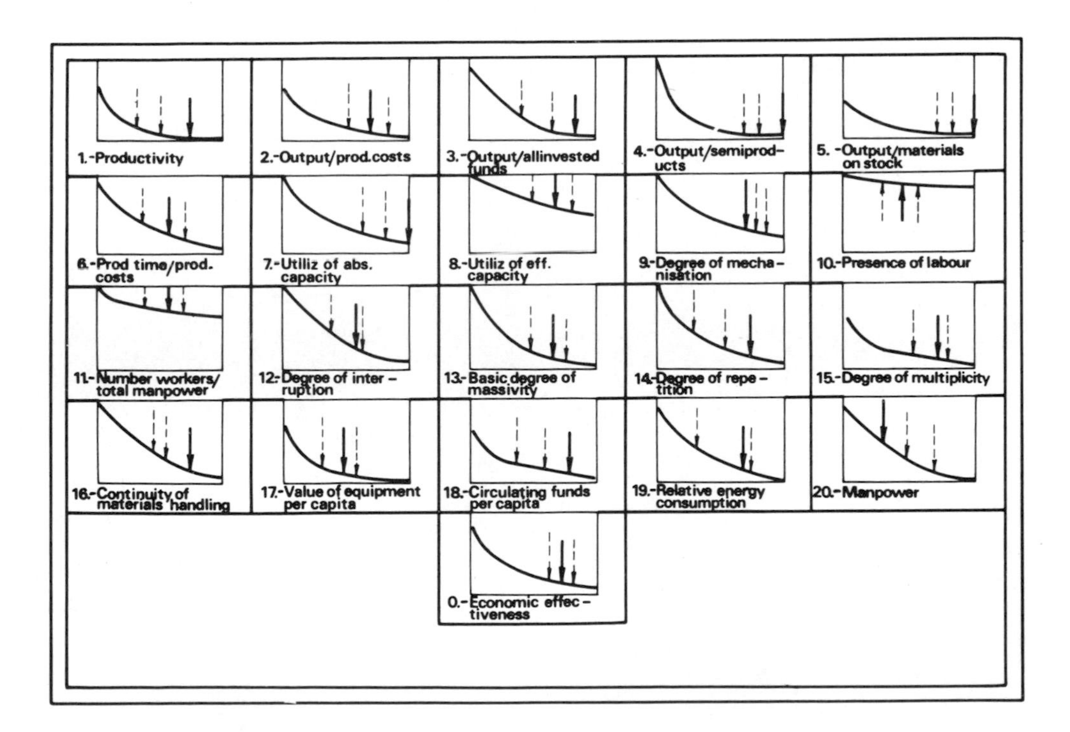

Fig. 3. Intervariogram. Example of principle with estimated curves. Intervariogram composed of intervariational distribution curves of several production indicators, taking into account in the intervariational analysis the type of production unit (factory) considered. Function of arrows explained in text.

Another important field of application should be that of optimisation of a great variety of industrial activities, for example of the operation or development of existing factories or the planning of new factories.

The x-weighting numbers should be applied with the aid of computers in the optimisation computation. This would be effected probably by the method of linear programming or some similar method.

Without mentioning other possibilities of application we should emphasise that even the mere knowledge of intervariational laws could promote the improvement of several engineering activities and decisions.

The author would welcome any proposal of co-operation and would answer any questions in connection with this paper.

THE DESIGN OF PRODUCTION SYSTEMS

A. S. Carrie
University of Strathclyde, Glasgow

SUMMARY

The paper reviews the problem of the design of production systems, and suggests that the philosophy of group technology is the correct approach. An essential requirement for any computer method is a means of providing the data in a suitable form, but the coding and classification techniques of group technology are inadequate for this purpose. Taxonomy, the science of classification, is introduced. The application of numerical taxonomy to the design of production systems is discussed, and promising results from its use in plant layout are given.

The paper then proceeds to consider work being carried out on the development of computer methods for the design of production systems. The first requirement is for a workpiece description system, and the outline specification of such a system based on the numerical control languages is suggested. This description would be analysed and an operation list determined. Alternatively, the operation list could be determined externally and specified in a similar language. The operation list is then analysed and times and costs estimated. Various plant layouts can be drawn up and their relative merits assessed by computer simulation and a form of discounted cash flow analysis.

1.0 INTRODUCTION

The problem of designing a production system can, in its most general sense, be very simply stated:

Given a set of products, their design specifications and estimated future sales requirements, what system of production should be set up to produce them?

Such a statement of the problem raises other questions: is this a realistic statement? How often is the question asked in industrial practice? How often should it be asked? What about the continually changing conditions in industry?

Such a question may seem academic, especially in such general terms, and virtually insoluble. The academic approach to such a general problem is to adopt a heuristic procedure whereby:

(1) Areas of a large general problem are identified which are sufficiently specific in themselves to be amenable to solution;

(2) These sub-problems are solved;

(3) The solutions are brought together with whatever adjustments are necessary to provide an aggregate solution to the general problem.

Sub-problems of the production system design problem are to determine:

- The operations required on each component
- The sequence of these operations
- The machines to be used
- The layout of the machines
- The handling methods

and several others.

Thus, at this level, an apparently academic question is found to be intensely practical, engaging many thousands of technicians and engineers.

It is the purpose of this paper to examine traditional approaches to the production system design problem and to outline a programme of research being carried out at Strathclyde University leading to the development of a methodology for the solution of the overall problem.

2.0 TRADITIONAL APPROACHES TO PRODUCTION SYSTEM DESIGN

A large part of the problem is concerned with plant layout. Traditional approaches tend to consider the layout in the context of two principal layout types:

(1) The *product line layout*, in which the plant required for the manufacture of a specific product is laid out in a line. The order of the machines is determined by the operation sequence of the product, and only that product is produced by that line.

(2) The *functional layout* in which all machines of similar type are located together and any of the components can be produced by any of the machines of each of the types required in the component's operation sequence.

The two layout types tend to contrast in terms of fast throughput versus long delivery times; low against high work in progress; high against low utilisation of facilities, for the product and functional layouts respectively. Real layouts tend to be between these two extremes, and the system design problem becomes one of finding the most suitable compromise, thus, reverting to the general problem outlined above. This necessitates a detailed analysis of the shape, surface features, accuracy requirements, material specification and other details of each component and of the tools and machines available and capable of producing the required component features.

3.0 GROUP TECHNOLOGY

In recent years group technology has emerged as a technique for achieving highly efficient production systems for the manufacture of selected groups of components. The achievements have been well publicised[1,2]. Group technology represents the very essence of the heuristic approach to the production design problem. The overall problem is broken down into one sub-problem for each group of components. Within each group the requirements and solutions can be highly developed. The aggregation of the group solutions into one whole solution becomes the responsibility of the higher levels of production management.

Group technology comprises two levels of techniques: techniques for finding groups of components for which an efficient production system should be possible, and techniques for designing the production system for the group of components. Because of this heuristic approach the group technology philosophy represents an ideal method of tackling the general production system design problem.

Group technology relies heavily on component coding or classification systems. The two best-known of these are the Opitz[3] and Brisch[4] systems. These two systems represent opposites in approach to coding and classification.

The Opitz system provides a nine-digit code number. Each digit refers to specific aspects of the component and the significance of each digit is always the same, independent of the company or type of product. These nine digits are respectively:

Geometrical Code (digits 1-5)

1	Component Class, e.g.	1 = rotational part with length: diameter ratio between 0·5 and 3·0
		8 = non-rotational component with length: width ratio less than 3·0 and length: thickness ratio greater than 4·0
2	Overall or Main Shape, e.g.	2 = external shape stepped to one end or smooth, with screw thread
3	Rotational Surface Machining, e.g.	5 = internal shape, with screw threads
4	Plane Surface Machining	
5	Auxiliary Holes, Gear Teeth, Forming	

Supplementary Code (digits 6-9)

6	Dimensions; diamter D or edge A, e.g.	3 = greater than 50 mm and less than 100 mm.
7	Material, e.g.	5 = alloy steel, not heat treated
8	Initial Form, e.g.	0 = Round Bar, black
9	Accuracy, e.g.	1 = in digit 2 (main shape)
		4 = in digit 5 (gear teeth or holes)

The inadequacy of this system, such as the inability to uniquely specify certain workpiece characteristics, such as internal or external grooves, splines and slots, and the 'crudity' with which certain aspects can be specified (e.g. alloy steel not heat treated) have been pointed out by several writers especially Gombinski[5]. The Opitz system is, therefore, inadequate for the detailed analysis of workpiece accuracy, shape, etc. referred to above.

The Brisch system, by contrast, is tailor-made to the particular company and its products, and consists of an hierarchical five-digit monocode in which each successive digit provides a sub-division of the component class previously specified, and a polycode which can be any number of digits in length and in which each digit has some specified significance. For example, digit 6 might refer to the type of heat treatment to be used, digits 23 and 24 might be the EN numbers of the material type. Since the polycode can be of any length, a large amount of information can be recorded in it. However, it has the serious disadvantage that if it has more than about a dozen digits, a 'code-book' must continually be referred to and the code system becomes unmanageable. With a large number of digits, as would be required for the production design problem, the polycode becomes virtually a workpiece description system.

It is, therefore, concluded that neither of these two principal group technology systems is capable of transmitting, in adequate detail, the information needed for the general production systems design problem.

4.0 NUMERICAL TAXONOMY

Taxonomy, the science of biological classification, has existed for several years, and it seemed appropriate to examine its potentialities for workpiece classification.

Taxonomy involves objects to be classified called 'operational taxonomic units' or OTUs for short, which possess 'characters'. The OTUs are sorted into classes or 'taxa' based on their possession or lack of these characters.

Since the advent of the electronic computer, numerical taxonomy has been developed. Numerical taxonomy[6] can be defined as the systematic numerical evaluation of the affinity or similarity between taxonomic units and the grouping of these units into taxa on the basis of their affinities.

Three stages are involved in numerical taxonomy:

(1) Prepare a data matrix. A matrix is prepared specifying a value for each of many characters possessed or lacked by each object to be classified. These values may be either binary, indicating the level at which the character is possessed by the object, e.g. 0 = no, 1 = yes, or numeric, indicating a measurement of the character on the object, e.g. 24·2, 45·8, – 1·0 or 4·2. The matrix is of size $N \times M$ where there are N objects, and they are to be classified on the basis of M characters.

(2) Calculate a similarity matrix. The $N \times M$ data matrix is analysed, by any of several techniques, and a triangular, $\frac{1}{2}N(N-1)$, similarity matrix is calculated. Its entries give a numerical value of the similarity between each pair of objects.

(3) Cluster analysis. Similarity, and its converse dissimilarity, can be considered as nearness and distance respectively in an M-dimensional space. Cluster analysis seeks to find clusters of points in this space which are close together, but reasonably far from any others. If the objects to be classified fall naturally into a few distinct types then there will be a distinct cluster for each type. On the other hand, if no distinct clusters exist the analysis will show this fact. In this way numerical taxonomy possesses greater flexibility than many analytical techniques in that it does not force the data into a pattern, willy-nilly, but shows how well they fit the pattern. The result of cluster analysis is a 'dendogram', rather like a family tree, showing the construction of the clusters.

5.0 NUMERICAL TAXONOMY APPLIED TO PRODUCTION SYSTEMS

The potentialities of applying numerical taxonomy to the design of production systems are being examined in the Department of Production Engineering at the University of Strathclyde. This is being done in two ways.

(1) Methods of specifying the appropriate characteristics of the workpieces.

(2) Development of the data processing procedures.

Since the effort in item (1) would be fruitless if the procedures in item (2) were unsuccessful, the development and potential of data processing procedures has been examined first. As stated above, the first stage in the numerical taxonomy problem is the preparation of a data matrix. Several problems in production engineering are naturally expressed in this way, and two of them have been examined, using a taxonomy computer program which was available. These two problems were:

5.1 *Numerical taxonomy in production flow analysis*

Production flow analysis, Burbidge's[7] method of group technology, involves a component-machine matrix, whose entries show the machines required to produce each component. Entry (I,J) is 1 if component I has an operation on machine J, or 0 otherwise. This matrix has been analysed by a student of operational research[8] and numerical taxonomy has been found much faster than the laborious trial and error sorting process and is much more consistent.

While this method is suitable for finding groups of components for group production, the organisation of the machines within the group cannot be indicated by the application of numerical taxonomy to the matrix used in the problem. A suitable matrix for this problem would be a component-operation matrix in which the entries are 1,2,3, etc. according to the number in which the operation appears in the operation sequence for the component, but, unfortunately, the necessary data processing is more complicated and has not been attempted.

5.2 *Numerical taxonomy and travel chart analysis in plant layout*

Nevertheless, the layout of the machines within a group, or in any general lay-out problem can be tackled by means of the travel chart[9]. This is a machine-machine matrix in which the entries give the amount of material to be moved from one machine to another. Thus entry (I,J) specifies the movment from I to J, and entry (J,I) specifies movement in the reverse direction. Entries $(I,\widehat{I})$, (J,J), etc. on the main diagonal are undefined. The application of numerical taxonomy to this matrix has been examined by an M.Sc. student[10] of production engineering.

Existing algorithms for plant layout are designed to find a suitable layout in one of two forms only. One form of algorithm attempts to find the best order for placing the machines in a single production line, such as the methods of Hollier[11]. The result of this exercise is an order of machines which, if the machines were entered in that order in the travel chart, would cause the non-zero entries to be concentrated above and as near to the main diagonal as possible. The other form of algorithm finds an arrangement of departments each having some floor space requirement, within an overall rectangular factory area. CRAFT[12] and CORELAP[13] are of this type. Neither of these approaches seems totally satisfactory in that they assume that either a line or a functional layout is to be developed without first providing an indication of which will provide best results. In addition, the CRAFT approach is highly unsatisfactory in that the resulting department areas may be of irregular shape, and also the problem of plant layout of machines within each department still remains to be tackled.

In applying numerical taxonomy to plant layout our approach has been to start with the premises that real plant layouts tend to consist of machines laid out in aisles, which will normally be straight and parallel, and that movement will be relatively easy between machines in the same aisle, but more difficult between machines situated on separate aisles. The first stage of a plant layout algorithm should, therefore, seek to find various self-contained groups of machines which can be assigned to each of the aisles. The second stage would be to organise the machines along each aisle, and to decide the order of the aisles to which the machine groups should be assigned. Numerical taxonomy, in particular cluster analysis, is ideal for finding these groups of machines, and in addition will show the extent to which these groups are self-contained, thereby indicating whether a functional, single line is, or several group lines are, liable to be the most suitable type of layout.

Employing this approach, the numerical taxonomy program was applied to the travel charts used in two previously published problems. The first is from the problem of Singleton[14], which Hollier[11] used in developing his multi-product line layout algorithms. The travel chart, Table 1, involves 27 machines, and shows the movements involved in the manufacture of 12 products. The ordering of the machines suggested by Singleton's algorithm and by Hollier's several algorithms are shown in Table 2.

Both these writers pointed out that machines 8 and 26 deliver material to and receive

TABLE 1. SINGLETON'S TRAVEL CHART

	1	2	3	4	5	6	7	8	9	10	11	12	13	14	15	16	17	18	19	20	21	22	23	24	25	26	27
1								11																			
2								59							4			5	16					4		32	
3				29				23			8															68	
4								11	29																		
5								5																		4	
6								24							7		37										
7								5		4							16	24									
8		120	21	11			49			7					19				20			12				64	24
9			4								24															28	
10												72															
11													28								60						
12																										32	40
13																					28						
14																									88		
15						11		16														34					
16																				16			56				
17	11							40																		14	
18										29																	
19															31											16	
20						57		12									8		11							4	
21													12														
22								67																			
23																				56							
24										4																	
25																72				16							
26			67			9		69	27	16	56						4			4		21					5
27			36					5		12																16	

TABLE 2. RESULTS DUE TO SINGLETON AND HOLLIER

a: 14 25 16 23 20 6 17 1 5 19 15 22 2 8 7 18 10 24 26 12 27 3 4 9 11 13 21

b: 14 25 16 23 20 6 17 19 15 1 22 5 2 8 7 18 10 12 24 26 27 3 4 9 11 13 21

c: 14 25 16 23 20 6 17 1 18 10 12 5 19 15 22 8 2 26 24 7 27 3 4 9 11 13 21

d: 14 25 16 23 20 6 17 1 7 18 10 12 27 3 26 8 2 19 15 22 4 9 11 21 13 5 24

Key a: Minimising transport distance (Singleton)
b: Minimising transport distance (Hollier)
c: Minimising backtracking movements (Hollier)
d: Maximising in-sequence movements (Hollier)

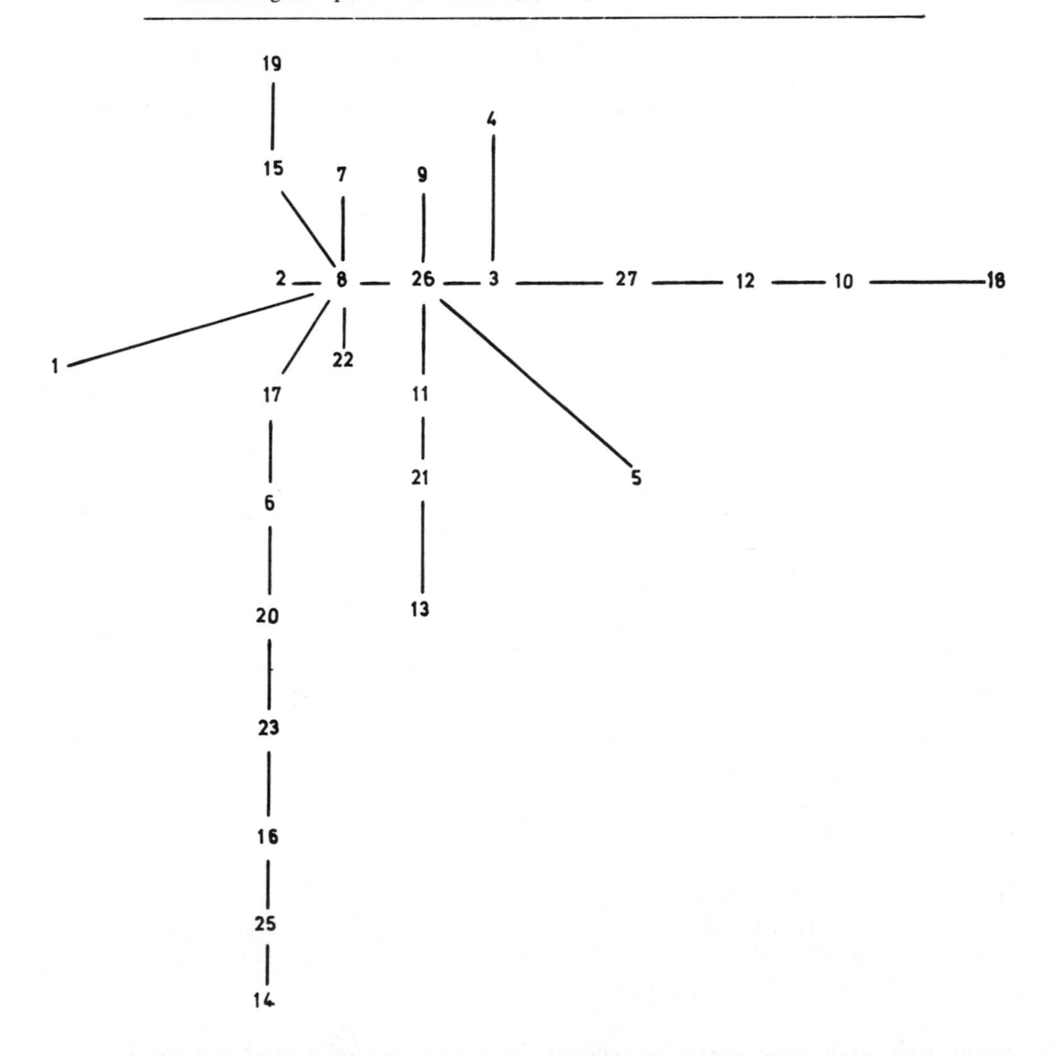

Fig. 1. Cluster analysis linkage chart based on total material movements. (Distances between operations are inversely proportional to similarities.)

material from nearly all other machines, and that this made finding the optional order very difficult and that perhaps a single line was not the best type of layout. The linkage chart produced by cluster analysis, with the similarity matrix based on the total movements between pairs of machines, shown in Fig. 1, clearly indicates the central importance of machines 8 and 26. The long arms 27-12-10-18, 14-25-16-23-20-6-17 and 11-21-13 are clearly candidates for assigning to separate aisles. When the similarity matrix is calculated on relative movement between pairs of machines, i.e. the ratio of the movement between two machines to the total movement between these machines and all other machines, a simpler linkage chart, shown in Fig. 2 is obtained. This could almost be used directly as a plant layout diagram. When the

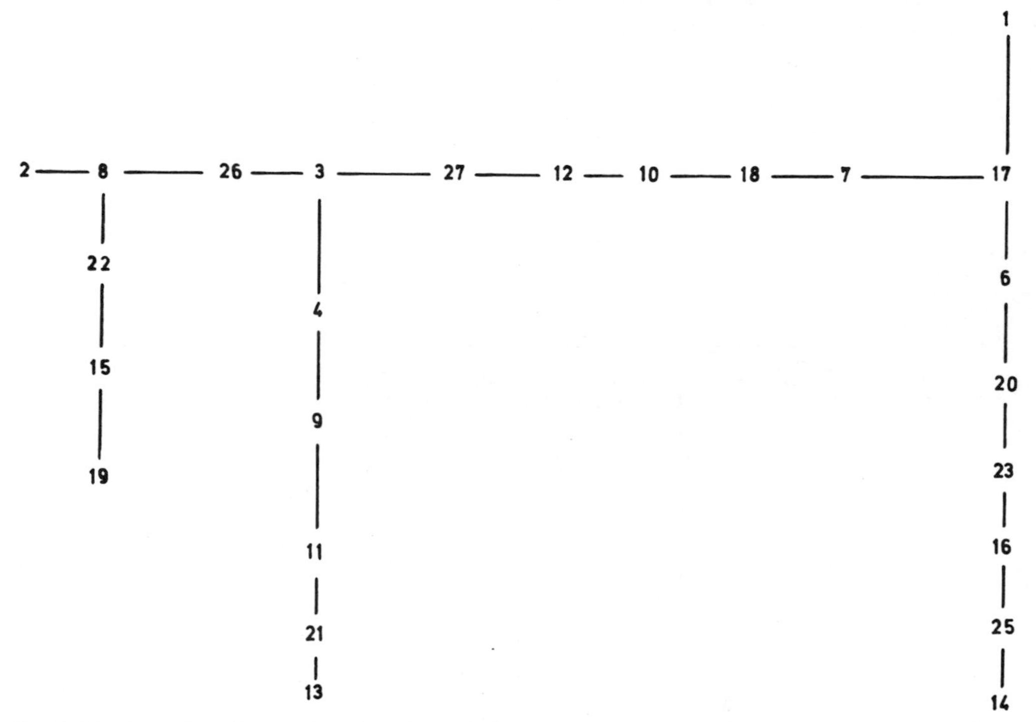

Fig. 2. Cluster analysis linkage chart based on relative material movements. (Distances between operations are inversely proportional to similarities.)

computer program is instructed to find clusters of machines with not more than six machines per cluster, and using relative movements, the following groups are produced.

Cluster	Machines
A	11, 13, 21
B	19, 15, 22
C	4, 9, 3, 26, 2, 8
D	24, 7, 18, 10, 12, 27
E	17, 1, 5
F	14, 25, 16, 23, 20, 6

An examination of the sequences determined by Singleton and Hollier, and of these clusters, will show that the sequences represent virtually an ordering of these clusters with some mingling of clusters C and D. Despite the fact that numerical taxonomy was applied with the purpose of finding independent groups, it appears to be effective also for determining the sequence in which to place machines in a single line layout.

The second travel chart problem to which numerical taxonomy was applied is that quoted by Buffa[12] in his discussion on the CRAFT program. His travel chart, with the entries rounded off to nearest integers, is shown in Table 3 and involves 20 departments.

TABLE 3. BUFFA'S TRAVEL CHART

	A	B	C	D	E	F	G	H	J	K	L	M	N	P	R	S	T	U	V	W
A		2	1							1	1			1						
B	2		1	24	1		14		1	1									7	
C	1	1				2			3	4				13					14	
D		24			1	6	8		2			1						2	16	
E		1		1			2	1		2					1					
F			2	6			6											1		
G		14		8	2	6		24		2				1				2		4
H					1		24						1						8	33
J		1	3	2										8			8			
K	1	1	4		2		2					12		19	2				5	
L	1											2		3	1	22				
M				1						12	2				2		15		8	
N								1						8	1	6				
P	1		13				1		8	19	3		8		10			1		
R					1					2	1	2	1	10			5			
S											22		6				12			
T									8			15			5	12			8	
U				2		1	2							1					5	
V		7	14	16				8		5		8					8	5		
W							4													

The solution achieved by CRAFT is shown in Table 4 and exhibits irregular department shapes, for example departments V,S,W. The linkage chart resulting from cluster analysis is shown in Fig. 3. The dotted lines indicate that the points J,A and E did not become 'dense', having only very small business with other departments. Nearly all the points lie on a single continuous line. This strongly suggests that a line would be a better type of layout than a process layout.

CORELAP[13] is another program which produces a process layout. It uses as input a chart showing, in terms of the letters A (absolutely), E (especially), I (important), O (ordinary), U (unimportant), X (not desirable), the importance of closeness between pairs of departments. This is very similar to a crude similarity matrix.

It can be concluded from these results that numerical taxonomy can be usefully applied to plant layout problems. Moreover, the logic of Hollier's methods can be incorporated in cluster analysis by imposing additional restrictions on the criteria by which points join clusters. Therefor the previous algorithms for generating either a single line by Hollier's method or a

TABLE 4. LOCATION PATTERN PRODUCED BY CRAFT FOR A 30 × 20 FACTORY FLOOR SPACE

	1	2	3	4	5	6	7	8	9	10	11	12	13	14	15	16	17	18	19	20	21	22	23	24	25	26	27	28	29	30
1	E	E	E	E	E	E	F	F	F	L	L	L	L	L	L	L	S	S	S	S	S	S	S	S	S	U	U	U	U	U
2	E	E	E	E	E	E	F	F	F	L	L	L	L	L	L	L	S	S	S	S	S	S	S	S	S	U	U	U	U	U
3	E	E	E	E	E	F	F	F	F	L	L	L	L	L	L	L	S	S	S	S	S	S	S	S	S	U	U	U	U	U
4	E	C	C	C	C	F	F	F	F	L	L	L	L	L	L	L	S	S	S	S	S	S	S	S	S	U	U	U	U	U
5	C	C	C	C	C	F	F	F	F	L	L	L	L	L	L	L	S	S	S	S	S	S	S	S	S	U	U	U	U	U
6	C	C	C	C	C	C	D	D	D	L	L	L	L	L	L	L	S	S	S	S	S	S	S	S	S	U	U	U	U	U
7	C	C	C	C	C	C	D	D	D	L	L	L	L	L	L	G	G	G	S	S	S	S	S	S	S	U	U	U	U	U
8	C	C	C	C	C	D	D	D	D	L	L	L	L	L	L	G	G	G	S	S	S	S	S	S	S	U	U	U	U	U
9	C	V	V	V	V	D	D	D	D	L	L	L	L	L	L	G	G	G	S	S	S	S	S	S	S	W	W	W	W	U
10	V	V	V	V	V	D	D	D	D	N	N	N	N	N	N	H	H	H	T	T	T	T	W	W	W	W	W	W	W	W
11	V	V	V	V	V	V	B	B	B	N	N	N	N	N	N	H	H	H	T	T	T	T	W	W	W	W	W	W	W	W
12	V	V	V	V	V	V	B	B	B	N	N	N	N	N	N	H	H	H	T	T	T	T	T	W	W	W	W	W	W	W
13	V	V	V	V	V	V	B	B	B	P	P	P	P	P	P	J	J	J	T	T	T	T	T	W	W	W	W	W	W	W
14	K	K	K	K	K	K	B	B	B	P	P	P	P	P	P	J	J	J	T	T	T	T	T	W	W	W	W	W	W	W
15	K	K	K	K	K	K	B	B	B	P	P	P	P	P	P	J	J	J	T	T	T	T	T	T	W	W	W	W	A	A
16	K	K	K	K	K	K	B	B	B	P	P	P	P	P	P	R	R	R	T	T	T	T	T	T	T	A	A	A	A	A
17	K	K	K	K	K	K	M	M	M	M	M	M	R	R	R	R	R	R	T	T	T	T	T	T	T	A	A	A	A	A
18	M	M	M	M	M	M	M	M	M	M	M	M	R	R	R	R	R	R	T	T	T	T	T	T	T	A	A	A	A	A
19	M	M	M	M	M	M	M	M	M	M	M	M	R	R	R	R	R	R	T	T	T	T	T	T	T	A	A	A	A	A
20	M	M	M	M	M	M	M	M	M	M	M	M	R	R	R	R	R	R	T	T	T	T	T	T	T	A	A	A	A	A

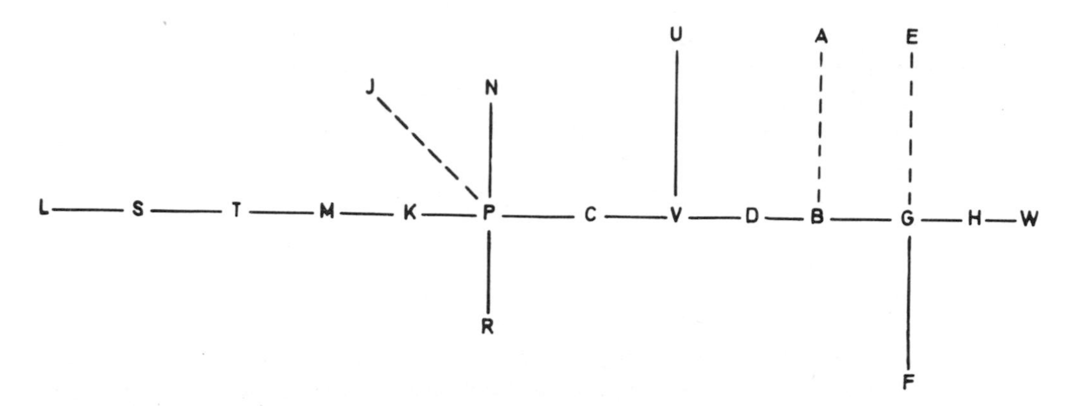

Fig. 3. Cluster analysis linkage chart for Buffa's data. (Distances are inversely proportional to similarities.)

functional layout by CORELAP represent different special cases of the cluster analysis procedures contained within numerical taxonomy. Numerical taxonomy is, therefore, a more general technique than the previous ones.

A large-scale project is being carried out during April and May 1971 in a Scottish company, whereby the layout plans recently drawn up for a factory extension will be compared to those produced by numerical taxonomy, providing a thorough testing of the technique.

6.0 DEVELOPMENT OF COMPUTER METHODS FOR THE DESIGN OF PRODUCTION SYSTEMS

The classification of workpieces would normally be the first step in designing the production system, and yet classification techniques (numerical taxonomy) have been shown to be useful in the later stages, namely the development of the layout. A programme of research is under way in the Department of Production Engineering to explore their full potentialities. The travel chart used as input in the examples quoted can only be compiled after the operations have been determined and the machines selected for each component, and these data collected for all the components. Research activities are progressing along two lines:

(1) To determine the desirability and capability of using numerical taxonomy as a means of component classification and the methods for automatically determining the operation sequences and the machines to be used so as to compile the travel chart.

(2) Developing and extending the use of numerical taxonomy in determining the most suitable plant layout.

7.0 COMPONENT CLASSIFICATION AND DESCRIPTION

Any computer method requires a means of providing the data to be analysed in some digital form. In the design of production systems the data to be analysed are the workpieces, so that a workpiece description system must be available. Only once the description is inside the computer can classification and operation planning techniques be applied. Consequently, a workpiece description system is being developed.

In marked contrast to the numerical digit code used in group technology, inadequate for this purpose, is the symbolic language used by the numerical control languages.

The sequence of statements:

```
P1 = POINT/20, 10
C1 = CIRCLE/CENTER, P1, RADIUS, 2·5
PAT1 = PATERN/ARC, C1, 0, CLW, 6
DRILL1 = DRILL/SO, DIAMET, 0·75, DEPTH, 1·5
ACT/DRILL1
GOTO/PAT1
```

defines a pattern of points around a pitch circle, a drilling operation and the instructions to carry out the operation at each point on the pitch circle.

The principal advantage of this language is that, being English-like, it is easy to read and write, and can be learned very quickly. By simply writing enough lines any amount of information can be conveyed in a form which would enable a computer to perform the required calculations. Hence this type of language is preferable to the numeric digit codes method of group technology.

The numerical control language structure has, therefore, been adopted for the proposed workpiece description system. Another reason for adopting this method is that the types of information being handled and the calculations to be undertaken are closely similar to those in NC languages, and as the technological capability of these languages develops the same filing methods for material, machine and tool information can be used, thereby achieving compatibility at all stages.

The structure of the proposed language is:

(1) Grammar, vocabulary and structure, largely compatible with the APT system.
(2) An arithmetic capability equal to that of the NEL-NC system, but excluding macros and looping.
(3) A geometric capability, equal to that of 2PL or EXAPT 1 including MATRIX and TRASYS but excluding nesting facilities.
(4) A surface element definition capability.

The relative position of each surface element would be specified by the geometric definitions accompanying its surface definition. For example, the part illustrated in Fig. 4 could be defined as follows:

```
PARTNO    EXAMPLE WORKPIECE DESCRIPTION
PART/MATERL, EN24, BAR, DIAMET 4·5, LENGTH, 4·5
ZSURF/0
F1 = FACE/OD, 4·0
D1 = CYLNDR/DIAMET, 4·0, LENGTH, 0·75, MAXDIA
B1 = BORE/DIAMET, 1·5, LENGTH, 1·0, STEPTO, 1·0
C1 = CIRCLE/0, 0, 3·25
PATI = PATERN/ARC, C1, 0, CLW, 6
H1 = HOLE/THREAD, 0·375, 16, UNC, DEPTH, 0·75, THRU, AT, PAT1
ZSURF/1·0
B2 = BORE/DIAMET, 1·0, LENGTH, 3·0, THRU, MINDIA
M1 = MATRIX/ZXROT, 180, TRANSL, 0, 0, 4·0
TRASYS/M1
ZSURF/0
```

```
F2 = FACE/OD, 2·5
D2 = CYLNDR/DIAMET, 2·5, LENGTH, 3·25, RAD, 0·19, STEPTO, 4·0, CHAMF, $
   45, 0·125
TRASYS/NOMORE
FINI
```

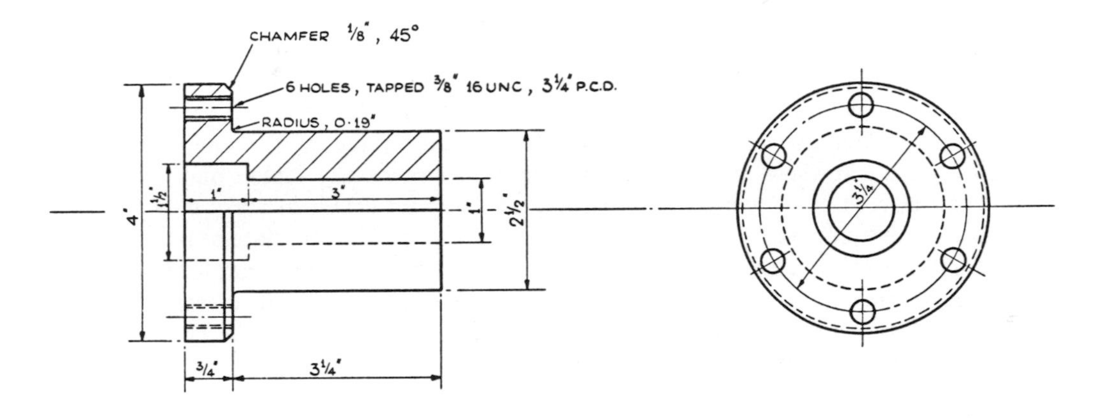

Fig. 4. Component used as an example for workpiece description system.

This description would be read into the computer where each statement would be decoded and the surface element definitions translated into a standard form, equivalent to the canonical forms used for geometric definitions.

Each surface element definition attempts to define the element as completely as possible, so that the machining operations required for its generation can be determined with the minimum of cross-reference to other definitions. The CONTUR type of definition in 2C or EXAPT 2 has been avoided, since this would require motion statements to be incorporated in the program, greatly increasing the amount of analysis needed to determine the required operations.

The grammar of the numerical control languages of the form.

SYMBOL = MAJOR WORD/MODIFIERS

also has advantages from the classification aspect for two reasons. Firstly, a scan of the major words gives an indication of the nature of the part. For example, they show that this example workpiece is:

(a) rotational
(b) has stepped external rotational machining;
(c) has stepped internal elements;
(d) the only planar elements are the end faces;
(e) has holes on a pitch circle.

Consequently its Opitz code number would be 01102, 1102, or 21102, and an examination of the modifiers shows that the L/D ratio is 1·0 so that the actual code number is 11102. Secondly, in relation to the taxonomic classification data matrix, the major words, CYLNDR, BORE, HOLE, etc. represent binary characteristics which are either absent or present, indicating the types of machining operation required and the modifiers represent numeric characteristics, whose values will determine the details of the operations, the machines and tools to be used and the time and cost of producing each surface element.

The development of this language is being carried out in co-operation with a local engineering company.

8.0 PROCESS PLANNING

The workpiece description statements, having been decoded and translated into a standard form within the computer, can then be acted upon by process planning routines which determine how each BORE, HOLE, etc. is to be produced. The particular routines could be selected, on a taxonomic basis, depending on the similarity of the specific component to some standard component whose detail planning had been done previously. Unlike numerical control, where the objective is to punch on to paper tape the exact details of every movement, tool change etc., the purpose of these routines in the design of production systems is to enable less detailed analysis to be undertaken, such as the selection of the machine and determination of sequences of operations and the time and cost of each operation with a certain degree of accuracy. In common with numerical control, however, the routines can be as simple or complex as necessary depending on the objective and could follow the pattern suggested by De Palo[15,16] in his flexible method for determination of feeds and speeds. For the present exercise the routines are being kept as simple as possible, being acceptable within the context of individual cases, thereby removing the necessity for storing large quantities of data concerning machinability, tool files, machine files, etc. It is planned to insert dummy routines in the procedures whose functions will be to provide whatever data of this type are necessary in a form compatible with that used in the numerical control programs, thereby enabling the same filing system to be used at a later date.

The output of these process planning routines would be an operation list, stating for each operation the machine, speed, feed, tool and other relevant data.

9.0 TIME AND COST ESTIMATING PROCEDURES

The next stage, after specifying the operations to be performed, is to determine the time and cost involved in their execution. In addition, in several recent student projects on such topics as group technology, plant layout, and capital investment appraisal, the need has arisen for a quick method of estimating the time and cost of operations on components. The number of components involved in these projects may range from a dozen to several hundred. This has often been impossible in the time available, and rough approximations have had to be used. Nevertheless, the logic in the calculations is very simple, involving elementary formulae and data obtained from time standards. Consequently it is proposed to develop computer routines for carrying out these estimating tasks. These routines can be applied to the output of the process planning stage of the design problem.

However, it is also proposed to provide a language whereby externally determined operation plans can be estimated. An operation sheet for a part taken from a local firm contains the line:

6: f. bore 6 in dia to 6·000/6·003 in dia 110 r.p.m. 0·016 in/rev

An APT-like statement for this operation could be:

OP6 = BORE/DIAMET, 6·000, 6·003, LENGTH, 1·25, SPEED, 110, FEED, 0·016

This closely resembles the EXAPT 1 bore instruction, and, in fact, the list of operations on a part could be expressed as a list of such statements, together with some additional information concerning machines, tools, etc.

```
                PARTNO EXAMPLE OPERATION LIST
                PART/MATERL, 4
                MACHIN/LATHE, PREOPT, TURRET, COLLET, MANUAL
                TOOLST/FRONT, 2, 1001, 1002, TURRET, 3, 2001, 2002, 2003
        OP1  =  TURN/CROSS, DIAMET, 2·5, FEED, 0·25, SPEED, 200, TOOL, 1001
        OP2  =  TURN/LONG, LENGTH, 10, DIAMET, 2·5, FEED, 0·25, SPEED, 200, $
                TOOL, 1002
        OP3  =  DRILL/DIAMET, 0·5, DEPTH, 1·5, TOOL, 2001, FEED, 0·15, SPEED, $
                250
        ....
        ....
        ....
                MACHIN/CYLGRD, CHBCHI
        OP10 =  GRIND/EXTRL, FEED, 0·001, TRAV, 0·4, SPEED, 80, PASSES, 10
        ....
        ....
        FINI
```

These statements would correspond closely to the entries contained in the normal operation sheet so that they could be coded directly from it.

When the operations to be performed are specified to this level of detail the time and cost of each operation can be easily calculated. In addition, the times of other activities such as load, unload, set or reset tools, can be determined, based on knowledge of the machine, method of workholding or type of tool, from standard times commonly existing in industry, or from predetermined motion times.

A project is being carried out in co-operation with the same local firm, in which the details of the language will be worked out and the estimating routines developed. It is hoped that this language and program will be useful in many current projects, such as group technology based on production flow analysis.

10.0 DESIGN AND EVALUATION OF THE SYSTEM

Once the list of operations is specified in this way and the times and costs analysed for all the components concerned, the total loads on machines can be determined and capacities fixed. The travel chart matrix can be computed and numerical taxonomy used to indicate the most suitable plant layout. In interpreting the linkage chart it may be desirable to consider more than one possible arrangement of the plant. The final evaluation of these layouts can be undertaken by computer simulation and a form of discounted cash flow analysis to assess the economic worth of alternative production systems. Various formulations of such an assessment have been considered, and a program written based on the following principle. The layouts are simulated so that an assessment may be made of the throughput times of the products, of the level of investment in work-in-progress, and of the running costs of the systems. The capital cost of the equipment and inventory and the manufacturing lines are then supplied to a discounted cash flow analysis which spreads these capital costs over the items being produced, and computes the direct manufacturing cost which would have to be charged to the products so as to produce a specified annual rate of return on the system.

11.0 CONCLUSION

This paper, and the activities discussed in it, are motivated by two considerations. On the one hand it has reviewed the problem of the design of production systems and put forward an approach to developing a general computer program for its solution. On the other it has been concerned with the various levels at which the problem is encountered in normal industrial practice. A computer program is being developed, in co-operation with engineering firms to tackle the problem.

The proposed program is designed to be as similar as possible to the APT-based numerical control languages, thereby achieving compatibility with data files created for numerical control purposes, and simplifying its introduction to, and use in, industry. The program is designed to be usable at various levels, depending on the extent to which the process planning has already been done. Particularly at the operation sequence language level the program should be a ready tool for analysing existing or proposed systems encountered in many industrial projects and provide assistance to the many production engineers who think they could make good use of a computer if the systems people would only spend some time examining their requirements.

REFERENCES

1 NATIONAL ECONOMIC DEVELOPMENT OFFICE, '*Production Planning and Control*', London, H.M.S.O., 1966

2 DURIE, F. R. E., 'A Survey of Group Technology and its Potential for User Application in the U.K.', *The Production Engineer*, February 1970.

3 OPITZ, H., 'A Workpiece Coding and Classification System', London, Pergamon Press, 1970.

4 SIDDERS, P. A., 'Flow Production of Parts in Small Batches', *Machinery*, 31st January and 25th April 1962.

5 GOMBINSKI, J., 'Fundamental Principles of Component Classification', *Ann. of C.I.R.P.* 7, 1969.

6 SOKAL, R. R., 'Numerical Taxonomy', *Scientific American,* **215**, (6) December 1966.

7 BURBIDGE, J. L., 'Production Flow Analysis', *Seminar on Group Technology,* Birniehill Institute, East Kilbride, January 1970.

8 McAULEY, J., 'A Study of Machine Grouping', *Diploma Thesis,* Operational Research Department, University of Strathclyde 1970.

9 SMITH, W. P., 'Travel Charting – First Aid for Plant Layout' *J. Ind. Engineering,* **6** (1) 1955.

10 CHOUDHURY, S., 'Numerical Taxonomy Applied to Plant Layout, *M.Sc. Thesis* Production Engineering Department, University of Strathclyde, 1970.

11 HOLLIER, R. H., 'The Layout of Multi-product lines', *Int. J. Prod. Research,* **2** (1) 1962

12 BUFFA, E. S., ARMOUR, G. C., and VOLLMAN, T. E., 'Allocating Facilities with CRAFT', *Harvard Business Review,* March-April 1964.

13 LEE, R. C., and MOORE J. M., 'CORELAP – Computerised Relationship Layout Planning', *J. Ind. Engineering* **18** (3) 1967.

14 SINGLETON, W. T., 'Optimum Sequencing of Operations for Batch Production', *Work Study and Industrial Engineering* **6**, (3) 1962.
15 DE PALO, P., 'A Flexible Method of feed and speed selection for Machining Operations', *NEL Memo X5/206,* 1969.
16 DE PALO, P., 'An NC Program Module for the calculation of feed and speed', *NEL Memo X5/219*, 1970.

INDUSTRIAL APPLICATIONS OF SMALL COMPUTERS

Bryan Davies
National Engineering Laboratory

Note: The paper is presented with permission of The Director, National Engineering Laboratory, East Kilbride, Glasgow. The views expressed by the author do not necessarily represent those of the Department of Trade and Industry.

SUMMARY

The potential market and uses of the small cheap computer are described whilst the lack of engineering applications packages is pointed out. Possible uses of small computers in design are discussed, together with work done in this field at the Birniehill Institute. Present and future developments in computerized quality control and inspection are outlined and trends in machine control and the automatic control of continuous and discontinuous processes are also discussed. Future computer trends in industrial research and development and engineering maintenance are forecast. The author is of the opinion that the small cheap computer with its appropriate software is likely to be used extensively in industry in the future enabling the United Kingdom Industry to remain internationally competitive.

1.0 INTRODUCTION

In recent years, there has been a large increase in the numbers and varieties of small computers produced, both within the UK and overseas, as a consequence of break-throughs in electronic component technology leading to reductions in cost – to the point where the small computer can economically duplicate tasks of much larger computers, but in a much smaller scale and at a much lower cost [1,2]. For example, a company with as few as 25 to 100 employees can afford a computerised payroll or job-costing system because of lower hardware cost, the development of user-orientated computer programming languages (e.g. BASIC and FOCAL) and packages (a few) for the small computer, though the last are still limited in general availability. Further reductions in cost with commensurate increase in power can be expected. The large degree of flexibility obtained from a high degree of modularity in both hardware and software must be also emphasised.

The potential importance of the cheap, small computer as a means of increasing productivity and profitability is seen when it is appreciated that about 90% of UK manufacturing companies employ less than 250 persons, and that they produce about 50% of the manufactured goods by value.

But there is a very serious gap in awareness of what can be done, and what is available, between the computer manufacturer and the practising engineer in the UK, e.g. there is no rationalised stockpile of engineering applications packages, even for the large computers which are now readily available in the UK.

Various agencies such as The National Computing Centre (NCC), The Computer Aided Design Centre, Cambridge (CADC), The Aldermaston Project for the Application of Computers to Engineering (APACE), The National Engineering Laboratory (NEL), The Machine Tool Industries Research Association (MTIRA), The Atomic Energy Research Establishment (AERE), The Production Engineering Research Association (PERA) and some consultancies, and others, offer packages or list the source of programs in addition to the computer manufacturers and their bureaux. These vary considerably in practicality and completeness – adoption, test, modification and maintenance are very real problems – as every NC machine tool user will know. The situation is much worse for small computers, where few applications packages exist. One major related problem must be that, generally speaking, engineering education has, until recent years, virtually ignored the computer, so that middle and senior management will not be attuned to it – they may even be hostile to it. The 'new maths' now being taught in schools must eventually lead to a computer awareness amongst future generations of engineers that is unknown today – more training at primary and secondary school level in computer programming is needed.

Currently, an opinion is being expressed, that the role of the small computer will by 1975 be limited to on-line data collection and validation as a peripheral to a larger processor – in spite of existing examples, particularly in production control and automated warehousing, being to the contrary. Other comments in the same vein are that the price and the limitations of software will also weigh against use of the small computer in the stand-alone mode[3,4]. The opinion is of considerable interest, particularly as it may stimulate small computer manufacturers into refutation.

In the UK we have about 30 manufacturers marketing some 50 varieties of small computer of size less than 16kword storage at 16 bit word size, say. The estimated deliveries for 1969 are 1250 at a value of £12·5M, or about 10% of the market. Of these about 70% are absorbed in commercial applications, with educational and scientific applications taking up a further substantial proportion, leaving quite a small number actually applied in industry to manufacture and design. Of these it can be assumed from previous experience that the major application is to process control, and furthermore that the principal use of the small computer is as an integral part of an electromechanical system produced by an OEM (original equipment manufacturer). The OEM can negotiate specially favourable prices with particular computer manufacturers if he uses that manufacturer's products alone. This is of importance in terms of the total product price, and eventual shake-out in the small computer industry.

The potential benefits to be derived from the use of the small computer are enormous, and it is important for UK industry generally to become aware of this potential. Some examples from current practice are given below.

2.0 DESIGN

In the past, even in big companies, the design engineer has been something of a Cinderella when it came to capital invested in him as a person – and largely this is still true. Most commercial mechanical engineering design is still a matter of pencil on paper, catalogues, British Standards, intuition, and rule-of-thumb. The slide rule, log book, pencil and paper, and, occasionally, a battered old mechanical or electromechanical calculator are the usual Drawing

Office (D.O.) tools. Once in a while a planimeter appears, but this is obsolete.

A recently published survey[5] on the scope for computer aids to design in the engineering industry concluded that future use of computer aided design (CAD) by small firms does not lie with expensive hardware and sophisticated techniques. Where there is scope for CAD it was thought that small firms need outside help to identify the opportunities for CAD, which should lead to reduced lead time and improved design, and in particular reduced time for the preparation of estimates and their improved accuracy.

Somewhat in contrast is one of the conclusions of a long-range technological forecasting study by the Birniehill Institute[35] which indicates the use of small computers in CAD electronics as the most immediate and likely application of the small computer. To this must be added the report on the U.S. semiconductor industry[20] which concludes that where customers do not have an extensive design capability the IC (integrated circuit) house must provide it, and as a result there will emerge a design industry complete with CAD facilities which will parallel the growth in the computer software industry, which today in the US is beginning to rival the hardware industry in size.

Only the computer is seen as being able to meet the scarcity in design talent, by reducing the manual effort required, and storing the complex successive stages of large-scale integrated circuit design. Motorola, Microsystems International (Canada), and Texas Industries (Europe) have or are in a position to create CAD centres. In the U.K. we already have talented software organisations in this area e.g. APACE and REDAC. The CAD Centre, Cambridge, is also likely to be involved. During 1971, the Birniehill Institute, NEL will present seminars on CAD, specifically aimed at designers new to computer methods, and other organisations may join this activity. The need for information storage and retrieval systems for parts already designed

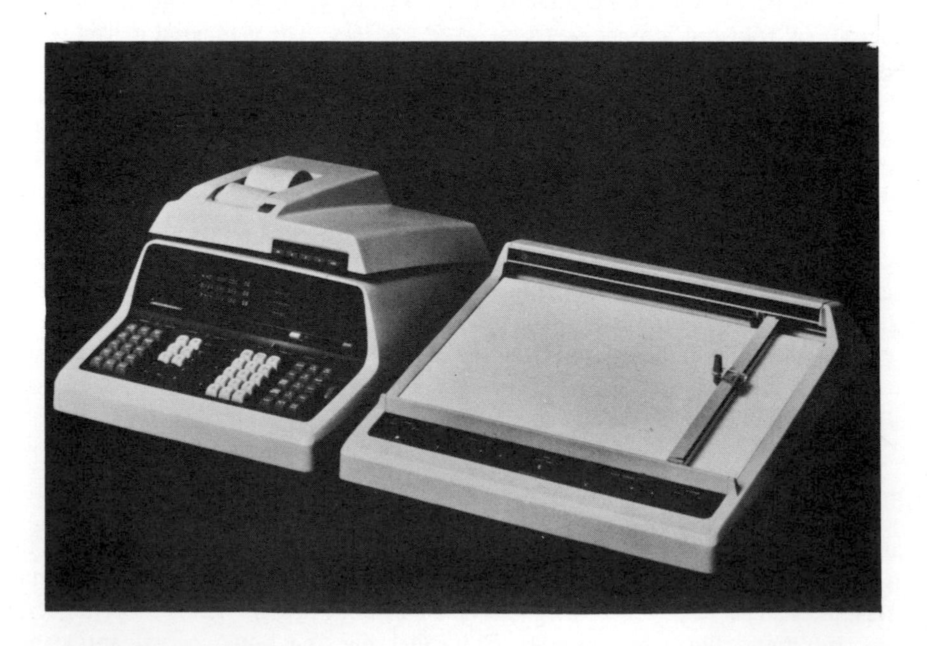

Fig. 1. Hewlett-Packard Model 9100A/9125A computing calculator and x-y recorder.

and for bought-out materials and components was also emphasised in the MinTech survey[5], but surprisingly not in terms of computer usage. It was implied that the high cost of automatic draughting equipment in terms of a small computer application, and of interactive graphics, is restricting their widespread acceptance. This seems to be a generally accepted view. Other effects such as availability of software must influence the situation as well.

A great deal of day-to-day DO calculation can be carried out by the relatively cheap desk-top programmable computers recently offered, such as the Hewlett-Packard 9100A (Fig. 1–with 9125A plotter)[6]. The relatively large storage/calculator capacity of such machines, and the access to trigonometric functions by a single key stroke allow calculations of some elaboration to be done in minutes, and programs written on magnetic card may be stored and the recalculation done in a few seconds. Use of small desk-top computers in the DO will save draughtsmen's time, and with errors in calculation virtually eliminated will clearly save money. Stored program libraries on magnetic cards will increase DO efficiency by gradually reducing the need to write new programs.

An interesting example of a fairly trivial but tedious and lengthy formula evaluation has been given by Texas Instruments[7]. Fig. 2 shows a flexivity equation for a three-layer thermo-

$$\frac{1}{\rho} = \frac{12E_1a_1E_2a_2E_3a_3\left[\left(\frac{a_1+2a_2+a_3}{2E_2a_2}\right)(\alpha_3-\alpha_1)+\left(\frac{a_2+a_2}{2E_1a_1}\right)(\alpha_3-\alpha_2)+\left(\frac{a_1+a_2}{2E_3a_3}\right)(\alpha_2-\alpha_1)\right](t-t_0)}{E_1^2a_1^4+E_2^2a_2^4+E_3^2a_3^4+2E_2a_2E_3a_3(2a_2^2+3a_2a_3+2a_3^2)+2E_1a_1E_3a_3(2a_1^2+2a_3^2+6a_2^2+6a_2a_1+6a_3a_2+3a_1a_3)+2E_1a_1E_2a_2(2a_1^2+3a_1a_2+2a_2^2)}$$

Where ρ = radius of curvature; α_1, α_2 and α_3 = coefficients of thermal expansion; E_1, E_2, and E_3 = moduli of elasticity; a_1, a_2 and a_3 = thickness; t = final temperature; t_0 = initial temperature.

Fig. 2, Flexivity equation for a three-layer thermostat meter. (With acknowledgement to *Product Design Engineering*).

stat meter evaluated on an IBM 1620 computer. It is evident that errors could quite easily arise in computations of this nature if done manually. The calculation and mathematics are tedious and, typically, of such low intellectual content as not to maintain the interest of a designer once the equation has been derived symbolically. Nevertheless, the evaluation is important to the design itself. It would be a waste of a big computer to put it to work on jobs like this (as is often done), which may be regarded as typical day-to-day DO calculations. Surely it wastes a designer's time and the firm's investment in him to make him enumerate such a formula?

Our own experience at the Birniehill Institute in experimenting in a DO environment with a small computer (an 8k Marconi-Elliott 905 with 928 display and Calcomp drum-plotter with tape reader and teletype) confirmed our expectancy of difficulty in achieving serious CAD (or CAM), with core-size limitations and no disk or tape back-up. The CAD work that has been done could probably have been performed just as well with a desk-top calculator. Regarding applications to production engineering, we have performed a useful demonstration of the potential speeding-up of production flow analysis using computer graphics interactively but for a very small sample of a machine shop. The larger analysis requires a big computer and this has been done using the Univac 1108 at NEL with a teletype as the input device. The length of time for print-out on a teletype is restrictive. Sophisticated analytical design requires access to a large computer for which the small DO computer may be a useful peripheral.

It was interesting to see how the staff concerned reverted to past DO experience in seeking out applications areas for experimentation – in this case, gear design. A comment was made that lengthy calculations of considerable tedium are sometimes performed manually in industry, even when computer programs exist for their evaluation, because of the lengthy turn-around times with computer bureaux. This problem, if it exists, is resolvable by using remote access with conversational programming, or, for simple calculations, a small, local, stand-alone system. It is essential that this be provided with an adequate back-up store. The general implementation of small, local, stand-alone DO systems (in the UK typically a 16k computer with back-up disk and graph-plotter) seems to await price reductions – which ought to result from further anticipated reductions in the price of electronics. However, even a total system cost of £20–£50k (say) does not represent a large investment per designer, when discounted over the life of the equipment.

It is evident from the most recent developments in the Birniehill experiment with a more ambitious program (for the design of two mating gears) that the earlier, more elementary, programming stage of learning was essential. The real DO past experience and irritation at having to do laborious, tedious, routine calculations strongly influences, psychologically, subject selection. We suspect that the draughtsman will see and use these new tools readily once he has been made aware of their potential and is permitted to use them. Their immediate, local accessibility in the DO may be significant but we cannot yet conclude on this aspect. As an experiment in the possible effect of introducing such a system into a DO, we believe the effort to have been worthwhile, but unfortunately biased from being relieved of commercial pressures.

Experience in engaging DO staff at O.N.C., H.N.C. and unqualified levels on sophisticated analytical design using very large computers shows that a certain element does respond with enthusiasm and increasing expertise. Thus, there is no reason to suppose that the response to smaller computers in the DO will be any less enthusiastic provided the education and training are adequate.

There is a growing awareness of all kinds of innovative concepts in mechanical engineering at present and one that is being spoken of more frequently is 'added value'. Perhaps this comes from the value engineers and from advocates of Problem Analysis by Logical Approach (PABLA) qualitative design systems. Molins, in advocating their System 24 on film, made use of the added value concept[21]. A recent review[19] of work done at the Lockheed Missile and Space Company regarding the designer's influence on manufacturing costs quotes the result that the designer can vary the amount of direct manufacturing labour by 35%, or, the 'least costly' design can be made for 65% of the manufacturing cost of the 'most costly' design. The designer can vary indirect labour costs, procurement, planning, scheduling, dispatching, order control and stocking by 65%. These figures are in sharp contrast with those for the manufacturing operation itself, which Lockheed found could only influence its own costs by about 10%. Whilst the 'value' of a product may be used as a measure of its earning capacity, it is paradoxical that the designer must try to minimise the manufacturing cost whilst at the same time – by increasing its complexity – attempting to maximise its 'value'.

Analysis of the total design situation at Lockheed led to the conclusion that the real problem was the lack of readily available information that could aid the designer in his selection of the characteristics of the design, and presumably also the lack of an information storage and retrieval system in which components data had been rationalised, e.g. by group technology. This is a situation which is made for small and large computer intervention, as no doubt Lockheed Missile and Space Company, with its vast expertise in computer applications, recognised.

3.0 QUALITY ASSURANCE AND QUALITY CONTROL

Quality assurance is a term currently used to embrace a range of activities aimed at ensuring that the design specification is properly interpreted during manufacture. In the UK the emphasis now being placed on 'design' rather than manufacture demonstrates the awareness of the quality assurance man in that he closes the loop in the iterative process between DO and shop-floor.

Watkins[22] identifies two pressures on factory quality control/inspection departments: the collection of more data, more rapidly; and the requirement to reduce the direct expenditure on inspection, or quality control, the aim being to obtain a higher degree of adherence to design specification, leading to less scrap and rework. To meet these pressures at Raleigh Industries, a small computer-based system (an ARCH 102) has been combined with multi-dimensioning gauging fixtures using inductance displacement gauges and analogue-to-digital converters, to monitor the manufacture of components on the shop floor on an inter-machine basis, thereby providing quality control of the machines. The system, installed in August 1968, had to obtain savings to cover costs of £25 000 for the electronics and £10 000 for the installation and the gauging equipment less the development grants, which left £26 000 to be accounted for. By the end of 1968, it was evident that the savings would be made at the rate of £8000 per year, reducing the number of persons employed on a sample audit. Other benefits accruing from the almost total elimination of scrap and rework have not been costed. Inspection times per batch have been considerably reduced, typically by an order of magnitude, whilst the total number of parts inspected has been increased.

Hewlett-Packard have described[6] calculations on the 9100A desk top computer to determine the measurement over rollers for both gear wheel and pinion, which, with an input of the number of teeth, the diametrical pitches, the pressure angle, the circular tooth thickness at pitch circle, and the roller diameter, took only about a minute to compute. This would have taken a skilled draughtsman about an hour, and a less skilled inspector even longer. These are typical of the many day-to-day calculations now carried out manually in U.K. inspection departments.

The development of huge electronic, mechanical and electromechanical systems (e.g. computers, missiles and rockets, military systems) has meant that their testing is commensurately sophisticated. Testing covers, of course, both development and production, and it has now been recognised that the test apparatus must be capable of change to meet the unique requirements of particular projects, and that it must be capable of use by semi-skilled personnel with little or no fault-finding skill of their own [8,13] . Test must ensure that production meets specification: they must yield statistical data from which parameters such as mean-time-between-failures may be calculated[9] .

Many of the larger firms in the U.K. have established quality assurance laboratories and test houses, not only for their own purposes but also to provide a service to others. A typical example of the use of a small computer is that provided by the Hawker Siddeley Dynamics TRACE System, which is in day-to-day use with airline operators. It uses a Honeywell 316 small computer to check out avionics equipment. Honeywell markets an equivalent TITAN tester of its own, which can be tailored to customers' requirements and is highly modular in concept.

Several examples of different modes of computerised inspection and in-process monitoring were shown at the 1970 Chicago Machine Tool Exhibition[10]. These included a Browne and

Sharpe set-up of a Hydrocut vertical machining centre combined with a Validator 100-co-ordinate measuring machine for the checking of non-mathematical contours, using a PDP8/L computer to optimize incremental data (including conversion to linear or circular interpolation for the particular NC system used), which is then part-processed and transmitted to the machine tool controller. Feeds, speeds, tool selection and other functions are input by teletype. The system can be used to inspect one workpiece whilst the next is being machined from a stored program. A similar system for the digitising-contouring of surfaces not easily described mathematically was shown by Cincinnati Milacron. It used a 16 in Hydro-Tel profile miller with Acramatic IV NC system, an electronic tracer and a CIP small computer with the usual I/O devices. The tracer probes the prototype surface and the machine tool slide position sensors measure position co-ordinates back through the NC system to the computer, which produces a paper tape. The Model 3000 Sheffield Cordax measuring machine is computer-controlled and produces NC tapes; the other Cordax machines are merely computer-assisted – the distinction being on computer or manual control. Among future developments proposed were systems based on a lasers to measure linear dimensions, and optical rotary encoders to measure angular dimensions, the output being fed to a computer-controlled inspection machine. Such an arrangement would be used for the precision checking of cams, gears, index plates, aerofoils, and other complex shapes. Presumably, the whole system is to be built around a small computer.

Ferranti have since the early sixties developed inspection machines which have progressed co-ordinate measuring machine for the checking of non-mathematical contours, using a PDP8/L computer. The latest machines from Ferranti can be used to obtain co-ordinate data by automatically scanning 2D and 3D contours, and then to generate tapes for use on NC machine tools, by combination with numerical tool code information[23].

Ferranti have found savings to accrue from, amongst other things, a reduction in the documentation of instructions that detail set-up and inspection. These can now be incorporated into the tape containing the inspection program.

Sophisticated R & D into computer-controlled inspection using small computers is being carried out at the Cranfield Unit for Precision Engineering (CUPE)[24], which has recently specified the design of an NC measuring centre with computer control, and has suggested as a future possibility for the UK, the use of the measuring centre as part of a closed loop embracing the NC machine tool as an in-process controller, e.g. as in the Browne & Sharpe set-up referred to earlier. Adaptive control is envisaged for the future.

4.0 MACHINE TOOLS

The trend in machine tool technology is towards increased automation, and to a lessening of direct human involvement in the actual cutting process. How long will tape control for NC machine tools last? This question is clearly of importance in the UK, where industry has yet to take up NC in a big way. Are we in a privileged position – in that our belatedness enables us to make the step from conventional machining to complete numerical control (D.N.C.), bypassing NC on the way? It has been suggested[11] that in the US in five years time as much as 50% of NC will be computer-controlled, and if the machines on show at the Chicago 1970 Machine Tool Exhibition are indicative, the process has in fact begun[10].

The variety of small computers used was dedicated to one or two machines, or used as multiplex message-switching devices to many machines. Some firms producing D.N.C. systems have now developed their own small computers e.g. the GEPAC 30 is in use in the GE CommanDir system and Cincinnati Milacron have produced the CIP/2100. Conversely, the Digital Equipment Corporation, (DEC), whose PDP8 is widely applied, has produced its own machine control system. Plug-in units are on offer as CRT/keyboard I/O stations for local program updating, debugging or reprogramming. The small computer may be used to input data on job status for workshop loading programs at a central processor as a part of a management information/manufacturing system. 'Soft-wired' control systems are available in which logic functions are contained in a stored program rather than 'hard-wired' into unchangeable circuitry. These include integral small computers usually dedicated to the control of a single machine tool.

The highly sophisticated American D.N.C. (direct numerical control) systems operating, such as System Gemini (Kearney and Trecker) are expensive and large-scale, and are not yet mirrored in the UK, apart from the Molins System 24, which was once the most advanced of all. In these, the small computer represents a small fraction of total cost. There were a few dedicated small-computer approaches to CNC, e.g. that of Warner and Swasey, aimed at the US small- and medium-sized factory, in which a PDP-8/I with back up disc is used to replace a conventional NC system. The basic DEC PDP-8/I is 8k (12 bit words), core expansible to 32k and the disc expansible to 512k. The first implementations are to two turret lathes, the SC25, four-axis turret lathe, and the SC16, two-axis turret lathe. The computer is used for coarse interpolation and the fine interpolation is done at the machine. For smaller machines, the logic will be built into the headstock (including a typewriter keyboard); on larger machines it will be in a separate console. An interesting comment taken from the Westinghouse Machine Tool Electrification Forum[26] is that the use of a computer in the NC system will give little benefit in terms of machine tool maintenance where the problems are often mechanical e.g. servo-drives, limit switches etc.

5.0 AUTOMATIC CONTROL OF CONTINUOUS PROCESSES

The computer has been applied to continuous process control – the measurement and control of the manufacture and flow of materials – in large industries such as steel, petro-chemicals[18], smelting, paper, sugar and cotton for some time. It has been argued by Thomas[12] that with reduced prices and increased reliability the application of small process control computers in small industries will be equally profitable, and, moreover, will provide indispensible management information.

Thomas describes the types of computer control as:

- Supervisory – information from the process is used by the computer to issue printed instructions for manual adjustment of plant.
- Analogue with set-point adjustment – in which the computer adjusts the set points of the analogue process controllers.
- Direct – in which signals are provided to adjust controllers directly, this type is highly dependent on the quality of control instrumentation available (Fig. 3).

The programmable computer, as distinct from the special-purpose controller, should be adaptable to changes in requirements, and permit some degree of process optimisation and simulation at the design stage. By incorporation in larger systems it also permits integration of the control of several closed-loop processes leading to a management information system. The

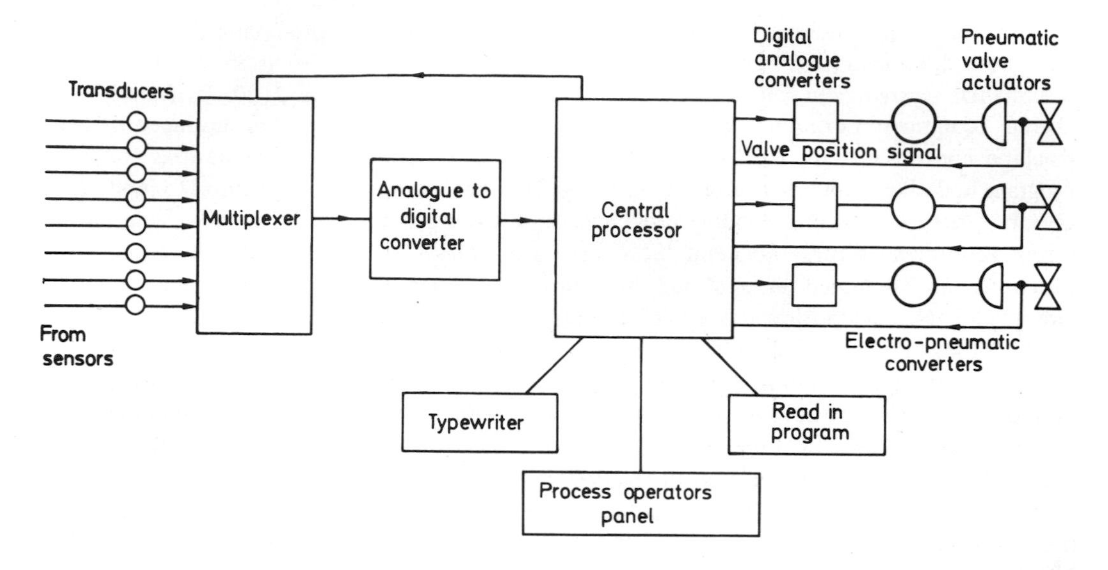

Fig. 3. Direct control of a process by computer. (With acknowledgement to *Automation for Management*).

extent of permissible management decision-making is one of the fascinating aspects of integrated process control – when and by how much does a manager exercise control on the system as distinct from needing information?

The Ferranti Argus 600 represents what the small firm or the system designer would be looking for, i.e. a low-cost, programmable computer that is more than a simple sequence controller, suitable for on-line operation. At £1700 it approaches the '£1000 computer' that hard-headed mechanical engineers dream about, and should be achievable by 1972/73. In the control sense, the ability to program gives increased scope – particularly for mechanical-assembly-type discrete process control, which we expect will increase in importance in the next few years. However, the factory environment may cause severe implementation problems for solid-state electronics due to industrial electrical noise[35].

Industrial applications occur in quite different environments to scientific applications, generally more severe. Some of the interface problems for industry have been described by Groves[29]. He points out that facilities are often required to control equipment in plants which are separated by considerable distances, that industrial 'noise' can be significant, and that the design requirements include high levels of reliability and protection. Groves suggests that the hardware design criterion to minimise noise should be that noise be eliminated after it has passed through a circuit – attenuation is not enough. Other electrical requirements are modularity, flexibility and versatility, which seem obvious enough. Modularity is obtained by a range of plug-in logic modules identical with those in the computer.

In the UK today the range of continuous-process computerised control is extensive in sophistication (and cost) – ranging from simple data-logging of performance to embryonic integrated systems. The use of the computer itself to simulate the plant at the conceptual stage is likely to be extensive[13] and when national expertise is more readily to hand, computer software applications availability is essential. This area of engineering is not likely to be less demanding in spite of its historical start on others.

6.0 AUTOMATIC CONTROL OF DISCONTINUOUS PROCESSES

One of the potentially most applicable areas for small computers is in automated mechanical assembly lines. There are a few examples to date and these include the application of the computer to both monitoring and control. One interesting example is of a system designed by Entrekin Computers Inc. (USA)[34] to monitor disk brake assembly at Cross Corp, using a PDP8/I, 12k (12 bit) words of core, 65k (12 bit) words of random access disk (see Fig. 4). The computer logs operating data, reporting out any exceptions to standard. It checks every station's cycle time, notes stoppages, checks on component position at each station and whether station mode is manual or automatic. It keeps a running inventory of parts in and out, of rejects and of parts repaired, and monitors stock levels in feeders and signals for refills. Automatic inspection is aided by computer records of each part. What is interesting is that the automatic assembly aspect and the full managerial control systems are capable of addition to the very open-ended computer packages – neither have been fully realised in this case yet. It is worthy of mention as it exemplifies the step-by-step approach to a totally integrated manufacture/management/design system.

7.0 SOFTWARE FOR SMALL COMPUTERS

Applications software and executive programs are not always readily available for small computers because the computers themselves are relatively recent in origin. Economics may not permit manufacturers of relatively cheap machines to develop and supply standard software. Furthermore, the variety of applications and configurations tends to militate against standard packages[14]. The functions of a small computer cover DDC (direct digital control), supervisory control, sequence control, data acquisition, operator communication and computation, for which are needed basic support software, operating system software and applications software. With regard to the latter, in the event of his supplier failing him, the industrial newcomer can obviously either write his own program to solve particular problems, use a bureau, or purchase an existing program. The lesson for both the user and the computer manufacturer is to choose the organisation that has the most to offer.

8.0 THE MANAGEMENT INFORMATION SYSTEM (MIS)

The number of examples of factory management information systems that use a small computer is small. Nonetheless they are growing in number. In the UK we are fortunate in having readily available the ICL PROMPT program, which can be run on a small 1901 16k processor with tape back-up. A typical installation which is used for production control is that at Serck Audco Valves, a firm perhaps best known for its pioneering work in the UK on group technology[16]. This is a typical UK small-to-medium-sized firm of about 1000 employees, with a turnover of £4M, producing some 2000 finished products using 15 000 individual piece parts and raw materials. The number of different parts per product varies between 5 and 200. Serck Audco's ICL 1901 with four 20kc tape decks covers the functions of stock control (up-dated per fortnight per machining cell), material control, foundry stock and production program purchasing reports, and listings for control committee decisions.

It is an interesting confirmatory fact of the theory of each firm's uniqueness that the PROMPT package has to be tailored by Serck Audco to its own unique requirements. Serck

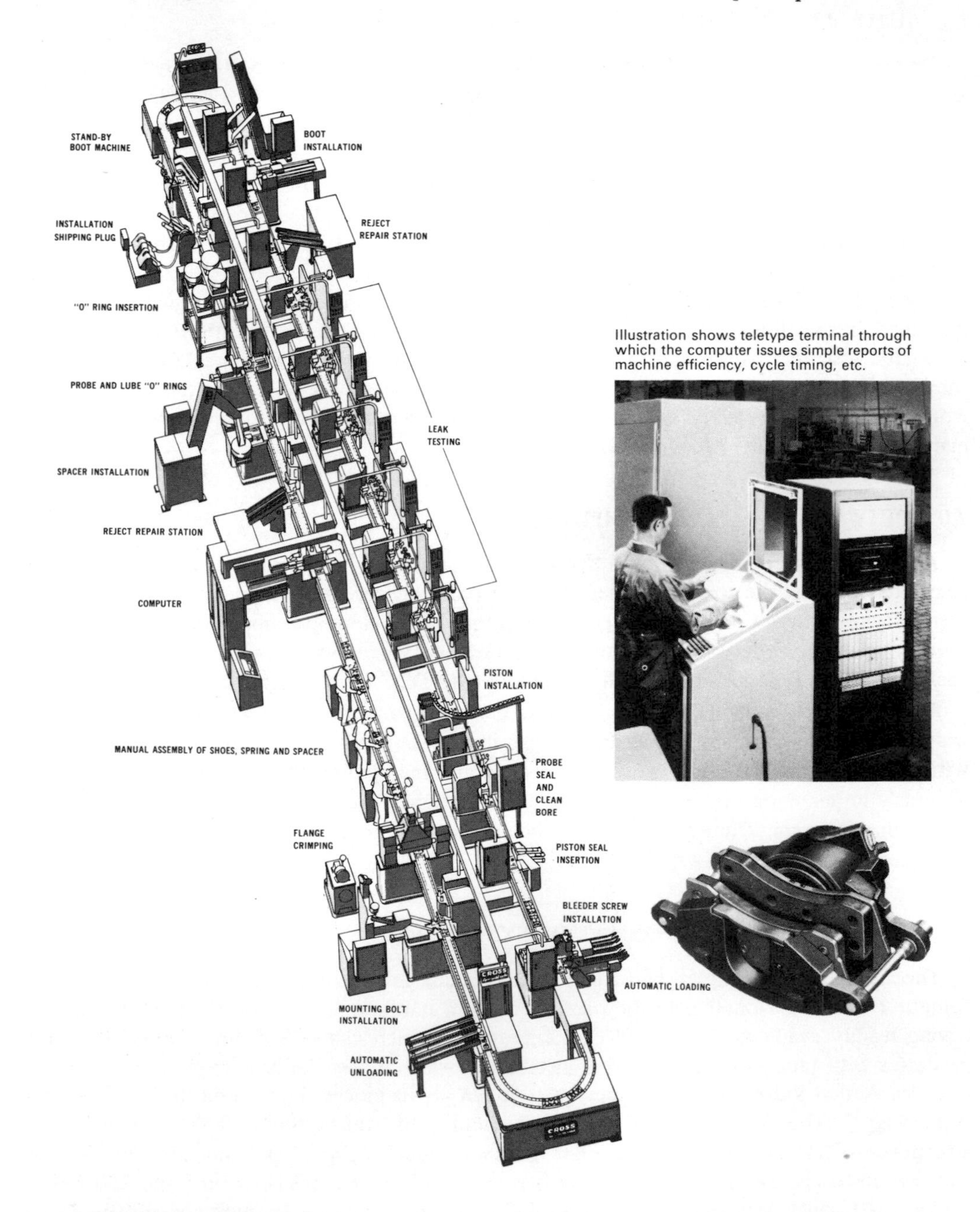

Fig. 4. Entrekin computer monitoring system on assembly line for disk brake calipers. (With acknowledgement to Entrekin Computers Inc., Michigan).

Audco's objectives initially were simply to reduce the volume of clerical work and to cut down the operating cycle, by using a combination of computerised production control and group technology to casting, machining and assembly processes. The firm has found that output has nearly doubled whilst the number of employees has decreased, with a corresponding reduction in stocks and work-in-progress, followed by reductions in the manufacturing cycle time from 5 to 12 weeks on average. (See Table 1, reproduced from ref. 16 by permission of Serck Audco Valves Limited.)

TABLE 1. SERCK AUDCO PERFORMANCE RECORD

(1) Year	(2) Net despatches £M	(3) Value added £M	(4) Average no. employed	(5) Total wages and salaries £M	(6) Despatches per employee £	(7) Value added per employee £	(8) Average income per employee £
1961/62	2·220	1·615	1001	·714	2218	1613	714
1962/63	2·184	1·580	952	·706	2294	1660	742
1963/64	2·585	1·872	903	·752	2863	2073	833
1964/65	2·922	2·007	941	·844	3105	2133	897
1965/66	3·363	2·303	992	·951	3390	2320	953
1966/67	3·768	2·646	987	·979	3818	2681	992
1967/68	3·576	2·519	906	·951	3947	2780	1050
1968/69	4·984	3·117	1130	1·338	4411	2758	1184
1969/70	6·047	3·810	1175	1·625	5146	3242	1385

During this period the progress in Column (7) was retarded by the problems associated with the transfer of £1·3M of ball valve production to Newport.

Based on 10 months' actual – year ending 26 September 1970.

A computer-based management information system has been introduced into the small company of J Goulder and Sons Limited, which employs 180 people. The system is based upon a small local Friden computer at the firm and a larger computer at Huddersfield Technical College – the system could however be accommodated on a small computer. Comments made by the firm were that its approach to computerised production control was to set-up a Working Party (with delegated tasks) comprising Managing Director, Works Director, Financial Director, Production Controller, Office Manager, Production Manager, representatives of Huddersfield Technical College and Elliott and Friden. All who had to work the system were involved from the outset, including the trade unions. Initiation and continuing support came from top management, surely enlightened, not from O and M or systems analysts. Implementation of such systems involves new thinking at the top of the management structure[30,31].

9.0 FACTORY DATA COLLECTION

It is a prerequisite to the use of a small or large computer in a factory environment that there be a collection system that feeds it 'clean' data. Application studies have shown that straightforward mechanisation of existing procedures will not provide the best economic solution, furthermore, once established, the data collection system must be allowed to deal with exceptions. These should not be dealt with manually[17]. It is unlikely that any standard systems application package will ideally satisfy a user – it has to be tailored to the individual firm's requirement. Reference 17 is a valuable comprehensive statement of the method of analysis that should be used prior to setting up a factory data collection system, e.g. Cornelius states that by queueing theory it can be shown that for up to 70% of maximum average switching rate, outstation (shop floor report station) queueing times will be insignificant, thus giving a basic figure for system design, and indicating when the designed system will experience significant delays.

A typical shop-floor example of a data collection terminal is shown in Fig. 5, (reproduced by permission of Feedback Limited).

Fig. 5. Feedback data collection terminal.

10.0 FUTURE TRENDS

The Birniehill Institute in conjunction with its October 1970 conference on the Industrial Applications of Small Computers conducted a survey of small computer hardware/software for the UK, and a Delphi experiment into future applications[33]. Some results of these experiments, which were only partially available at that time and are still incomplete at the time of writing[15], follow.

Taking the survey as a whole, the average small computer hardware/software works out as:

Computer	
Cost	£3000 – 6000, limit £20,000
Memory size	4-32k
Word length	16 bits
Cycle time	900 nanoseconds
Add time	2 microseconds
Multiply time	5 microseconds
Peripherals	
Card reader	300 c/min
Line printer	300 lines/min
Paper tape reader	300 c/s

There is an increasing tendency to include ROMs (read-only memories) particularly for executive operations, thereby realising an increase in operating speed. Some important control applications of small computers require the capability of automatic recovery e.g. after power failure.

The software available for small computers apart from the essential assembler is generally the IFIPS sub-set of ALGOL, ASA FORTRAN IV and ASA Basic FORTRAN. Apart from those supplied by the older-established manufacturers like DEC, IBM, ICL and Honeywell few applications packages are available yet. This fact may be significant in relation to the small-computer manufacturers' individual survival.

A price reduction of about 40% per generation, each generation remaining active for about 2 years, will lead us to a small average computer priced about £1000 in 1972/73.

The common small computer language seems to be tied up with the development of time-sharing, pointing towards JOSS-type languages like BASIC and FOCAL (DEC).

11.0 INDUSTRIAL R & D

A fascinating use of the small computer is that in a mobile instrumentation laboratory. (Reference has already been made to the use of the computer in test equipment.) An application by the British Non-Ferrous Metals Research Association combines studies of process control, design and quality assurance. A mobile Ferranti Argus 500 computer is taken to foundries where measurements are obtained on pressures, piston and metal movements, and metal and die temperature. The computer collects, analyses and obtains stress data: during the injection stroke of the die-casting machine, 4000 readings are taken and put directly into store for subsequent processing during the slow part of the cycle. The total cycle time is about a quarter of a minute. One long-term aim is to study the feasibility of reactive control on various parts of the cycle, using the flexibility of the computer to explore ideas which would ultimately be carried out in production with much simpler control devices[25].

CUPE also have a mobile computing system based on a Marconi-Elliott 903 computer (an 8k core, 18 bit word length), a paper tape reader, paper tape punch, teletype and digital graph plotter, a general-purpose interface, 12 input-output channels multiplexed to the computer and a real-time clock. This machine is intended for shop-floor use, e.g. to optimise digital servos and to calibrate machine-tool slide-ways.

12.0 TEROTECHNOLOGY

Terotechnology, a term for engineering maintenance, is claimed to be an area where considerable cost savings can be effected[32]. CEGB's approach to preventive scheduled maintenance by computer[31] has been to use the computer available rather than to purchase a new small computer for scheduling maintenance. However, what has been done on the ICL 1905 in the Midlands Region could have been achieved on a small computer – a complete system to provide schedules for preventive maintenance activities. The initial work will be followed up by production of work order cards, and later plant history catalogues and integration generally into a management information system (MIS), e.g. cost analysis, resource utilization analysis, simulation etc.

13.0 THE TOTAL FACTORY

Fig. 6 is an attempt to list the potential applications of the small computer in small- to medium-sized factories. It is surely in the nature of things that applications will still sometimes be arbitrary and even expediently introduced. We should expect, however, with increased use of simulation techniques, possibly a customer service offered by the vendor, and that the actual selection of the computer and its peripherals will be reasonably scientifically based. UK consulting engineers invariably approach an assignment by taking a very common-sense, production engineer's look at the system as a whole – Hoskyns Systems Management, Herbert-Ingersoll/W. S. Atkins are typical organisations which combine production engineering and computer applications in the best type of systems engineering approach.

14.0 COMPUTER-ASSISTED INSTRUCTION

Computer-assisted instruction (CAI) must be regarded as extremely important to all engaged in training, industry and academics alike.

The size of the UK education bill has been predicted by Vaizey[27] as reaching £6000M by the year 2000 – three times the current level of £2200M per annum (6·25% of the G.N.P.). By 1980 this bill will rise to £5000M or 6% of the G.N.P.

Also, the numbers of students will increase from 443 000 (220 000 in universities) at present to about 727 000 (450 000 in universities) by 1981, with a further 98 000 in Scotland. However, the expanded university capacity is only expected to be 400 000. What is to be done with the 50 000 extra? New universities or a new approach? The question[8] is posed: do we have the resources to meet the explosive increase in demand for higher education? Does the current

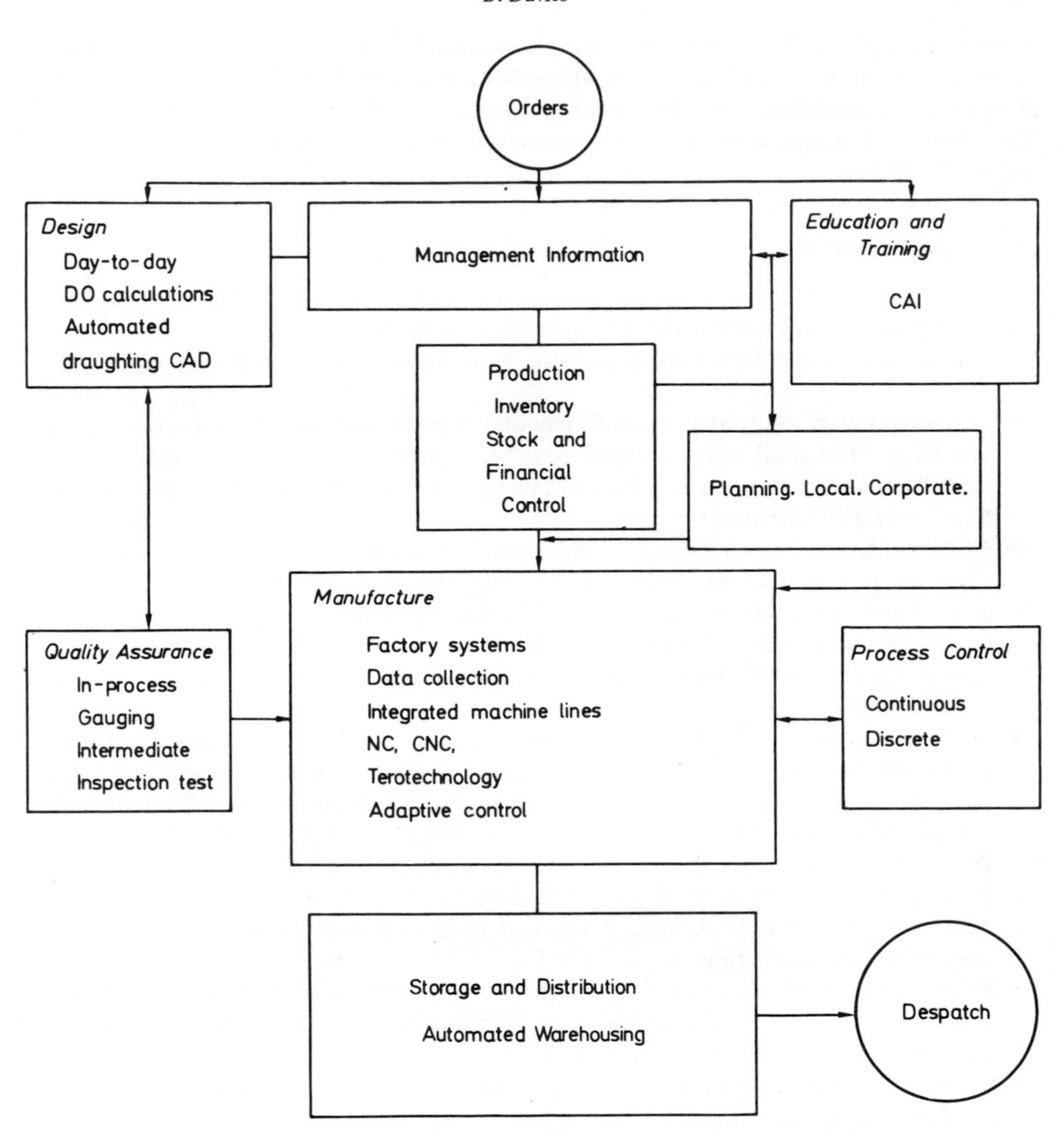

Fig. 6. Potential applications of small computers in the small firm.

university thinking provide a good enough base for the professional engineer who may opt for production engineering, management, administration? Can we extrapolate perhaps from the work of Leeds University[28] and, regarding it as embryonic, see the small computer as providing both the range asked for by the industrialist and the academic rigour of the educationalist, also perhaps coping by new teaching methods with the mass-education demands of the future? It is interesting to conjecture that the continuing survey of students' progress with CAI eliminates, and make impractical, the written examination.

The demand from industry for more education and training will grow. This will be due partly to the influx of better-educated engineers, and partly to a growing awareness generally in the engineering fraternity. Clearly, firms will be faced with work-load problems, if long periods of absence are permitted to employees for education. In-works instruction for a period, each day or week, using an in-house terminal connected to an educational centre which offers CAI, may be the answer.

15.0 CONCLUSIONS

The following conclusions come partly from the Conference on Industrial Applications of Small Computers held at the Birniehill Institute, October 26-30, 1970.

- The modern concept of the highly modular, cheap, small computer is only about two years old.
- The computer generally and the small computer in particular have left the era of mysticism and magic. The small computer must now be regarded just as any other manufacturing tool. The production engineer is slow in taking it up, but this is no fault of his – he must be re-educated in the use of computers.
- Whilst hardware costs are rapidly decreasing to the point where the minimal cost of a small computer in 1980 may be £20, the '£1000 computer' should be with us by 1972/3. The cost of software (particularly for compilers and operating systems) is escalating. It is possible that a software industry comparable in size and value to the hardware industry will arise in the U.K., particularly – and hopefully – with regard to selling services overseas as an invisible export.
- The non-availability of applications software packages, representing as it does a real lack of joint effort between the computer house and the practising industrial engineer, is critical and must be overcome if U.K. industry is to take full advantage of the small computer and remain competitive internationally. There are notable exceptions but these are few. Without the applications software, the hardware is of little value.
- It is now cheaper in a total control system to use small programmable computers, duplicated (for safety), than to design and produce special-purpose, hard-wired logic because of mass production.
- Small computers are characterised by a large number of options. This modularity, combined with the influence of rapidly falling prices, is chiefly responsible for promoting the use of small computers in industrial equipment.
- The OEM (original equipment manufacturer) tends to favour one computer manufacturer, on pure technological grounds. This is likely to emphasise the position of the better-known, longer-established firms and reduce the numbers of small computer manufacturers in the U.K.

REFERENCES

1. KOTSAFTIS, C. J., 'Mini-computerised Managing', *American Machinist,* Nov. 3, 1969, 108.
2. NRDC., *Inventions for Industry,* No. 36, Oct. 1970: *Small Computers – The Coming Revolution.*
3. HOSKYNS GROUP LIMITED, *U.K. Computer Industry Trends 1970 to 1980,* Oct. 1969.
4. *Univac Users Association Conference,* Amsterdam, Apr. 1970.

5. MINTECH, *The Scope for Computer Aids to Design in the Engineering Industry,* Aug. 1970: Report by Urwick Technology Management Limited.
6. TIDD, J., 'Speed up Gear Design Calculations', *Product Design Engineering,* Sep. 1970, **50.**
7. ORNSTEIN, J. L., 'Computer speeds selection of thermostat metals', *Product Design Engineering,* Sep. 1970, **47.**
8. WHITMAN, B., 'The importance of quality control', *Electronics Today, Electroniscope,* Oct. 1970.
9. SMITH, C. S., *Quality and Reliability: An Integrated Approach,* Pitman.
10. HATSCHEK, R. L., 'Computers take control', *American Machinist,* Aug. 10, 1970, 85.
11. 'Wide Role for Computers Seen', *American Machinist,* Jun. 15, 1970, 85.
12. THOMAS, H. A., *Automation for Management.*
13. LEWIS, A., 'Automatic Testing', *Kent Technical Review,* (2), Oct. 1970.
14. TEMPLE, R. H., and DANIEL, R. E., 'Status of Mini-computer Software', *Control Engineering,* Jul. 1970, 61.
15. *Industrial Applications of Small Computers Conference,* Birniehill Institute, Oct. 26-30, 1970.
16. WOOLLEY, V. S., 'Production Planning and Control in a Group Technology Environment', *Group Technology Seminar,* Birniehill Institute, Oct. 14, 1970.
17. CORNELIUS, V., 'The Application of a Data Collection System in Machine Shops and Assembly Shops', *Industrial Applications of Small Computers Conference,* Birniehill Institute, Oct. 26-30, 1970.
18. 'Computer Control for New Petrochemical Complex', *Automation,* Jul. 1970.
19. POSSER, F. H., 'Designing Labour out of the Product', *Design and Components in Engineering,* Oct. 7, 1970.
20. SMITH, K., 'The Frontier Spirit Makes Fortunes Overnight in American Electronics Gold Rush', *Electronics Weekly,* Aug. 5, 1970, 7.
21. MOLINS LTD., Film: 'Beyond the Steel Barrier'.
22. WATKINS, C., 'Quality Control by Computer', *Industrial Applications of Small Computers Conference,* Birniehill Institute, Oct. 26-30, 1970.
23. FRETWELL, F., 'Computer Facilities in the Inspection Machine Field', *Industrial Applications of Small Computers Conference,* Birniehill Institute, Oct. 26-30, 1970.
24. DINSDALE, J., 'The NC Measuring Centre: Computer Controlled Inspection', *Industrial Applications of Small Computers Conference,* Birniehill Institute, Oct. 26-30, 1970.
25. WATTS, G. A., 'A Small Computer Applied to the Study of Pressure Die Casting', *Industrial Applications of Small Computers Conference,* Birniehill Institute, Oct. 26-30, 1970.
26. *Westinghouse Machine Tool Electrification Forum,* Pittsburgh, May 26-27, 1970.
27. '£6000 m education bill by year 2000', *New Electronics,* Aug. 4, 1970.
28. HARTLEY, J. R., SLEEMAN, D. H., WOODS, P., 'Assisting the Computer to Assist Instruction', *Industrial Applications of Small Computers Conference,* Birniehill Institute, Oct. 26-30, 1970.
29. GROVES, R. D., 'Real-time Computer Interfaces', *Control and Instrumentation,* May 1970, 50.

30. GOULDER, B., MOSS, A., and SHAW, R., 'Computer Based Management Information in a Small Company', *Proc. I.Mech.E.,* **184**, Part 1, (22), 1969, 397-461. Also discussion on a paper based on above at Birniehill Institute, East Kilbride, Oct. 27, 1970.
31. ROBERTS, G., 'Computer Monitored Maintenance Scheduling of Plant', *Industrial Applications of Small Computers Conference,* Birniehill Institute, Oct. 26-30, 1970.
32. 'Terotechnology', *Target,* Jul/Aug. 1970.
33. DAVIES, B., and BRUCE, J. W., 'Summary of Results of a Long Range Technological Forecasting Study', *Industrial Applications of Small Computers Conference,* Birniehill Institute, Oct. 26-30, 1970.
34. ENTREKIN, D. A., 'Computer Control – Why, How', *American Machinist,* Special Report No. 641, Mar. 9, 1970, 87.
35. WILLARD, F. G., 'Transient Noise Suppression in Control Systems', *Control Engineering,* Sep. 1970, 59.
36. MORRIS, J. J., 'What to Expect when You Scale Down to a Minicomputer', *Control Engineering,* Sep. 1970, 65.